Discrete Cosine Transform

Second Edition

Discrete Cosine Transform

Second Edition

Humberto Ochoa-Domínguez
K. R. Rao

CRC Press
Taylor & Francis Group
Boca Raton London New York

CRC Press is an imprint of the
Taylor & Francis Group, an **informa** business

CRC Press
Taylor & Francis Group
6000 Broken Sound Parkway NW, Suite 300
Boca Raton, FL 33487-2742

International Standard Book Number-13: 978-1-138-30472-7 (Hardback)

Library of Congress Cataloging-in-Publication Data

Names: Rao, K. Ramamohan (Kamisetty Ramamohan), author. | Ochoa-Domínguez, Humberto, author.
Title: Discrete cosine transform / Humberto Ochoa-Domínguez and K. R. Rao.
Description: Second edition. | Boca Raton, FL : Taylor & Francis Group, CRC Press, 2019. | Revised edition of: Discrete cosine transform : algorithms, advntages, applications / K. R. Rao, P. Yip. 1990. | Includes bibliographical references and index.
Identifiers: LCCN 2019000195 | ISBN 9781138304727 (hardback : alk. paper)
Subjects: LCSH: Signal processing--Mathematics. | Discrete cosine transforms.
Classification: LCC TK5102.5 .R335 2019 | DDC 621.382/23--dc23
LC record available at https://lccn.loc.gov/2019000195

Visit the Taylor & Francis Web site at
http://www.taylorandfrancis.com

and the CRC Press Web site at
http://www.crcpress.com

Acknowledgments

This book is the result of the longterm association of the authors, Humberto Ochoa-Domínguez and K. R. Rao. Dr. Ochoa's significant and sustained contributions made this book possible. Special thanks go to their respective families for their support, perseverance and understanding. Dr. Ochoa would like to express his gratitude to Claudia Valenzuela, for her continuous support during the writing of this book, and to the Universidad Autónoma de Ciudad Juárez (UACJ). Dr. Rao would like to acknowledge the support provided in various forms by Dr. Crouch, Dean of the College of Engineering, and Dr. J. W. Bredow, Chair, Department of Electrical Engineering, both at UTA. Dr. G. J. Sullivan, Microsoft, Dr. H. Kalva, Florida Atlantic University and Dr. W.-K. Cham, Chinese University of Hong Kong provided various resources in this regard. Shiba Kuanar, Harsha Nagathihalli Jagadish, Venkata Siva Reddy Anusha and Swaroop Krishna Rao at UTA contributed countless hours in tying up all loose ends (references, copyright releases, proofreading and a million other details). The graduate students and alumnae in the Multimedia Processing Lab (MPL) at UTA in various ways have made constructive comments.

Contents

Preface

After the original book, *Discrete Cosine Transform (DCT),* by K. R. Rao and P. Yip was published by Academic Press (now Elsevier) in 1990, several variants to the DCT have been developed, not to mention the widespread applications in the international and national standards, such as H.264/AVC, AVS China, AV1 and the HEVC. Versatile Video Coding (VVC), projected to be the new video coding standard by 2020, has multiple size transforms, including DCTs and DSTs. Some of these are integer DCTs (IntDCTs), low-complexity ICT (LCICT), modified DCT (MDCT), integer MDCT, etc. Others are directional DCT (DDCT), steerable DCT, etc. Other groups, such as SMPTE, VP8-9 by Google, ATSC, AV1 by Alliance for Open Media (AOM), have also incorporated IntDCTs in their video coding specifications. Both AVS China and HEVC adopted IntDCTs of size 32x32 in which smaller sizes (16x16, 8x8 and 4x4) are embedded. Suggestions for developing IntDCT of size 64x64 (preserving the embedding property) for possible adoption in future standards are made. Order-64 LCICT has been developed and its application in HEVC is analyzed. In fact, in the next generation video coding, VVC, transform size of 128x128 is considered. The IntDCTs are nearly orthogonal and the basis vectors are not of constant length. Their inner (scalar) products vary slightly. Interpolation filters for evaluating the fractional pels from full pels for application in motion compensation are based on DCT.

Newer applications of DCT include mirroring/rotation. An extension to DCT, called steerable DCT (SDCT) and its properties are also described. Some recent applications such as focus and blurriness measurements using reorganized DCT coefficients for auto focus and DCT-based quality degradation metrics are also emphasized. Other applications include corner and edge detections in the DCT domain, visual recovery by multi-scale DCT regularizer, reversible data hiding, DCT-inspired feature transform for image retrieval and reconstruction, face recognition, corner detection directional block-wise image representation, DCT-based watermarking, spatio-temporal and foveated masking effects, fingerprint orientation, image fusion and face detection. VVC also considers multiple size DCTs/DSTs. Other applications include palm-print/fingerprint recognition, video extraction, image/video watermarking, forgery detection, quality degradation metrics, signal- and noise-level estimation, multiview image and video compression, trans-multiplexers, motion analysis/estimation, phoneme recognition, steganalysis, copyright protection, 3D-video watermarking, biometric watermarking, extraction of ROI, artist identification, audio coding, progressive image transmission, vector cardiograms, face-scene analysis, visual recovery, moving object detection, steganography, image fusion, motion estimation, inpainting, medical image processing / compression, Wiener filter, etc. These applications are extensive and evolving. While DCT is not a panacea to all problems, it has its place – a major one – in the general signal processing arena.

Educational tools such as ftp/websites, open source software, software manuals, source codes, image/video test sequences, books, overview papers, etc. are provided. The overall objective is to make this book user friendly from teaching, academic and research perspectives. Image-coding standards such as JPEG, JPEG XS, JPEG XT and JPEG XR, wherein DCT and its variants are used, are described.

This book focusses on these new IntDCTs and their recent applications. References related to the IntDCTs and their hardware realization are also listed. Open source software for their efficient implementation is provided. Using the encoder/decoder block diagrams, their role in video coding standards is furnished. Fast implementation of these DCTs, based on the sparse matrix factorization, is provided. The principal objective of this revised edition is to acquaint the academic, industrial and research community with the latest developments in DCT and in all of its variants. The spectrum of additional applications after 1990 is extensive and continuously evolving. All these developments justify the need for the second edition. Comments, suggestions and reviews (also criticisms) from the academic/research/industrial community are especially welcome.

The textbook consists of eight chapters, organized as follows:

Chapter 1: Highlights the importance of DCT and presents some applications areas.

Chapter 2: Covers the Karhunen–Loève transform (KLT) to perform signal analysis in the KLT domain, examples, applications, problems and projects.

Chapter 3: Provides a thorough analysis of concepts, properties, definitions of the discrete cosine transform and provides its relation to the KLT, fast algorithms and implementation via other transforms. Our coverage includes examples, applications, problems and projects using DCT.

Chapter 4: Treats the analysis of the discrete sine transform. We carry out a detailed discussion about the DST via other transforms and provide fast algorithms. The chapter includes applications, problems and projects.

Chapter 5: Is concerned with the analysis of the modified discrete cosine transform, its properties and definitions. We discuss the evenly and oddly stacked modified discrete cosine transform and block transform. We also cover applications, problems and projects of this transform.

Chapter 6: Discusses the integer discrete cosine transform and their matrices used in modern video coding. We carry out a detailed discussion of different techniques to obtain the integer matrices including the integer Tchebichef transform. We define the different metrics to evaluate a transform and give applications, problems and projects that can be implemented using the integer transform.

Chapter 7: Introduces the directional discrete cosine transform and the steerable DCT. We carry out a detailed analysis of the transforms. The chapter provides examples of use and ends with applications, problems and suggested projects using directional transform.

Chapter 8: Covers transform mirroring and rotation. The chapter provides examples of use, applications, problems and projects.

MATLAB® programs are listed wherever they are possible.

Appendix A: Reviews the DCT in applications such as the different image coding standards, (i.e., JPEG, JPEG XR, JPEG XT, JPEG Pleno, JPEG AIC and JPEG XS). The appendix includes projects for the different JPEG standards focussed to the DCT.

Appendix B: Reviews the DCT in applications such as the different video coding standards and provides the Bjøntegaard metric used to compare the coding efficiency between two different compression algorithms. The appendix includes projects focused on the DCT for the different video coding standards.

Appendix C: Offers general definitions and mathematical formulas. Topics include the cosine modulated filter bank and flowgraphs of fast discrete cosine transform algorithms.

Acronyms

AAC Advanced Audio Coding
AAC-SSR AAC Scalable Sample Rate
AC Non-Zero Frequency Coefficient
AI All-Intra
AIC Advance Image Coding
ALLR Alternative Local Linear Regression
AMT Adaptive Multiple Core Transform
AOM Alliance for Open Media
ASIC Application-Specific Integrated Circuit
AVC Advanced Video Coding
AVS Audio and Video Coding Standard Workgroup of China
AVS1 First Generation Audio Video Coding Standard of China
AVS2 Second Generation Audio Video Coding Standard of China
AZB All Zero Block
BD-PSNR Bjøntegaard Delta Peak-Signal-to-Noise Ratio
BD-rate Bjøntegaard Delta Rate
BPNN Back Propagation Neural Network
BRAM Block RAM
BRO Bit-Reversed Order
CABAC Context Adaptive Binary Arithmetic Coding
CB Coding Block
CIF Common Intermediate Format
CR Compression Ratio
CTU Coding Tree Unit
CU Coding Unit
DC Zero Frequency Coefficient
DCT Discrete Cosine Transform
DCT-I Type-I Discrete Cosine Transform
DCT-II Type-II Discrete Cosine Transform
DCT-III Type-III Discrete Cosine Transform
DCT-IV Type-IV Discrete Cosine Transform
DCT-VIII Type-VIII Discrete Cosine Transform
DCTIF Discrete Cosine Transform Interpolation Filter
DDCT Directional Discrete Cosine Transform
DFT Discrete Fourier Transform
DHT Discrete Hartley Transform
DIF Decimation in Frequency
DIR Directional Features

DIT Decimation in Time
DLT Discrete Legendre Transform
DPCM Differential Pulse Code Modulation
DPU Data Processing Unit
DSIS Double Stimulus Impairment Scale
DST Discrete Sine Transform
DST-I Type-I Discrete Sine Transform
DST-II Type-II Discrete Sine Transform
DST-IV Type-IV Discrete Sine Transform
DST-VII Type-VII Discrete Sine Transform
DT-CWT Dual-Tree Complex Wavelet Transform
DTT Discrete Tchebichef Transform
DWT Discrete Wavelet Transform
EBCOT Embedded Block Coding with Optimized Truncation
EOT Even/Odd Transform
EOTF Electro-Optical Transfer Function
EPE Energy Packing Efficiency
FC Fractional Correlation
FCT Fourier Cosine Transform
FDCT Fast Discrete Cosine Transform
FDIS Final Draft International Standard
FDM Frequency-Division Multiplexing
FFT Fast Fourier Transform
FHT Fast Hartley Transform
FIN Fineness of a Subimage
FPGA Field Programmable Gate Array
fps Frames Per Second
FST Fourier Sine Transform
FT Fourier Transform
FVC Future Video Coding
GPU Graphics Processing Unit
H.264/AVC MPEG-4 Part 10
Hcub Heuristic Cumulative Benefit
HD High Definition
HDL Hardware Description Language
HDR High Dynamic Range
HEVC High Efficiency Video Coding
HFR High Frame Rate
HVS Human Visual System
IAU Input Adder Unit
ICT Information and Communications Technology
ICT Integer Cosine Transform
IDCT Inverse Discrete Cosine Transform
IDFT Inverse Discrete Fourier Transform

IDST Inverse Discrete Sine Transform
IEC International Electrotechnical Commission
IF Interpolation Filter
IFFT Inverse Fast Fourier Transform
IMDCT Inverse Modified Discrete Cosine Transform
IMDST Inverse Modified Discrete Sine Transform
IntDCT Integer Discrete Cosine Transform
IntDST Integer Discrete Sine Transform
IntDTT Integer Discrete Tchebichef Transform
ISDN Integrated Services Digital Network
ISO International Organization for Standardization
IST Integer Sine Transform
IT Information Technology
JEM Joint Exploration Model
JFIF JPEG File Interchange Format
JND Just-Noticeable Distortion
JPEG Joint Photographic Experts Group
JPEG XT JPEG eXTension
JVET Joint Video Experts Team
KLT Karhunen-Loève Transform
LCICT Low-Complexity Integer Cosine Transform
LD Low Delay
LDDP Local Distortion Detection Probability
LOT Lapped Orthogonal Transform
MACE Middle Frequency AC Energy
MC Motion Compensation
MCM Multiplierless Constant Multiplication
MDCT Modified Discrete Cosine Transform
MDDT Mode Dependent Directional Transforms
MDNSST Mode-Dependent Non-Separable Secondary Transforms
MDST Modified Discrete Sine Transform
ME Motion Estimation
MOS Mean Opinion Score
MP3 Moving Picture Experts Group Layer-3 Audio
MPEG Moving Picture Experts Group
MSE Mean Squared Error
MSP Main Still Picture
MSSIM Multi-Scale Structural Similarity Index
MUX Multiplexer
MV Motion Vector
MV-HEVC Multiview High Efficiency Video Coding
MVC Multiview Video Coding
NO Natural Order
NSST Non-Separable Secondary Transform

OA Overlap-Add
OAU Output Adder Unit
PCA Principal Component Analysis
PCM Pulse Code Modulation
PET Positron Emission Tomography
PFA Prime Factor Algorithm
PIT Progressive Image Transmission
PIT Pre-Scaled Integer Transform
PQF Polyphase Quadrature Filter Bank
PSNR Peak Signal-to-Noise Ratio
PVC Perceptual Video Coding
QMF Quadrature Mirror Filter Bank
QP Quantization Parameter
QR Quick Response
RA Random Access
RBC Recursive Block Coding
RC Register Cells
RDO Rate-Distortion Optimization
RICT Recursive Integer Cosine Transform
RO Reversed Order
ROI Region of Interest
ROM Read Only Memory
ROT Rotational Transform
RT Rapid Transform
SA-DCT Shape-Adaptive DCT
SARFT State Administration of Radio, Film, and Television
SAU Shift-Add Unit
SCC Screen Content Coding
SCT Symmetric Cosine Transform
SD Standard Definition
SDCT Steerable Discrete Cosine Transform
SDFT Shifted Discrete Fourier Transform
SDT Signal-Dependent Transform
SMF Sparse Matrix Factor
SNR Signal-to-Noise Ratio
SPECT Single-Photon Emission Computed Tomography
SR-SIM Spectral Residual-Based Similarity
SSB Single-Sideband
SSIM Structural Similarity Index
SST Symmetric Sine Transform
SVD Singular Value Decomposition
TB Transform Block
TDAC Time-Domain Aliasing Cancellation
TDM Time–Division Multiplexing

TU Transform Unit
UHD Ultra High Definition
VHDL Very High Speed Integrated Circuits Hardware Design Language
VLC Variable Length Coding
VLSI Very-Large-Scale Integration
VR Virtual Reality
VVC Versatile Video Coding
WCG Wide Color Gamut
WHT Walsh-Hadamard Transform
WMV Windows Media Video
WFTA Winograd Fourier Transform Algorithm

Contributors

Humberto Ochoa-Domínguez
Universidad Autónoma de Ciudad Juárez, Departamento de Ingeniería Eléctrica y Computación
Ciudad Juárez, Chihuahua, México

K. R. Rao
The University of Texas at Arlington, Department of Electrical Engineering
Arlington, Texas, USA

List of Figures

List of Tables

1 Introduction

Transform coding is an integral component of image and video processing applications. The main goal is to map a set of pixel values into a few transform coefficients with certain desired properties. These properties help reconstruct the pixels with the least loss possible. The mapping process, or transform, must fulfill two important features: The coefficients in the transform domain should be decorrelated and the total energy must be conveyed by only a few coefficients with the remaining coefficients carrying insignificant energy.

A number of transformation techniques exist that fulfill the above properties, although their energy packing capabilities vary. The discrete cosine transform (DCT) is widely used because its strong energy packing and decorrelation properties makes it comparable to the Karhunen–Loève transform (KLT), which is an optimal transform in terms of retained transform coefficients and decorrelation because it completely removes the statistical dependence between transformation coefficients.

The disadvantage of the KLT is that its transformation process depends on the input signal and does not have a fast algorithm so that the computational cost is high, especially in areas of transform coding. Sinusoidal transforms, like the discrete cosine transforms and discrete Fourier transforms (DFT) use image-independent transformations. Moreover, fast algorithms and architectures are available for DCT and DFT. However, DCT results in less blocking artifacts due to the even-symmetric extension properties of DCT and it is real, unlike complex computations used in DFT.

The DCT is preferred over the KLT because it is separable, real, orthogonal and has a fast algorithm. There are 8 types of DCT. Nevertheless, the DCT-II and DCT-IV are constantly applied in image processing. However, recently, other types of DCT have found applications in versatile video coding (VVC) such as DCT-VIII. The following list outlines the common applications of the DCT in different areas.

APPLICATIONS

1. 3D IDCT for video coding standards [163].
2. 3D video watermarking [21].
3. Acoustic echo and feedback cancellation [157].
4. Artist identification [233].
5. Audio coding [153], [154], [155], [157], [160], [280], [546], [564].
6. AV1 codec [64].
7. Biometric watermarking [492].
8. Brain tumor compression classification [448].
9. Complex cepstra feature analysis [336].
10. Compressive sampling [14], [31], [79], [182], [258], [530].
11. Consumer electronics.

   a. 3D high efficiency video coding [478].
   b. DCT pyramid for error concealment [572].
12. Copyright protection [539].
13. Corner detection method [271].
14. DCT-based features [392].
15. DCT-based watermarking [291], [295], [479].
16. DCT filtering [82], [349].
17. DCT-inspired feature transform for image retrieval and reconstruction [514].
18. DCT QM DCT quality degradation metric (QM quality metric) [29].
19. DCT-regularized extreme visual recovery [516].
20. Digital watermarking survey [22], [438].
21. Directional block-wise image representation [198].
22. Directional textures auto inspection [370].
23. Directional transforms for video coding [20], [39], [52].
24. Downsampling and upsampling [117].
25. Edge detection method [28], [382].
26. Encryption [568].
27. Extraction of ROI [86].
28. Extreme visual recovery [516], [515].
29. Face detection [9], [22].
30. Face recognition/description [26], [44], [80], [176], [188], [218], [224], [230].
31. Face scene analysis [224].
32. Feature extraction [25], [114].
33. Fingerprint orientation [286]
34. Fingerprint-based biometric watermarking [492].
35. Focus/blurriness measure [582].
36. Forgery detection [180], [299].
37. Future video coding -- Multiple size DCTs/DSTs [23], [62], [63], [67], [324].
38. Geophysical transient EM method [586].
39. Image analysis [65].
40. Image enhancement [72].
41. Image forgery [15], [141].
42. Image fusion [79], [339], [365].
43. Image interpolation [4], [12], [84], [117], [318].
44. Image noise-level estimation [474], [551], [553], [565].
45. Image-quality assessment [10], [15], [69], [71].
46. Image retrieval and reconstruction [11].
47. Image rotation [144].
48. Image watermarking [124], [125], [293][298], [311], [331], [410], [461], [577].
49. Inpainting [99], [281], [363], [585].
50. Interpolation filters for fractional pel motion estimation in HEVC are based on DCT [28].
51. JPEG XT [19], [20], [40], [399], [400], [401], [402], [403], [404], [405], [406], [407].

52. Just-noticeable distortion (JND) profile [25], [30].
53. Local distortion detection probability model [28].
54. Lossless audio compression [564].
55. Medical image compression [77].
56. Medical image fusion [59], [164].
57. Medical image watermarking [68].
58. Medical imaging [182], [411], [473].
59. Mirroring/rotation [250], [383] and [430].
60. Motion analysis [46], [256].
61. Motion analysis in 3D-DCT domain [46].
62. Motion estimation [256].
63. Moving object detection [472], [479].
64. Multiview image and video compression [275].
65. Palm print identification/recognition [16], [126], [260], [574].
66. Pattern recognition [587].
67. Perceptual audio coding [160].
68. PET-image reconstruction [251], [484], [485], [486].
69. Phoneme recognition [336].
70. Progressive image transmission [83].
71. Reversible data hiding [191].
72. RF multi beam digital aperture-array [376].
73. Segmentation [171], [186], [392].
74. Signal and noise-level estimation [553].
75. Spatio-temporal and foveated masking effects [30].
76. Speech recognition [422].
77. Steerable DCT [137], [138], [139].
78. Steganography [181], [194], [333], [334], [446].
79. Steganalysis [181].
80. Surveillance
    a. DCT background modeling [472], [512].
    b. DCT-based high-resolution image reconstruction (i.e., Hallucinating face) [579].
    c. Tracking [32], [113], [334].
81. TDAC analysis/synthesis [51], [259], [379], [380].
82. Transmultiplexers [341], [342], [344].
83. Vector cardiograms [120].
84. Vehicular black box camera [252].
85. Video extraction [355].
86. Video watermarking [21], [22], [74], [128], [194], [244], [321], [576].
87. Voice-activity detection [152].
88. Wireless-sensor networks [384].
89. Watermarking detection [40].
90. Wiener filter [555].
91. Wireless acoustic sensor networks [37].

2 The Karhunen–Loève Transform

INTRODUCTION

In the signal-processing field, many discrete random processes are correlated. In other words, the processes are associated with a certain degree or they are somewhat similar. The Karhunen–Loève transform (KLT) is a well-established unitary optimal transform in terms of energy compaction and decorrelation because it packs the energy of a signal in a few coefficients and diagonalizes the covariance or correlation matrix, indicating that the coefficients are completely decorrelated [315], [394]. This property is necessary for processing (quantization, coding, etc.) of any coefficient in the KLT domain, which has no direct relation with the others. The shortcoming of the KLT is that it is signal dependent. Hence, there is no fast algorithm to implement it. Nevertheless, the KLT has been used as a benchmark to evaluate the performance of other transforms such as the discrete cosine transform (DCT). Even though it is a suboptimal transform, it is extremely popular in video coding because it approaches the KLT for a Markov-1 signal with large correlation coefficient, it is signal independent and it has fast algorithms resulting in efficient implementations (see Chapter 3).

The KLT has also provided an incentive to the researchers to develop other signal independent transforms that approach the performance of KLT and have fast algorithms. Because of the importance of this transform, this chapter defines and develops the KLT for one- and two-dimensional random signals and lists the performance criteria. It concludes with applications that illustrate the decorrelation property and its significance in image compression.

THE KARHUNEN–LOÈVE TRANSFORM FOR A 1D RANDOM FIELD

Let $x = (x_1, x_2, \ldots, x_N)^T$ be a random complex sequence, x^H the Hermitian vector equal to its conjugate transpose $x^H = (x^*)^T$ and R the $(N \times N)$ correlation matrix given by,

$$R = E[xx^H] = E\left[\begin{pmatrix} x_1 \\ x_2 \\ \vdots \\ x_N \end{pmatrix} (x_1^*, x_2^*, ..., x_N^*)\right] \tag{2.1}$$

$$= E\begin{bmatrix} x_1x_1^* & x_1x_2^* & \dots & x_1x_N^* \\ x_2x_1^* & x_2x_2^* & \dots & x_2x_N^* \\ \dots & \dots & \dots & \dots \\ x_Nx_1^* & x_Nx_2^* & \dots & x_Nx_N^* \end{bmatrix} \tag{2.2}$$

$$= \begin{bmatrix} E[x_1x_1^*] & E[x_1x_2^*] & \dots & E[x_1x_N^*] \\ E[x_2x_1^*] & E[x_2x_2^*] & \dots & E[x_2x_N^*] \\ \dots & \dots & \dots & \dots \\ E[x_Nx_1^*] & E[x_Nx_2^*] & \dots & E[x_Nx_N^*], \end{bmatrix},$$

where $E[\cdot]$ is the expectation operator, $E[x_px_q^*]$ is the autocorrelation of x_p when $p = q$ and the cross correlation when $p \neq q$. Let Φ be the unitary matrix which diagonalizes R such that,

$$\begin{gathered} \Phi^{-1} = \Phi^H, \Phi\Phi^H = I, \\ \Phi^H R\Phi = \Phi^{-1}R\Phi = \Lambda \\ \Lambda = Diag\,[\lambda_1, \lambda_2, \dots, \lambda_N]\,. \end{gathered} \tag{2.3}$$

Here, $\lambda_i, i = 1, 2, \dots, N$ are the eigenvalues of R, I is the identity matrix, and Φ is called the KLT matrix and decorrelates the random sequence x. This can be seen when the forward and inverse KLT are considered. For example, let $y = (y_1, y_2, ..., y_N)^T$ be the forward transform of x, and it represents the random sequence in the transform domain. Therefore,

$$y = \Phi^{-1}x = \Phi^H x,$$

and let the inverse transform of **y** be

$$x = \Phi y. \tag{2.4}$$

The correlation matrix of y is then,

$$E\left[yy^H\right] = E\left[\Phi^H xx^H \Phi\right] = \Phi^H E\left[xx^H\right]\Phi = \Phi^H R\Phi = \Lambda. \tag{2.5}$$

It is clear from (2.5) that the random sequence y has no cross-correlation. In other words, x is decorrelated by the KLT matrix Φ. Such a transform is also called principal component or Hoteling transform and it is a statistically optimal transform. The performance of all other transforms (suboptimal) is evaluated against this benchmark, KLT.

EIGENVALUES AND EIGENVECTORS OF *R*

To show the signal dependence of KLT, the eigenvalues and eigenvectors of R are considered. Multiplying both sides of (2.3) by Φ

$$\begin{aligned} \Phi\Phi^H R\Phi &= \Phi\Lambda \\ R\Phi &= \Phi\Lambda, \end{aligned} \tag{2.6}$$

where $\Phi = [\phi_1, \phi_2, \ldots, \phi_N]$ and $\phi_i = [\phi_{i1}, \phi_{i2}, \ldots, \phi_{iN}]^T$, $i = 1, 2, \ldots, N$ being the ith column of Φ. Writing the right-hand side of (2.6) in full, we have,

$$\begin{bmatrix} \phi_{11} & \phi_{21} & \cdots & \phi_{N1} \\ \phi_{12} & \phi_{22} & \cdots & \phi_{N2} \\ \cdots & \cdots & \cdots & \cdots \\ \phi_{1N} & \phi_{2N} & \cdots & \phi_{NN} \end{bmatrix} \begin{bmatrix} \lambda_1 & 0 & \cdots & 0 \\ 0 & \lambda_2 & \cdots & 0 \\ \cdots & \cdots & \cdots & \cdots \\ 0 & 0 & \cdots & \lambda_N \end{bmatrix} = \tag{2.7}$$

$$= \left[\lambda_1 \begin{pmatrix} \phi_{11} \\ \phi_{12} \\ \vdots \\ \phi_{1N} \end{pmatrix}, \lambda_2 \begin{pmatrix} \phi_{21} \\ \phi_{22} \\ \vdots \\ \phi_{2N} \end{pmatrix}, \ldots, \lambda_N \begin{pmatrix} \phi_{N1} \\ \phi_{N2} \\ \vdots \\ \phi_{NN} \end{pmatrix} \right],$$

or

$$R\phi_i = \lambda_i \phi_i, \qquad i = 1, 2, \ldots, N. \tag{2.8}$$

From (2.8) we can see that $\lambda_i s$ (the eigenvalues of R) are real and positive and $\phi_i s$ are the corresponding eigenvectors. When the eigenvalues are arranged in non-increasing order, that is $\lambda_1 \geq \lambda_2, \ldots, \geq \lambda_N$, the auto-correlations of the transformed signal vector are arranged in descending order.

Similar considerations, based on the diagonalization of the covariance matrix of a random vector x, produce a KLT based on the covariance matrix.

Let $\mu_y = E[y] = (E[y_1], E[y_2], \ldots, E[y_N])^T = (\mu_{y1}, \mu_{y2}, \ldots, \mu_{yN})^T$ be the mean of the random vector y and $E[(y - \mu_y)(y - \mu_y)^H]$ the covariance matrix in KLT domain. Then,

$$E\left(\begin{pmatrix} y_1 - \mu_{y1} \\ y_2 - \mu_{y2} \\ \cdots \\ y_N - \mu_{yN} \end{pmatrix} \left[(y_1 - \mu_{y1})^* (y_2 - \mu_{y2})^* \ldots (y_N - \mu_{yN})^* \right] \right). \tag{2.9}$$

It is desired that (2.9) be a diagonal matrix, i.e.,

$$E[(y - \mu_y)(y - \mu_y)^H] = \Lambda' = Diag[\lambda_1', \lambda_2', \ldots, \lambda_N'].$$

This diagonalization can be achieved if the vector $[y - \mu_y]$ is related to the signal vector $[x - \mu_x]$ by a unitary transform Ψ such that,

$$[y-\mu_y] = \Psi^{-1}[x-\mu_x],$$

where $\mu_x = E[x]$. As it can be seen, Ψ is made up of the eigenvectors for the data covariance matrix $E[(x-\mu_x)(x-\mu_x)^H] = R'$. Therefore,

$$R' = \Psi\Lambda',$$

or

$$R'\psi_i = \lambda_i'\psi_i, i = 1,2,\ldots,N,$$

where $\Lambda' = \Psi^{-1}R'\Psi$ then,

$$E\left[(y-\mu_y)(y-\mu_y)^H\right] = E\left[\Psi^{-1}(x-\mu_x)(x-\mu_x)^H\Psi\right]$$

$$= \Psi^{-1}E[(x-\mu_x)(x-\mu_x)^H]\Psi = \Psi^{-1}R'\Psi = \Lambda'. \quad (2.10)$$

Observe that although Ψ and Φ play the same role of diagonalization, they are in general different unless x is a zero mean random vector.

Figure 2.1 shows the basis functions for the KL expansion for N= 16 and sample signal assumed to have Markov-1 statistics with adjacent correlation coefficient $\rho = 0.95$. (See Prob. 2.5.)

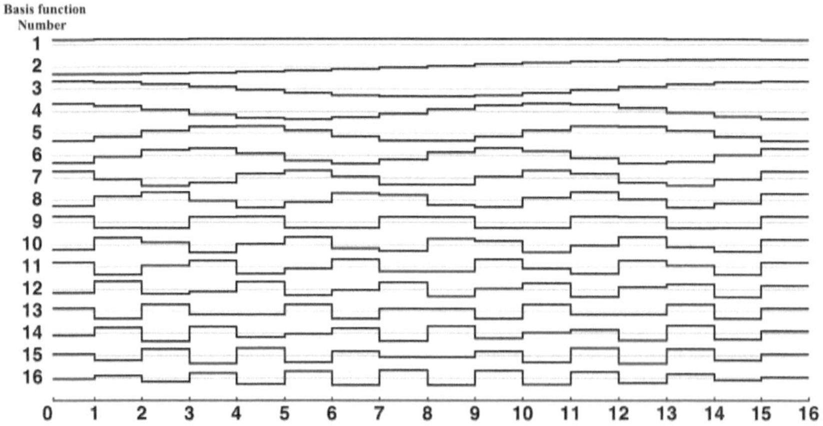

Figure 2.1: KLT basis functions for N= 16 and ρ = 0.95 for a Markov-1 signal.

APPLICATION OF KLT TO DATA COMPRESSION

Figure 2.2 shows the role of KLT in data compression in a general variance zonal filter. The data vector x is transformed by the operation block A into the transform domain vector y which undergoes the operator block $\hat{I}_m$ and is then transformed

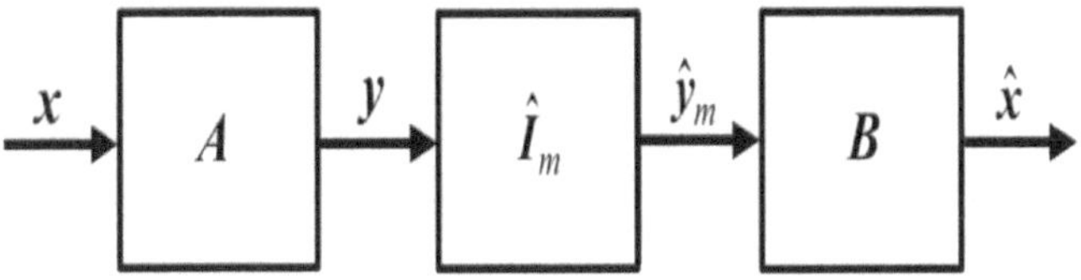

Figure 2.2: Maximum variance zonal filter.

back by the block B into the data domain $\hat{x}$. The blocks A, $\hat{I}_m$ and B are selected to minimize the mean square error (MSE) of $\hat{x}$ with respect to x.

As the KLT diagonalizes the correlation or covariance matrix, it is possible to represent an N-dimensional random vector x by only some of its coefficients in the KLT domain by retaining m out of N coefficients that represent the m largest eigenvalues. The vector x can be reconstructed with minimal error by quantizing and coding these m coefficients. This is the essence of data compression or bandwidth reduction because the retained coefficients need a smaller number of bits compared to that required for x.

In the case of simple data compression the block $\hat{I}_m$ is a $(N \times N)$ diagonal $1 \leq m \leq N$ with the first m diagonal elements as ones and the remaining diagonal elements as zeros, i.e.,

$$\hat{I}_m = \text{Diag}[1,1,1,\ldots,1,0,\ldots,0].$$

The random vector $x = (x_1, x_2, \ldots, x_N)^T$ is mapped into y by the orthogonal matrix, i.e.,

$$y = Ax = (y_1, y_2, \ldots, y_N)^T \tag{2.11a}$$

$$\hat{y}_m = \hat{I}_m y = (y_1, y_2, \ldots, y_m, 0, 0, 0, \ldots, 0)^T \tag{2.11b}$$

$$\hat{x} = \hat{B}\hat{y}_m = (\hat{x}_1, \hat{x}_2, \ldots, \hat{x}_N)^T. \tag{2.11c}$$

Notice that B is another orthogonal matrix. It can be shown that the MSE between x and y is minimized when $A = \Phi^H$ and $B = A^{-1} = \Phi$ provided that the columns of Φ (eigenvectors) are arranged such that the corresponding eigenvalues are in non-increasing order $\lambda_1 \geq \lambda_2 \geq \ldots \geq \lambda_N$. This implies that A and B correspond to KLT and inverse KLT respectively. Hence, the MSE is given by,

$$MSE = \frac{1}{N} E\left[\sum_{k=1}^{N} (x_k - \hat{x}_k)^2\right] \tag{2.12}$$

$$= E\left[\sum_{k=m+1}^{N} y_k^2\right] = \sum_{k=m+1}^{N} \lambda_k,$$

since the MSE is invariant under orthogonal transformation. This implies that among all other orthogonal transforms, the KLT achieves the minimum MSE when only a subset of the *m* KLT coefficients is retained. The N-*m* smallest eigenvalues are set to zero. In general, the KLT is said to be the optimal transform because it has the following properties:

1. It packs most of the energy in the fewest number of KLT coefficients.
2. It minimizes the MSE between the original and the reconstructed signal for a given number of coefficients.
3. It achieves the minimum rate for rate-distortion function, among all unitary transforms.
4. It completely decorrelates the signal in the transform domain.

These properties are applicable to pattern recognition, classification and bit reduction (compression) among other areas.

Example 2.1. *Consider the problem of diagonalizing the following* 3×3 *matrix*

$$A = \begin{pmatrix} 2 & -1 & 0 \\ -1 & 2 & -1 \\ 0 & -1 & 2 \end{pmatrix}.$$

We know from (2.6) *that* $A\phi = \phi\Lambda$, *where* ϕ *and* Λ *are the eigenvalues and eigenvectors matrix respectively. Then, we can express the equation as,*

$$(A - \Lambda I)\,\phi = 0, \tag{2.13}$$

where I is the identity matrix. The characteristic equation is the determinant,

$$\det(A - \Lambda I) = 0$$

$$\det\begin{pmatrix} 2-\lambda & -1 & 0 \\ -1 & 2-\lambda & -1 \\ 0 & -1 & 2-\lambda \end{pmatrix} = \begin{pmatrix} 0 \\ 0 \\ 0 \end{pmatrix}.$$

Solving the determinant to find the $\lambda_i s$

$$\lambda^3 - 6\lambda^2 + 10\lambda - 4 = 0$$

$$(\lambda - (2+\sqrt{2}))(\lambda - 2)(\lambda - (2-\sqrt{2})) = 0$$

Therefore, $\lambda_1 = 2+\sqrt{2}$, $\lambda_2 = 2$, $\lambda_3 = 2-\sqrt{2}$. *Afterwards, the eigenvectors can be found as,* $\frac{1}{2}$

$$\phi_1 = \begin{pmatrix} -\frac{1}{2} \\ \frac{1}{\sqrt{2}} \\ -\frac{1}{2} \end{pmatrix}, \quad \phi_2 = \begin{pmatrix} -\frac{1}{\sqrt{2}} \\ 0 \\ \frac{1}{\sqrt{2}} \end{pmatrix}, \quad \phi_3 = \begin{pmatrix} \frac{1}{2} \\ \frac{1}{\sqrt{2}} \\ \frac{1}{2} \end{pmatrix}.$$

Finally, Λ is the diagonal matrix that contains the eigenvalues of A and Φ its eigenvectors,

$$\Lambda = \begin{pmatrix} 2+\sqrt{2} & 0 & 0 \\ 0 & 2 & 0 \\ 0 & 0 & 2-\sqrt{2} \end{pmatrix}, \quad \Phi = \begin{pmatrix} -\frac{1}{2} & -\frac{1}{\sqrt{2}} & \frac{1}{2} \\ \frac{1}{\sqrt{2}} & 0 & \frac{1}{\sqrt{2}} \\ -\frac{1}{2} & \frac{1}{\sqrt{2}} & \frac{1}{2} \end{pmatrix}.$$

Notice that Φ is an orthogonal matrix therefore $\Phi^{-1} = \Phi^T$. A has been diagonalized. The result can be verified by,

$$A = \Phi\Lambda\Phi^T.$$

The MATLAB code is the following

```
A=[2 -1 0; -1 2 -1; 0 -1 2];
[U D V] = eig(A);  % Also function svd() can be used
Arec = U * D * V';  % Matrix recovered
```

KLT FOR A 2D RANDOM FIELD

The KLT developed for a 1D random field can be extended to the 2D case. This has applications in processing of 2D signals such as multispectral imagery. For simplicity, assume the square random field to be real and represented by

$$\aleph = \begin{bmatrix} x_{11} & x_{12} & \cdots & x_{1N} \\ x_{21} & x_{22} & \cdots & x_{2N} \\ \cdots & \cdots & \cdots & \cdots \\ x_{N1} & x_{N2} & \cdots & x_{NN} \end{bmatrix}. \tag{2.14}$$

Define $\Re$ the correlation matrix as follows,

$$\Re = E[uu^T], \tag{2.15}$$

where u is the $\left(N^2 \times 1\right)$ column vector obtained from the lexicographic ordering of matrix $\aleph$, i.e.,

$$u = (x_{11}, x_{12} \ldots x_{1N}, x_{21}, x_{22} \ldots x_{2N}, x_{N1}, x_{N2} \ldots x_{NN})^T.$$

By using u, a one-dimensional random field to represent the $N \times N$ image, the diagonalization of its correlation matrix defined in (2.15) can be considered in Section 2. The problem, however, quickly becomes formidable for N of moderate size, since the diagonalization problem to produce the eigenvectors for the KLT is of dimension $N^2 \times N^2$. The relevant equation is

$$\Re\psi_i = \lambda_i\psi_i \quad i = 1, 2, \ldots, N^2, \tag{2.16}$$

where λ_i and ψ_i are the eigenvalues and eigenvectors of $\Re$. The ψ_is form the columns of the KLT matrix Ψ for this 2-dimensional random field. In terms of the pixels $x_{m,n}$ in the 2-dimensional image, (2.16) becomes

$$\sum_{m'=1}^{N}\sum_{n'=1}^{N} E\left[x_{m,n}, x_{m',n'}\right] \psi_{i,l}(m',n') = \lambda_{i,l}\psi_{i,l},$$

$$i,l,m,n = 1,2,\ldots,N. \tag{2.17}$$

By assuming separable statistics, the development of 2D KLT can be considerably simplified. Under this assumption, the row and column statistics are considered completely independent and identically distributed. Hence the correlation matrix and the eigenfunctions become separable as shown in the following,

$$E\left[x_{m,n}; x_{m',n'}\right] = E\left[x_{m,n}; x_{m',n}\right] E\left[x_{m,n}; x_{m,n'}\right], \tag{2.18}$$

and

$$\psi_{i,l}(m,n) = (\Phi_1)_{m,i}(\Phi_2)_{n,l} \qquad m,m',n,n',i,l = 1,2\ldots,N,$$

where Φ_1 and Φ_2 are factor matrices in the KLT matrix Ψ due to the separable statistics of rows and columns. In fact, the diagonalization problem of the $N^2 \times N^2$ correlation matrix $\Re$ can now be separated so that we have:

$$\Re = \Re_1 \bigotimes \Re_2, \tag{2.19a}$$

and

$$\Psi = \Phi_1 \bigotimes \Phi_2, \tag{2.19b}$$

where

$$\Phi_1 \Re_1 \Phi_1^H = \Lambda_1, \tag{2.20a}$$

and

$$\Phi_2 \Re_2 \Phi_2^H = \Lambda_2. \tag{2.20b}$$

Equations (2.19a) through (2.20b) represent two diagonalization problems. Λ_1 and Λ_2 are $N \times N$ diagonal matrices whose diagonal elements are the eigenvalues of the $N \times N$ correlation matrices $\Re_1$ (based on row statistics) and $\Re_2$ (based on columns statistics) respectively. The symbol $\bigotimes$ stands for Kronecker product of two matrices, i.e.,

$$A \bigotimes B = C = \begin{bmatrix} a_{11}B & a_{12}B & \ldots & a_{1n}B \\ a_{21}B & a_{22}B & \ldots & a_{2n}B \\ \ldots & \ldots & \ldots & \ldots \\ a_{n1}B & a_{n2}B & \ldots & a_{nn}B \end{bmatrix}, \tag{2.21}$$

where C is a matrix of size $(mp \times nq)$ when the matrices A and B are of sizes $(m \times n)$ and $(p \times q)$ respectively. The KLT of u is, i.e.,

$$v = \Psi^H u = \left(\Phi_1^H \bigotimes \Phi_2^H\right) u. \tag{2.22}$$

In conclusion, by modelling the image autocorrelation by a separable function (independent row and column statistics), diagonalization of an $N^2 \times N^2$ matrix $\Re$ is considerably simplified into diagonalization of two $N \times N$ matrices $\Re_1$ and $\Re_2$.

Example 2.2. *Obtain the 2D KLT basis functions for the stationary Markov-1 signal with correlation coefficient $\rho = 0.95$ and $N = 8$. Assuming that the statistical properties along row and along column are independent.*

A first-order Markov process is defined by,

$$R = \rho^{|i-j|} = \begin{bmatrix} 1 & \rho & \rho^2 & \dots & \rho^{N-1} \\ \rho & 1 & \rho & \dots & \rho^{N-2} \\ \rho^2 & \rho & 1 & \dots & \rho^{N-3} \\ \dots & \dots & \dots & \dots & \dots \\ \rho^{N-1} & \rho^{N-2} & \rho^{N-3} & \dots & 1 \end{bmatrix}. \tag{2.23}$$

Figure 2.3 shows the 2D KLT basis images when $\rho = 0.95$ and $N = 8$.

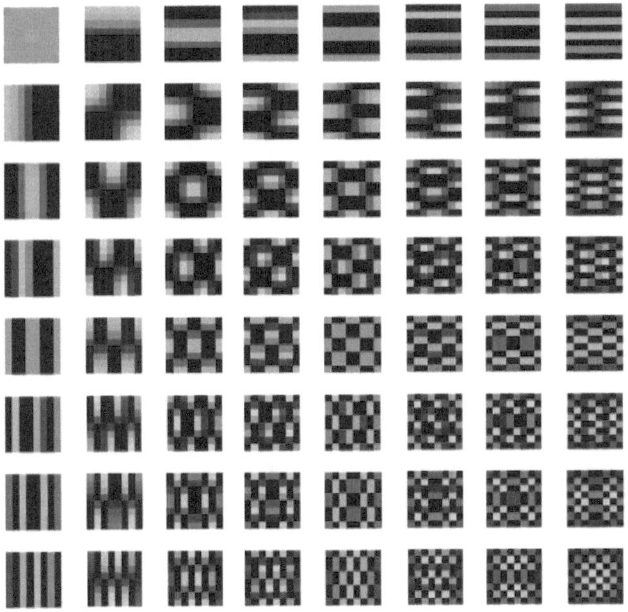

Figure 2.3: 2D KLT basis functions for the stationary Markov-1 signal with correlation coefficient $\rho = 0.95$ and $N = 8$.

The MATLAB code is shown following

```
ro =0.95;
N=8;

for i = 1 : N
    for k = 1 : N
        R(i , k)=ro ^( abs (k - i ));
    end
end

% Matrix R is symmetric

[V, D, U] = eig(R); % Also function svd() can be used

for i = 1 : N
    for k = 1 : N
        B = V(: , i )*V(: , k ) ';
        subplot( N, N, (k - 1) * 8 + i );
        imshow(B);
    end
end
```

Example 2.3. *Suppose that the points of Figure 2.4 correspond to positions* $(x_1, x_2)^T$ *of the pixels in an object inside an image. Determine the KL matrix* (Φ) *and the new coordinate system.*

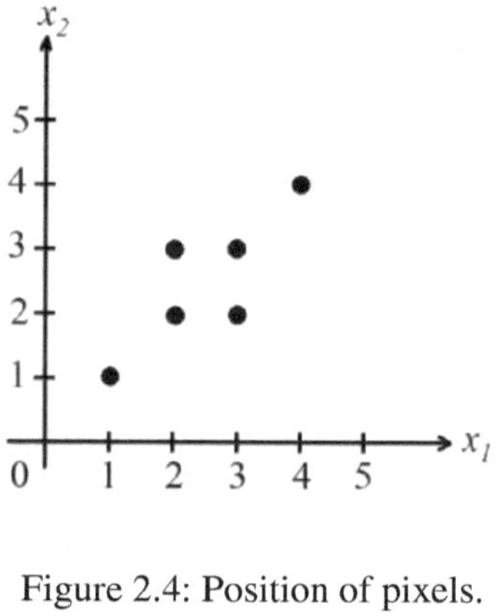

Figure 2.4: Position of pixels.

$$\text{Set of positions: } x = \left\{ \begin{pmatrix} 1 \\ 1 \end{pmatrix}, \begin{pmatrix} 2 \\ 2 \end{pmatrix}, \begin{pmatrix} 2 \\ 3 \end{pmatrix}, \begin{pmatrix} 3 \\ 2 \end{pmatrix}, \begin{pmatrix} 3 \\ 3 \end{pmatrix}, \begin{pmatrix} 4 \\ 4 \end{pmatrix} \right\},$$

with mean vector:

$$\mu_x = \begin{pmatrix} 2.5 \\ 2.5 \end{pmatrix},$$

and covariance matrix

$$C_x = E\left[(x-\mu_x)(x-\mu_x)^T\right].$$

Then,

$$
\begin{aligned}
(x_1-\mu_x) &= \begin{pmatrix}-1.5\\-1.5\end{pmatrix} \therefore (x_1-\mu_x)(x_1-\mu_x)^T = \begin{bmatrix}2.25 & 2.25\\2.25 & 2.25\end{bmatrix},\\
(x_2-\mu_x) &= \begin{pmatrix}-0.5\\-0.5\end{pmatrix} \therefore (x_2-\mu_x)(x_2-\mu_x)^T = \begin{bmatrix}0.25 & 0.25\\0.25 & 0.25\end{bmatrix},\\
(x_3-\mu_x) &= \begin{pmatrix}-0.5\\0.5\end{pmatrix} \therefore (x_3-\mu_x)(x_3-\mu_x)^T = \begin{bmatrix}0.25 & -0.25\\-0.25 & 0.25\end{bmatrix},\\
(x_4-\mu_x) &= \begin{pmatrix}0.5\\-0.5\end{pmatrix} \therefore (x_4-\mu_x)(x_4-\mu_x)^T = \begin{bmatrix}0.25 & -0.25\\-0.25 & 0.25\end{bmatrix},\\
(x_5-\mu_x) &= \begin{pmatrix}0.5\\0.5\end{pmatrix} \therefore (x_5-\mu_x)(x_5-\mu_x)^T = \begin{bmatrix}0.25 & 0.25\\0.25 & 0.25\end{bmatrix},\\
(x_6-\mu_x) &= \begin{pmatrix}1.5\\1.5\end{pmatrix} \therefore (x_6-\mu_x)(x_6-\mu_x)^T = \begin{bmatrix}2.25 & 2.25\\2.25 & 2.25\end{bmatrix}.
\end{aligned}
$$

Therefore

$$C_x = \begin{bmatrix}0.91666 & 0.75000\\0.75000 & 0.91666\end{bmatrix}.$$

The eigenvalues of C_x can be determined by solving the characteristic equation

$$
\begin{aligned}
\det\begin{pmatrix}0.91666-\lambda & 0.75\\0.75 & 0.91666-\lambda\end{pmatrix} &= 0\\
\lambda^2 - 1.8333\lambda + 0.2778 &= 0\\
\therefore \lambda_1 = 1.6666,\quad \lambda_2 &= 0.1666.
\end{aligned}
$$

The eigenvectors matrix is found by solving $(C_x - \Lambda I)\Phi = 0$ then

$$\lambda_1 \Rightarrow \phi_1 = \frac{1}{\sqrt{2}}\begin{pmatrix}1\\1\end{pmatrix},\quad \lambda_2 \Rightarrow \phi_2 = \frac{1}{\sqrt{2}}\begin{pmatrix}-1\\1\end{pmatrix}.$$

Finally,

$$\Lambda = \begin{pmatrix}1.66666 & 0\\0 & 0.166666\end{pmatrix},\quad \Phi = \frac{1}{\sqrt{2}}\begin{pmatrix}1 & -1\\1 & 1\end{pmatrix}.$$

The transformation output is $y = \Phi(x-\mu_x)$ and establishes a new coordinate system. The new center is the centroid of the object and the axes are parallel to the direction of the eigenvectors. This rotation aligns the data with the eigenvectors as shown in Figure 2.5. Because of this rotation, some elements of the image become uncorrelated and the eigenvalues are the variances of the components along each eigenvector.

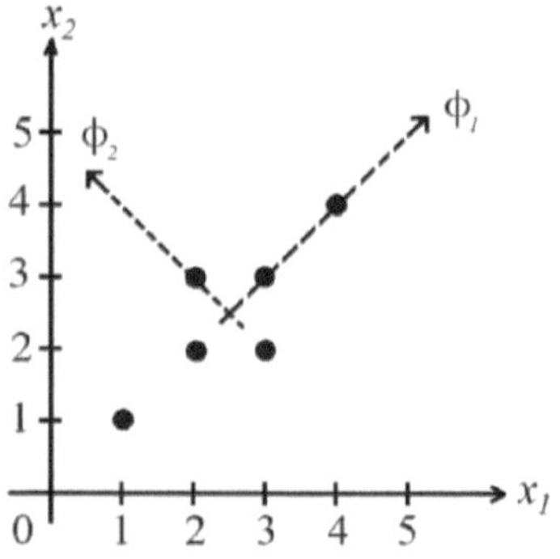

Figure 2.5: New coordinates system according to the eigenvectors (ϕ_1, ϕ_2).

Notice that four out of six points lie on the axis ϕ_1 *. Therefore, their projection (orthogonal projection) is zero and the projection of the remaining two points is minimum; in other words, the variance of the projected data is minimized. If we take only this axis, the dimensionality is reduced and the MSE is minimized. This axis is known also as principal subspace.*

The MATLAB code is shown following

```
% Set of coordinates
X =[[1;1],[2;2],[2;3],[3;2],[3;3],[4;4]];
N = size(X,2);

% Mean vector
Mu = [mean(X(1:1,:));mean(X(2:2,:))];

% x= X-Mu
x = [X(1:1,:)-Mu(1,1); X(2:2,:)-Mu(2,1)];

% Covariance matrix
Cx = zeros(2,2);
for i = 1:N
    Matrix = x(:,i:i)*x(:,i:i)';
    Cx = Cx+Matrix./N;
end
[Phi D]=svd(Cx) % Eigenvectors and eigenvalues

% Forward KLT
y = Phi*x ;

subplot(3,1,1);
plot(X(1:1, :), X(2:2,:),...
    'o', 'MarkerFaceColor', 'red');
```

```
title('Initial points');
xlabel('x_1');
ylabel('x_2');
axis([0 N 0 N]);

subplot(3,1,2);
plot(y(1:1, :), y(2:2, :),...
     'o', 'MarkerFaceColor', 'red');
title('Points after KL transformation');
xlabel('y_1'); ylabel('y_2');
axis([-floor(N/2) floor(N/2) -2 2]);

%Inverse KLT
xrec=Phi'*y; % Notice the transpose matrix
xrec = [xrec(1:1,:)+Mu(1,1); xrec(2:2,:)+Mu(2,1)];

subplot(3,1,3);
plot(xrec(1:1, :), xrec(2:2, :),...
     'o', 'MarkerFaceColor', 'red');
title('Points after inverse KL transformation');
xlabel('x_{1 rec}'); ylabel('x_{2 rec}');
axis([0 N 0 N]);
```

APPLICATIONS

Although the generation of KLT involves estimating correlation/covariance matrices with their diagonalization leading to eigenvalues and eigenvectors, a perusal of the references under KLT indicates that it has found applications in image compression [2], [394], multispectral image compression [414], [481], [493], image segmentation and indexing [171], recursive filtering [221], [60], image restoration [197], image representation, recovery and analysis [592], multilayer image coding [481], neural clustering [315], speech recognition [422], speaker recognition [69], speaker verification [347], feature selection [235], texture classification [477], image and video retrieval [11], [65], [88], [348], [470], [531], etc. Of particular significance is the paper by Saghri, Tescher and Reagan [414] wherein the KLT is applied in decorrelating across spectral bands followed by JPEG (Joint Photographic Experts Group) algorithm. They were able to produce a range of compression ratios (CR) starting with near lossless result at 5:1 CR to visually lossy results beginning at 50:1 CR. An adaptive approach in which the covariance matrix is periodically updated based on the terrain (water, forest, cloud, ice, desert, etc.) is utilized in achieving these high compression ratios. It is only appropriate at this stage to describe this application in detail.

Multispectral images (both satellite and airborne) exhibit a high degree of spatial and spectral correlations [234]. The proposed scheme (Fig. 2.6) involves 1D KLT to decorrelate across spectral bands followed by the JPEG algorithm (this involves 2D DCT of spectrally decorrelated images for spatial decorrelation).

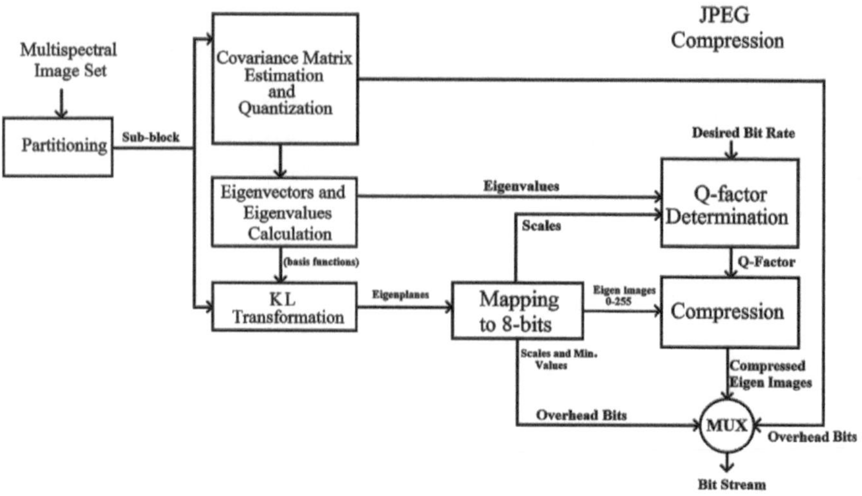

Figure 2.6: Terrain-adaptive compression block diagram [414].

KLT has been applied in video segmentation, classification and indexing [145], [171]. This is useful in random retrievals of video clips from large data bases. This approach for automatic video scene segmentation and content-based indexing is very robust and reduces the potential for fast scene change detection. The principal component analysis extracts effective discriminating features from the reduced data set that can be reliably used in video scene change detection, segmentation and indexing tasks [190], [253], [274].

The paper by Gastpar, Dragotti and Vetterly [151] has opened the possibility to apply more efficiently the KLT in the sensors networks area [34], [37], [89], [110], [115], [314], [384], [385], [420], [450], [544]. The transform considers local KLTs in turn at the various terminals of the network, leading to a locally convergent algorithm; depending on the overall correlation structure this iterative algorithm can reach the global maximum.

SUMMARY

Both 1D and 2D KLT are defined and developed. Their properties are outlined. By assuming row and column independent statistics, generation and implementation of 2D KLT are simplified into two 1D-KLTs. In spite of its computational complexity it is utilized in specific fields such as multispectral imaging. Also, KLT serves as a benchmark in evaluating other discrete transforms.

PROBLEMS

2.1. Show that when eigenvectors (columns) of Φ are rearranged, the corresponding eigenvalues rearrange accordingly.

2.2. See Figure 2.2. It is stated that the MSE between x and $\hat{x}$ is minimized when A and B correspond to KLT. Prove this. Show that this MSE is $\sum_{k=m+1}^{N} \lambda_k$.

2.3. Show that the sum of variances of an N-point signal under orthogonal transformation is invariant.
Hint: Given $y = Ax$, where $(A^*)^T = A^{-1}, x = (x_1, x_2, \ldots, x_N)^T$ the data vector and $y = (y_1, y_2, \ldots, y_N)^T$ the transform vector. Show that

$$\sum_{k=1}^{N} \sigma_{x_{kk}}^2 = \sum_{k=1}^{N} \sigma_{y_{kk}}^2,$$

where $\sigma_{x_{kk}}^2$ and $\sigma_{y_{kk}}^2$ are the variances of x_k and y_k respectively.

2.4. The rate-distortion function R_D in bits/sample for a specified distortion D is defined as

$$R_D = \frac{1}{N}\sum_{k=1}^{N} max\left(0, \frac{1}{2}log_2 \tilde{\sigma}_{kk}^2/\theta\right)$$

$$D = \frac{1}{N}\sum_{k=1}^{N} min\left(\theta, \tilde{\sigma}_{kk}^2\right).$$

The parameter θ is determined from D. $\tilde{\sigma}_{kk}^2$ is the variance of the kth coefficient in any orthonormal transform domain. Show that among all unitary transforms, KLT yields the minimum R_D.

2.5. For a first-order Markov process defined by,

$$R = \rho^{|i-j|} = \begin{bmatrix} 1 & \rho & \rho^2 & \cdots & \rho^{N-1} \\ \rho & 1 & \rho & \cdots & \rho^{N-2} \\ \rho^2 & \rho & 1 & \cdots & \rho^{N-3} \\ \cdots & \cdots & \cdots & \cdots & \cdots \\ \rho^{N-1} & \rho^{N-2} & \rho^{N-3} & \cdots & 1 \end{bmatrix},$$

where ρ is the adjacent correlation coefficient, show that the eigenvalues are

$$\lambda_k = \frac{1-\rho^2}{1-2\rho\cos(\omega_k)+\rho^2}, \qquad k = 1, 2, \ldots, N,$$

where ω_k are the real positive roots of the transcendental equation

$$\tan(N\omega) = \frac{(1-\rho^2)\sin(\omega)}{\cos(\omega)-2\rho+\rho^2\cos(\omega)}, \qquad \text{for } N = even.$$

Similar result is valid for N odd. Show that the mkth element of the KLT Φ matrix is

$$\phi_{mk} = \sqrt{\frac{2}{N+\lambda_k}} \sin\left(\omega_k \left[m - \frac{(N+1)\pi)}{2}\right] + \frac{k\pi}{2}\right), \qquad m,k = 1,2,\ldots,N.$$

2.6. Obtain the eigenvalues and sketch the eigenvectors for an order-1 Markov process with ρ=0.9 and N = 16.
2.7. Repeat Problem 2.6. for ρ = 0.85.
2.8. Repeat Example 2.1 for the following 3×3 matrix

$$A = \begin{pmatrix} 2 & -1 & -1 \\ -1 & 2 & -1 \\ -1 & -1 & 2 \end{pmatrix}.$$

2.9. Derive (2.22).
2.10. Repeat Example 2.3 for the set of points shown in Figure 2.7.

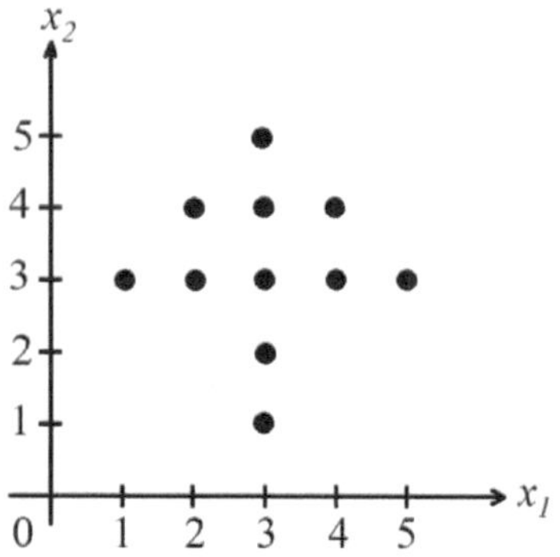

Figure 2.7: Position of pixels.

PROJECTS

P.2.1. In [452] a watermarking system with increasing concealing capacity using the KLT was proposed. Use similar ideas to implement a watermarking system for video sequences.
P.2.2. In [27] a method for removing Gaussian noise from digital images, based on the combination of the wavelet packet transform and the principal component analysis was proposed. Use this idea to denoise the coding blocks (CBs) at the input of the high efficiency video coding (HEVC) system.
P.2.3. In [231] the authors proposed an image enhancement algorithm for images compressed using the JPEG standard. The algorithm is based on a contrast measure defined within the discrete cosine transform (DCT) domain. Explore the possibility to implement a similar system based on KLT.
P.2.4. Implement a human face recognition system based on principal component analysis (PCA) [489].

P.2.5. In [345] the authors proposed two methods, based on KLT, for 4D reconstruction of cardiac gated single-photon emission computed tomography (SPECT) images. Explore the possibility to implement the two methods for positron emission tomography- (PET-) image reconstruction.

P.2.6. Use KLT for feature reduction intended for image segmentation [40].

P.2.7. Digital watermarking is a very common technique used against illicit copying and distribution of copyright material. Use KLT to implement a system for hiding information [452].

3 The Discrete Cosine Transform

INTRODUCTION

Transform coding has been widely used in image and video compression. There are several characteristics desirable for the purpose of data compression. Transforms are useful entities that encapsulate the (some/all) following characteristics:

Data decorrelation: The ideal transform completely decorrelates the data in a sequence/block; i.e., it packs the most amount of energy in the fewest number of coefficients. In this way, many coefficients can be discarded after quantization and prior to encoding. It is important to note that the transform operation itself does not achieve any compression. It aims at decorrelating the original data and compacting a large fraction of the signal energy into relatively few transform coefficients (see Chapter 2 on KLT).

Data-independent basis functions: Owing to the large statistical variations among data, the optimum transform usually depends on the data, and finding the basis functions of such transform is a computationally intensive task. This is particularly a problem if the data blocks are highly non-stationary, which necessitates the use of more than one set of basis functions to achieve high decorrelation. Therefore, it is desirable to trade optimum performance for a transform whose basis functions are data-independent.

Fast implementation: The number of operations required for an N-point transform is generally of the order $O(n^2)$. Some transforms have fast implementations, which reduce the number of operations to $O(n\ log_2 n)$. For a separable $n \times n$ 2D transform, performing the row and column 1D transforms successively reduces the number of operations from $O(n^4)$ to $O(2n^2\ log_2 n)$.

The discrete cosine transforms (DCT) and discrete sine transform (DST) are members of a family of sinusoidal unitary transforms. They are real, orthogonal, recursive and separable with fast algorithms for their computation. They have a great relevance to data compression. In discussing the DCT, we shall first consider the Fourier cosine transform (FCT), whose properties are well known. In this chapter, we treat the DCTs as discretized approximations of the continuous FCT. It would, however, be quite wrong. For in reality one has to deal with sample, measurements, and time instants. The continuum is merely an idealization to permit the use of calculus. Thus, in the remained of this chapter we will derive the various properties of the DCTs directly and attempt to derive properties similar to those possessed by the continuous FCT, which will serve as a convenient and familiar starting point.

THE FOURIER COSINE TRANSFORM

Definition: Given a function $x(t)$ for $-\infty < t < \infty$, its forward and inverse Fourier transform $F[\cdot]$, $F^{-1}[\cdot]$ respectively, subject to its existence condition for the integral, are given by,

$$X(\omega) \equiv F[x(t)] = \sqrt{\frac{1}{2\pi}} \int_{-\infty}^{\infty} x(t)e^{-j\omega t}dt \tag{3.1a}$$

$$x(t) \equiv F^{-1}[X(\omega)] = \sqrt{\frac{1}{2\pi}} \int_{-\infty}^{\infty} X(\omega)e^{j\omega t}d\omega. \tag{3.1b}$$

Here, $j = \sqrt{-1}$, $\omega = 2\pi f$; ω is the radian frequency and f the frequency in Hertz. If $x(t)$ is defined only for $t \geq 0$, we can construct a function $y(t)$ given by,

$$\begin{aligned} y(t) = &x(t) \quad t \geq 0 \\ &x(-t) \quad t \leq 0. \end{aligned}$$

$$\begin{aligned} F[y(t)] &= \sqrt{\frac{1}{2\pi}} \left[\int_{0}^{\infty} x(t)e^{-j\omega t}dt + \int_{-\infty}^{0} x(-t)e^{-j\omega t}dt \right] \\ &= \sqrt{\frac{1}{2\pi}} \int_{0}^{\infty} x(t)[e^{-j\omega t} + e^{j\omega t}]dt \\ &= \sqrt{\frac{2}{\pi}} \int_{0}^{\infty} x(t)\cos(\omega t)dt. \end{aligned} \tag{3.2}$$

We can now define this as the FCT of $x(t)$ given by,

$$X_c(\omega) \equiv F_c[x(t)] = \sqrt{\frac{2}{\pi}} \int_{0}^{\infty} x(t)\cos(\omega t)dt. \tag{3.3}$$

Note that $X_c(\omega)$ is an even function of ω. We can apply the Fourier inversion to (3.2) to obtain,

$$y(t) = x(t) \equiv F_c^{-1}[X_c(\omega)] = \sqrt{\frac{2}{\pi}} \int_{0}^{\infty} X_c(\omega)\cos(\omega t)d\omega \quad (t \geq 0). \tag{3.4}$$

Equations (3.3) and (3.4) define the FCT pair. Some of the properties are immediately clear.

PROPERTIES OF THE FCT

(a). *Inversion:*

$$F_c \equiv F_c^{-1}. \tag{3.5}$$

It is clear from (3.3) and (3.4) that

$$F_c^{-1}\{F_c[x(t)]\} = x(t) \qquad t \geq 0.$$

(b). *Linearity:*

$$F_c[\alpha x(t) + \beta y(t)] = \alpha X(\omega) + \beta Y(\omega), \tag{3.6}$$

where α and β are constants. F_c is clearly a linear operator.

(c). *Scaling in time:*

$$\begin{aligned} F_c[x(at)] &= \sqrt{\frac{2}{\pi}} \int_0^\infty x(at)\cos(\omega t)dt \\ &= a^{-1}\sqrt{\frac{2}{\pi}} \int_0^\infty x(r)\cos(\frac{\omega}{a}r)dr \\ &= \frac{1}{a}X_c\left(\frac{\omega}{a}\right) \qquad \text{for } a > 0, \end{aligned} \tag{3.7}$$

where $r = at$.

(d). *Shift in time:*

$$\begin{aligned} F_c[x(t-a)] &= \sqrt{\frac{2}{\pi}} \int_0^\infty x(t-a)\cos(\omega t)dt \\ &= \sqrt{\frac{2}{\pi}} \int_0^\infty x(r)\cos[\omega(r+a)]dr \\ &= \sqrt{\frac{2}{\pi}} \int_0^\infty x(r)\cos[\omega r + \omega a]dr \\ &= \sqrt{\frac{2}{\pi}} \left[\cos(\omega a)\int_0^\infty x(r)\cos(\omega r)dr - \sin(\omega a)\int_0^\infty x(r)\sin(\omega r)dr\right] \\ &= \cos(\omega a)F_c[x(t)] - \sin(\omega a)F_s[x(t)] \end{aligned} \tag{3.8a}$$

$$X_s(\omega) = F_s[x(t)] = \sqrt{\frac{2}{\pi}} \int_0^\infty x(t)\sin(\omega t)dt, \tag{3.8b}$$

where $r = (t-a)$ and F_s denotes the Fourier sine transform (FST). We also assumed that $x(t) = 0$ for $t < 0$.

(e). *Differentiation in time:*

$$\begin{aligned} F_c\left[\frac{d}{dt}x(t)\right] &= \sqrt{\frac{2}{\pi}} \int_0^\infty \left[\frac{d}{dt}x(t)\right]\cos(\omega t)dt \\ &= \sqrt{\frac{2}{\pi}} \left[x(t)cos(\omega t) + \omega\int x(t)\sin(\omega t)dt\right]_0^\infty. \end{aligned}$$

Assuming that $x(t)$ vanishes as t tends to infinity,

$$= -\sqrt{\frac{2}{\pi}}x(0) + \omega F_s[x(t)]. \tag{3.9}$$

Transforms of higher order derivatives may be obtained in a similar way (see Problem 3.1.).

(f). *The convolution property:* The convolution of the functions $x(t)$ and $y(t)$ is given by,

$$x(t) * y(t) = \int_0^\infty x(t-\tau)y(\tau)d\tau. \tag{3.10}$$

Taking the FST,

$$F_c[x(t) * y(t)] = \sqrt{\frac{2}{\pi}}\int_0^\infty \int_0^\infty x(t-\tau)y(\tau)\cos(\omega t)d\tau dt$$

$$F_c[x(t) * y(t)] = \sqrt{\frac{\pi}{2}}\left[X_c(\omega)Y_c(\omega) - X_s(\omega)Y_s(\omega)\right]. \tag{3.11}$$

Example 3.1. *The unit rectangular pulse (Heaviside unit step function) is defined as,*

$$x(t) = U(t) - U(t-1), \text{ where } U(t) = \begin{cases} 0 & \text{for } t < 0, \\ 1 & \text{for } t > 0. \end{cases}$$

The FCT is given by,

$$X_c(\omega) = \sqrt{\frac{2}{\pi}}\int_0^1 \cos\omega t dt = \sqrt{\frac{2}{\pi}}\left(\frac{\sin\omega}{\omega}\right). \tag{3.12}$$

This result is also known as the sinc function.

Example 3.2. *The exponential function is defined as,*

$$x(t) = e^{-j\alpha t}.$$

For $a \geq 0$, the FCT is given by,

$$X_c(\omega) = \sqrt{\frac{2}{\pi}}\int_0^\infty e^{-j\alpha t}\cos\omega t dt = \sqrt{\frac{2}{\pi}}\left(\frac{\alpha}{\alpha^2+\omega^2}\right). \tag{3.13}$$

Notice that the result is similar to the Laplace transform except for the scaling factor.

Example 3.3. *The Bessel function of the first kind,*

$$x(t) = J_0(\alpha t),$$

where $\alpha \geq 0$ and $J_0(\alpha t)$ denotes the zero-order Bessel function of the first kind.

The FCT is given by

$$X_c(\omega) = \sqrt{\frac{2}{\pi}} \int_0^{\infty} J_0(\alpha t) \cos \omega t dt = \begin{cases} \sqrt{\frac{2}{\pi}} \left(\frac{1}{\sqrt{\alpha^2 - \omega^2}} \right) & \textit{for } 0 < \omega < \alpha, \\ \infty & \textit{for } \omega = \alpha, \\ 0 & \textit{for } \omega > \alpha. \end{cases} \tag{3.14}$$

In the next section, we derive properties for the discrete cosine transform in parallel to the above properties for the Fourier cosine transform.

DEFINITIONS

From (3.3) we can see that the Fourier cosine transform has a kernel given by,

$$K_c(\omega, t) = \cos(\omega t), \tag{3.15a}$$

up to a normalization constant. (3.5) shows that this kernel is involutory if proper normalization is taken into account. Let $\omega_m = 2\pi m \delta f$ and $t_n = n \delta t$ be the sampled angular frequency and time, where δf and δt represent the unit sample intervals for frequency and time, respectively; m and n are integers. (3.15a) may now be written as,

$$K_c(\omega_m, t_n) = K_c(2\pi m \delta f, n \delta t) = \cos(2\pi m \delta f n \delta t) = K_c(m, n). \tag{3.15b}$$

If we let $\delta f \delta t = 1/(2N)$ with N an integer, we have,

$$K_c(m, n) = \cos\left(\frac{\pi mn}{N}\right). \tag{3.16}$$

Equation (3.16) represents a discretized Fourier cosine kernel. It is tempting to discretize the various properties from (a). to (f). in the previous section. However, it is far simpler, both mathematically and conceptually, to regard the discrete kernel (3.16) as elements in an $(N+1) \times (N+1)$ transform matrix denoted by $[M]$. Therefore, the mnth element of this matrix is,

$$[M]_{mn} = \cos\left(\frac{\pi mn}{N}\right) \qquad m, n = 0, 1, \ldots, N. \tag{3.17}$$

When we apply $[M]$ to a column vector $x = (x_0, x_1, \ldots, x_N)^T$ we obtain the vector $X = (X_0, X_1, \ldots, X_N)^T$ such that,

$$X = [M]x, \tag{3.18}$$

where,

$$X_m = \sum_{n=0}^{N} x_n \cos\left(\frac{\pi mn}{N}\right) \qquad m = 0, 1, \ldots, N. \tag{3.19}$$

The vector **x** is said to have undergone a discrete transform. Equation (3.19) was first reported by Kitajima in 1980 [255] and named the symmetric cosine transform (SCT).

Since inversion is part of transform processing in general, non-singular matrices are usually preferred. However, there are discrete transforms that are singular and non-invertible. An example of this is the rapid transform (RT) proposed by Reitboeck and Brody in [395]. In practice, if the non-singular matrix is real and orthogonal, its inverse is easily obtained as its transpose. Such unitary transform matrices (with the proper normalizations) have a preeminent place in the digital signal-processing field. We shall now present the definitions for the four discrete cosine transforms as classified by Wang [522].

(1). DCT-I. Defined for the order $N+1$.

$$\left[C_{N+1}^{I}\right]_{mn} = \sqrt{\tfrac{2}{N}}\left[k_m k_n \cos\left(\tfrac{\pi mn}{N}\right)\right] \qquad m, n = 0, 1, \ldots, N.$$

(2). DCT-II. Excellent energy compaction property and best approximation to KLT.

$$\left[C_N^{II}\right]_{mn} = \sqrt{\tfrac{2}{N}}\left[k_m \cos\left(\tfrac{m(n+0.5)\pi}{N}\right)\right] \qquad m, n = 0, 1, \ldots, N-1.$$

(3). DCT-III. Inverse of DCT-II.

$$\left[C_N^{III}\right]_{mn} = \sqrt{\tfrac{2}{N}}\left[k_n \cos\left(\tfrac{n(m+0.5)\pi}{N}\right)\right] \qquad m, n = 0, 1, \ldots, N-1.$$

(4). DCT-IV. Fast implementation of lapped orthogonal transform for the efficient transform-subband coding.

$$\left[C_N^{IV}\right]_{mn} = \sqrt{\tfrac{2}{N}}\left[\cos\left(\tfrac{(m+0.5)(n+0.5)\pi}{N}\right)\right] \qquad m, n = 0, 1, \ldots, N-1.$$

where,

$$k_p = \begin{cases} 1 & \text{if } p \neq 0 \text{ or } N, \\ \frac{1}{\sqrt{2}} & \text{if } p = 0 \text{ or } N. \end{cases} \tag{3.20}$$

Here, we note that DCT-II is the discrete cosine transform first developed by Ahmed, Natarajan and Rao [7]. The DCT-III is obviously the transpose of the DCT-II, and the DCT-IV is the shifted version of DCT-I. The basis functions for DCT-II for $N = 16$ are shown in Figure 3.1.

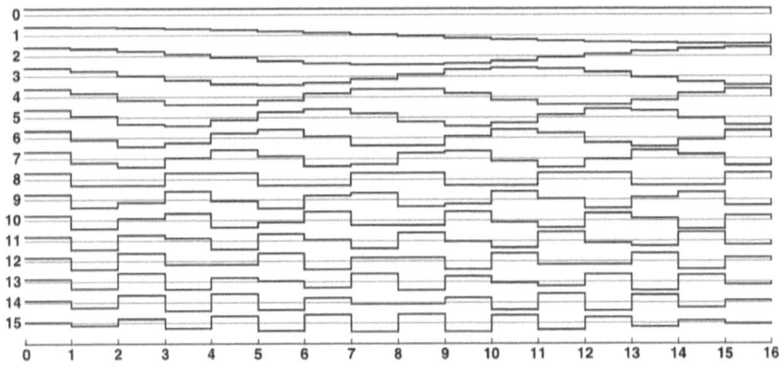

Figure 3.1: DCT-II Basis functions for $N = 16$.

THE UNITARITY OF DCTs

In the original derivations of these discrete transforms [7], [255], a diagonalization problem is considered and the unitarity of the transform matrices is assured, since they are the similarity transform matrices in the diagonalization problem.

Recall that the inner product of two N-dimensional real vectors $\mathbf{g}_k$ and $\mathbf{g}_m$ is:

$$< g_k, g_m >= \sum_{n=1}^{N} g_{nk} g_{nm}, \tag{3.21}$$

where $\mathbf{g}_k = [g_{1k}, g_{2k}, \dots, g_{Nk}]^T$. $\mathbf{g}_k$ and $\mathbf{g}_m$ are said to be orthogonal if $< g_k, g_m >= 0$ for $k \neq m$. In addition, if $< g_k, g_m >= 1$ for $k = m$, the vectors are said to be normalized, and the corresponding real transform matrix is then unitary. Instead of deriving the transforms by considering the diagonalization problem, we simply use the inner product concept (3.21) and investigate the unitary property of the discrete transforms directly.

(1). DCT-I. Let $\mathbf{g}_k$ denote the kth column vector in the $(N+1)$ by $(N+1)$ transform matrix C^I_{N+1}, then,

$$\begin{aligned} < g_k, g_m >&= \frac{2}{N} \sum_{n=0}^{N} k_n k_k \cos\left(\frac{nk\pi}{N}\right) k_n k_m \cos\left(\frac{nm\pi}{N}\right) \\ &= \left(\frac{k_k k_n}{N}\right) \Bigg\{ 1 + (-1)^{m+k} \\ &+ \sum_{n=1}^{N-1} \left(\cos\left[\frac{(m+k)n\pi}{N}\right] + \cos\left[\frac{(m-k)n\pi}{N}\right] \right) \Bigg\} \end{aligned}$$

$$
\begin{aligned}
&= \left(\frac{k_k k_n}{N}\right)\left\{\sum_{n=0}^{N-1}\cos\left[\frac{(m-k)n\pi}{N}\right]\right. \\
&\quad \left. + \sum_{n=1}^{N}\cos\left[\frac{(m+k)n\pi}{N}\right]\right\} \\
&= \left(\frac{k_k k_n}{N}\right)\mathrm{Re}\left\{\sum_{n=0}^{N-1} W_{2N}^{-n(m-k)}\right. \\
&\quad \left. + \sum_{n=1}^{N} W_{2N}^{-n(m+k)}\right\},
\end{aligned}
\tag{3.22}
$$

where we have defined W_{2N} as the primitive $2N$th root of 1, i.e.,

$$
W_{2N} = \exp\left(-\frac{j\pi}{N}\right) = \cos\left(\frac{\pi}{N}\right) - j\sin\left(\frac{\pi}{N}\right). \tag{3.23}
$$

To examine the unitary property, we consider,

$$
\begin{aligned}
\sum_{n=0}^{N-1}\left(W_{2N}^{-p}\right)^n &= \frac{1-W_{2N}^{-Np}}{1-W_{2N}^{-p}}, \qquad \text{where } p=(m-k) \\
&= \left[2\left(1-\cos\left(\frac{p\pi}{N}\right)\right)\right]^{-1} \\
&\quad \times\left\{1-W_{2N}^{-Np}-W_{2N}^{-p}+W_{2N}^{-(N-1)p}\right\}.
\end{aligned}
\tag{3.24}
$$

Similarly, for $q = m+k$,

$$
\begin{aligned}
\sum_{n=1}^{N}\left(W_{2N}^{-q}\right)^n &= \left[2\left(1-\cos\left(\frac{q\pi}{N}\right)\right)\right]^{-1} \\
&\quad \times\left\{W_{2N}^{-q}-W_{2N}^{-(N+1)q}-1+W_{2N}^{-Nq}\right\}.
\end{aligned}
\tag{3.25}
$$

Thus for $m \neq k$ and $p \neq 0$ we have

$$
Re\left\{\sum_{n=0}^{N-1} W_{2N}^{-np}\right\} = \frac{\left[1-(-1)^p\right]\left[1-\cos\left(\frac{p\pi}{N}\right)\right]}{2\left[1-\cos\left(\frac{p\pi}{N}\right)\right]} \tag{3.26}
$$

and

$$Re\left\{\sum_{n=1}^{N} W_{2N}^{-nq}\right\} = \frac{-\left[1-(-1)^q\right]\left[1-\cos\left(\frac{q\pi}{N}\right)\right]}{2\left[1-\cos\left(\frac{q\pi}{N}\right)\right]}. \tag{3.27}$$

Combining these two expressions, we see that since p and q differ by $2k$

$$<\mathbf{g}_k, \mathbf{g}_m> = 0 \text{ when } k \neq m \text{ (i.e., } p \neq 0). \tag{3.28}$$

For $k = m$, $k \neq 0$ or N, (3.22) reduces to

$$<\mathbf{g}_k, \mathbf{g}_m> = \frac{1}{N} Re\left\{\sum_{n=0}^{N-1} 1 + \sum_{n=1}^{N}\left(W_{2N}^{-2k}\right)^n\right\} = 1. \tag{3.29}$$

For $k = m$, $k = 0$ or N, we have

$$<\mathbf{g}_k, \mathbf{g}_m> = \frac{1}{2N} Re\left\{\sum_{0}^{N-1} 1 + \sum_{n=1}^{N} 1\right\} = 1 \tag{3.30}$$

Combining (3.28), (3.29) and (3.30), we can represent the unitarity property of DCT-I by,

$$<g_k, g_m> = \delta_{km}, \tag{3.31}$$

where,

$$\delta_{km} = \begin{cases} 1 & \text{for } m = k, \\ 0 & \text{for } m \neq k. \end{cases}$$

Thus

$$[C_{N+1}^{I}]^{-1} = [C_{N+1}^{I}]^{T} = [C_{N+1}^{I}]. \tag{3.32}$$

The procedure used for demonstrating the unitarity of DCT-I can be applied to the other DCTs with little or no modifications (see Problem 3.3.). We shall simply state the results:

(2). DCT-II:

$$[C_N^{II}]^{-1} = [C_N^{II}]^{-T} = [C_N^{III}]. \tag{3.33}$$

(3). DCT-III:

$$[C_N^{III}]^{-1} = [C_N^{III}]^{-T} = [C_N^{II}]. \tag{3.34}$$

(4). DCT-IV:

$$[C_N^{IV}]^{-1} = [C_N^{IV}]^{-T} = [C_N^{IV}]. \tag{3.35}$$

We note here that only DCT-I and DCT-IV are involutary, and thus property (3.5) of the FCT is retained only by these two DCTs. Therefore, the family of forward and inverse DCTs can be defined as follows:

(1). DCT-I:

Forward:

$$X_m^{C(1)} = \sqrt{\tfrac{2}{N}} k_m \textstyle\sum_{n=0}^{N} k_n x_n \cos\left(\tfrac{mn\pi}{N}\right), \qquad m = 0, 1, \ldots, N;$$

Inverse:

$$x_n = \sqrt{\tfrac{2}{N}} k_n \textstyle\sum_{m=0}^{N} k_m X_m^{C(1)} \cos\left(\tfrac{mn\pi}{N}\right), \qquad n = 0, 1, \ldots, N;$$

(2). DCT-II:

Forward:

$$X_m^{C(2)} = \sqrt{\tfrac{2}{N}} k_m \textstyle\sum_{n=0}^{N-1} x_n \cos\left(\tfrac{(2n+1)m\pi}{2N}\right), \qquad m = 0, 1, \ldots, N-1;$$

Inverse:

$$x_n = \sqrt{\tfrac{2}{N}} \textstyle\sum_{m=0}^{N-1} k_m X_m^{C(2)} \cos\left(\tfrac{(2n+1)m\pi}{2N}\right), \qquad n = 0, 1, \ldots, N-1;$$

(3). DCT-III:

Forward:

$$X_m^{C(3)} = \sqrt{\tfrac{2}{N}} \textstyle\sum_{n=0}^{N-1} k_n x_n \cos\left(\tfrac{(2m+1)n\pi}{2N}\right), \qquad m = 0, 1, \ldots, N-1;$$

Inverse:

$$x_n = \sqrt{\tfrac{2}{N}} k_n \textstyle\sum_{m=0}^{N-1} X_m^{C(3)} \cos\left(\tfrac{(2m+1)n\pi}{2N}\right), \qquad n = 0, 1, \ldots, N-1;$$

(4). DCT-IV:

Forward:

$$X_m^{C(4)} = \sqrt{\tfrac{2}{N}} \textstyle\sum_{n=0}^{N-1} x_n \cos\left(\tfrac{(2m+1)(2n+1)\pi}{4N}\right), \qquad m = 0, 1, \ldots, N-1;$$

Inverse:

$$x_n = \sqrt{\frac{2}{N}} \sum_{m=0}^{N-1} X_m^{C(4)} \cos\left(\frac{(2m+1)(2n+1)\pi}{4N}\right), \qquad n = 0, 1, \ldots, N-1;$$

where x_n is the input data, $X_m^{C(i)}$ is the transform sequence corresponding to DCT-i with $i = 1, 2, 3, 4$ and,

$$k_p = \begin{cases} \sqrt{\frac{1}{2}}, & p = 0 \text{ or } p = N, \\ 1, & \text{otherwise.} \end{cases} \tag{3.36}$$

Notice that the normalization factor $\sqrt{\frac{2}{N}}$ appears in both the forward and the inverse transforms and can be merged as $2/N$, and moved either to the forward or the inverse transform matrix. For example, the DCT-II pair can also be defined as,

Forward:

$$X_m^{C(2)} = \frac{2}{N} k_m \sum_{n=0}^{N-1} x_n \cos\left(\frac{(2n+1)m\pi}{2N}\right), \qquad m = 0, 1, \ldots, N-1;$$

Inverse:

$$x_n = \sum_{m=0}^{N-1} k_m X_m^{C(2)} \cos\left(\frac{(2n+1)m\pi}{2N}\right), \qquad n = 0, 1, \ldots, N-1; \tag{3.37}$$

It is evident that the terms $X_m^{C(2)}$ in (3.36) and (3.37) are related by the scale factor $\sqrt{\frac{2}{N}}$. By merging these normalization factors, the family of DCT matrices are no longer orthonormal. They are, however, orthogonal. The linearity property, as expressed by (3.6), for the Fourier cosine transform is easily extended to the DCTs. Since matrix multiplication is a linear operation, i.e.,

$$[\mathbf{M}]\{\alpha\mathbf{g} + \beta\mathbf{f}\} = \alpha[\mathbf{M}]\mathbf{g} + \beta[\mathbf{M}]\mathbf{f},$$

For the matrix $[\mathbf{M}]$, constants α and β, and vectors $\mathbf{g}$ and $\mathbf{f}$, all four DCTs are linear transforms, retaining property (3.6).

SCALING IN TIME

Recall that in the discretization of the Fourier cosine transform, we consider,

$$\delta f \delta t = \left(\frac{1}{2N}\right), \tag{3.38}$$

or

$$\delta f = \left(\frac{1}{2N}\right) / \delta t. \tag{3.39}$$

Since the DCTs deal with discrete sample points, a scaling in time has no effect in the transform, except in changing the unit frequency interval in the transform domain. Thus, as δt changes to $a\delta t$, δf changes to $\delta f/a$, provided N remains constant. Thus, a scaling in time leads to a scaling in frequency without scaling of the transform, and property (3.7) is retained, except for the $\frac{1}{a}$ factor, which is absent in the DCTs case.

Equation (3.39) may also be interpreted to produce the frequency resolution for a given sample length. Combining $N\delta t$ as T, we have

$$\delta f = \left(\frac{1}{2T}\right), \tag{3.40}$$

where T is the duration in time for a sequence of length N.

SHIFT IN TIME

Let $\mathbf{x} = [x_0, \ldots, x_N]^T$ and $\mathbf{x}_+ = [x_1, \ldots, x_{N+1}]^T$ be two sampled data vectors of dimension $N+1$. $\mathbf{x}_+$ is the data vector $\mathbf{x}$ shifted by one sample point. We shall detail here the shift property for the DCTs, corresponding to property (2.8) for the Fourier cosine transform. Again, we consider first the DCT-I. Recall that,

$$k_p = \begin{cases} 1 & \text{if } p \neq 0 \text{ or } N, \\ \frac{1}{\sqrt{2}} & \text{if } p = 0 \text{ or } N. \end{cases} \tag{3.41}$$

(1). DCT-I: Using (3.41), we can define the DCT-I of both $\mathbf{x}$ and $\mathbf{x}_+$ as

$$X_m^{C(1)} = \sqrt{\tfrac{2}{N}} \textstyle\sum_{n=0}^{N} k_m k_n \cos\left(\tfrac{mn\pi}{N}\right) x_n$$

and

$$X_{m+}^{C(1)} = \sqrt{\frac{2}{N}} \sum_{n=0}^{N} k_m k_n \cos\left(\frac{mn\pi}{N}\right) x_{n+1}, \tag{3.42}$$

where

$$\begin{aligned} \mathbf{X}^{C(1)} &= [X_0^{C(1)}, X_1^{C(1)}, \ldots, X_N^{C(1)}]^T \quad \text{and} \\ \mathbf{X}_+^{C(1)} &= [X_{0+}^{C(1)}, X_{1+}^{C(1)}, \ldots, X_{N+}^{C(1)}]^T, \end{aligned}$$

are, respectively, the DCT-I of the vectors $\mathbf{x}$ and $\mathbf{x}_+$. Taking the mth element of $\mathbf{X}_+^{C(1)}$ and changing the summation index n we obtain

$$\begin{aligned} X_{m+}^{C(1)} = \sqrt{\frac{2}{N}} \Bigg\{ & k_m \cos\left(\frac{m\pi}{N}\right) \sum_{n=0}^{N} k_n \cos\left[\frac{m(n+1)\pi}{N}\right] x_{n+1} \\ & + k_m \sin\left(\frac{m\pi}{N}\right) \sum_{n=0}^{N} k_n \sin\left[\frac{m(n+1)\pi}{N}\right] x_{n+1} \Bigg\}. \end{aligned} \tag{3.43}$$

We shall simplify this to

$$X_{m+}^{C(1)} = \sqrt{\frac{2}{N}} k_m \left\{ \cos\left(\frac{m\pi}{N}\right) S_1 + \sin\left(\frac{m\pi}{N}\right) S_2 \right\}, \tag{3.44}$$

where S_1 and S_2 can be further simplified

$$\begin{aligned} S_1 &= \sum_{n=0}^{N} k_n \cos\left[\frac{m(n+1)\pi}{N}\right] x_{n+1} \\ &= \sqrt{\frac{1}{2}} \cos\left(\frac{m\pi}{N}\right) x_1 + \sum_{n=2}^{N} \cos\left(\frac{mn\pi}{N}\right) x_n \\ &+ \sqrt{\frac{1}{2}} (-1)^m \cos\left(\frac{m\pi}{N}\right) x_{N+1}. \end{aligned}$$

The second term in the above expression can be made into the DCT-I of the vector **x** by introducing the proper term. With this, we obtain for S_1 the following result:

$$\begin{aligned} S_1 = \sqrt{\frac{N}{2}} \frac{X_m^{C(1)}}{k_m} - \frac{x_0}{\sqrt{2}} + \left(\frac{1}{\sqrt{2}} - 1\right) \cos\left(\frac{m\pi}{N}\right) x_1 \\ + \left(1 - \frac{1}{\sqrt{2}}\right) (-1)^m x_N + \left(\frac{1}{\sqrt{2}}\right) (-1)^m \cos\left(\frac{m\pi}{N}\right) x_{N+1}. \end{aligned} \tag{3.45}$$

Similarly, for S_2 we obtain

$$\begin{aligned} S_2 = \sqrt{\frac{N}{2}} X_m^{S(1)} + \left(\frac{1}{\sqrt{2}} - 1\right) \sin\left(\frac{m\pi}{N}\right) x_1 \\ + \frac{1}{\sqrt{2}} (-1)^m \sin\left(\frac{m\pi}{N}\right) x_{N+1}, \end{aligned} \tag{3.46}$$

where we have defined

$$X_m^{S(1)} = \sqrt{\frac{2}{N}} \sum_{n=1}^{N-1} \sin\left(\frac{mn\pi}{N}\right) x_n. \tag{3.47}$$

which is the discrete sine transform of type I (DST-I) (see Jain [220]). Substituting (3.45) and (3.46) into (3.44), we finally obtain the shift property for DCT-I:

$$\begin{aligned} X_{m+}^{C(1)} &= \cos\left(\frac{m\pi}{n}\right) X_m^{C(1)} + k_m \sin\left(\frac{m\pi}{N}\right) X_m^{S(1)} \\ &+ \sqrt{\frac{2}{N}} k_m \left\{ \left(\frac{-1}{\sqrt{2}}\right) \cos\left(\frac{m\pi}{N}\right) x_0 + \left(\frac{1}{\sqrt{2}} - 1\right) x_1 \right. \\ &\left. + (-1)^m \left(1 - \frac{1}{\sqrt{2}}\right) \cos\left(\frac{m\pi}{N}\right) x_N + (-1)^m \frac{x_{N+1}}{\sqrt{2}} \right\}. \end{aligned} \tag{3.48}$$

We note that the first two terms are similar to the expression in (3.8). The remaining terms make the shift property for the DCTs much more complicated compared with that for the Fourier cosine transform. However, it should be noted that when instantaneous DCTs have to be carried out on a continuous incoming data stream, (3.48) does represent a possible way of updating the instantaneous DCTs without having to perform complete DCTs at every instant. Such a situation may exist, for example, in the use of adaptive filtering in the transform domain. Similar relations are available for the remaining DCTs (see Problem 3.4.). We list them here without detailed derivation (see Yip and Rao [563]).

(2). DCT-II:

$$\begin{aligned} X_{m+}^{C(2)} &= \cos\left(\frac{m\pi}{N}\right) X_m^{C(2)} + \sin\left(\frac{m\pi}{N}\right) X_{m-}^{S(2)} \\ &\quad + \sqrt{\frac{2}{N}} k_m \left[(-1)^m x_N - x_0\right] \cos\left(\frac{m\pi}{2N}\right), \end{aligned} \tag{3.49}$$

where

$$\mathbf{X}^{C(2)} = \left[C_N^{II}\right] \mathbf{x}, \qquad \mathbf{X}_+^{C(2)} = \left[C_N^{II}\right] \mathbf{x}_+$$

are the DCT-II for the data vector $\mathbf{x}$ and its time-shifted version $\mathbf{x}_+$, respectively. Also, $X_{m-}^{S(2)}$ is used to denote the mth transform output of the DST (type II) for the sequence $x_0, x_1, \ldots, x_{N-1}$, as defined by Wang (see Wang [522]).

(3). DCT-III:

$$\begin{aligned} X_{m+}^{C(3)} &= \cos\left[\frac{(2m+1)\pi}{2N}\right] X_m^{C(3)} + \sin\left[\frac{(2m+1)\pi}{2N}\right] X_{m+1}^{S(3)} \\ &\quad + \sqrt{\frac{2}{N}} \left\{ \left(\frac{1}{\sqrt{2}} - 1\right) x_1 - \cos\left[\frac{(2m+1)\pi}{2N}\right] \frac{x_0}{\sqrt{2}} \right. \\ &\quad \left. + (-1)^m \left(1 - \frac{1}{\sqrt{2}}\right) \sin\left[\frac{(2m+1)\pi}{2N}\right] x_N \right\}, \end{aligned} \tag{3.50}$$

where

$$\mathbf{X}^{C(3)} = \left[C_N^{III}\right] \mathbf{x}, \qquad \mathbf{X}_+^{C(3)} = \left[C_N^{III}\right] \mathbf{x}_+$$

are the DCT-III of $\mathbf{x}$ and $\mathbf{x}_+$, respectively. Also, $\mathbf{X}^{S(3)}$ is the DST-III of the data vector (see (3.52)).

(4). DCT-IV:

$$\begin{aligned} X_{m+}^{C(4)} &= \cos\left[\frac{(2m+1)\pi}{2N}\right] X_m^{C(4)} + \sin\left[\frac{(2m+1)\pi}{2N}\right] X_m^{S(4)} \\ &\quad + \sqrt{\frac{2}{N}} \left\{ -\cos\left[\frac{(2m+1)\pi}{4N}\right] x_0 + (-1)^m \sin\left[\frac{(2m+1)\pi}{4N}\right] x_N \right\}, \end{aligned} \tag{3.51}$$

where

$$\mathbf{X}^{C(4)} = \left[C_N^{IV}\right]\mathbf{x}, \qquad \mathbf{X}_+^{C(4)} = \left[C_N^{IV}\right]\mathbf{x}_+$$

are the DCT-IV of the data vector and its time-shifted version, respectively. Also, $\mathbf{X}^{S(4)}$ is the DST-IV of the data vector (see (3.52)).

Equations (3.48) through (3.51) delineate the shift property of the four types of DCTs. All of them involve the corresponding discrete sine transforms, which are defined as follows [522]:

$$\text{DST-I:} \quad \left[S_{N-1}^{I}\right]_{mn} = \sqrt{\frac{2}{N}}\left\{\sin\left[\frac{mn\pi}{N}\right]\right\}, \quad m,n = 1,2,\ldots,N-1.$$

$$\text{DST-II:} \quad \left[S_{N}^{II}\right]_{mn} = \sqrt{\frac{2}{N}}\left\{k_m \sin\left[\frac{m(n-0.5)\pi}{N}\right]\right\}, \quad m,n = 1,2,\ldots,N.$$

$$\text{DST-III:} \quad \left[S_{N}^{III}\right]_{mn} = \sqrt{\frac{2}{N}}\left\{k_n \sin\left[\frac{(m-0.5)n\pi}{N}\right]\right\}, \quad m,n = 1,2,\ldots,N.$$

$$\text{DST-IV:} \quad \left[S_{N}^{IV}\right]_{mn} = \sqrt{\frac{2}{N}}\left\{\sin\left[\frac{(m+0.5)(n+0.5)\pi}{N}\right]\right\}, \quad m,n = 1,2,\ldots,N-1. \qquad (3.52)$$

where

$$k_p = \begin{cases} \frac{1}{\sqrt{2}} & p = N, \\ 1 & \text{otherwise.} \end{cases}$$

Similar to the DCT, the forward and inverse transforms for the DST family can be explicitly expressed. Also, the DST relations corresponding to (3.32) through (3.35) are [522]:

$$\begin{aligned}
&[\text{DST-I}]\text{: } [S_{N-1}^{I}]^{-1} = [S_{N-1}^{I}]^{T} = [S_{N-1}^{I}], \\
&[\text{DST-II}]\text{: } [S_{N}^{II}]^{-1} = [S_{N}^{II}]^{T} = [S_{N}^{III}], \\
&[\text{DST-III}]\text{: } [S_{N}^{III}]^{-1} = [S_{N}^{III}]^{T} = [S_{N}^{II}], \\
&[\text{DST-IV}]\text{: } [S_{N}^{IV}]^{-1} = [S_{N}^{IV}]^{T} = [S_{N}^{IV}].
\end{aligned} \qquad (3.53)$$

As in the case of the DCTs, the normalization factor $\sqrt{2/N}$, which appears in both the forward and the inverse transforms in (3.52) can be merged as $(2/N)$ and

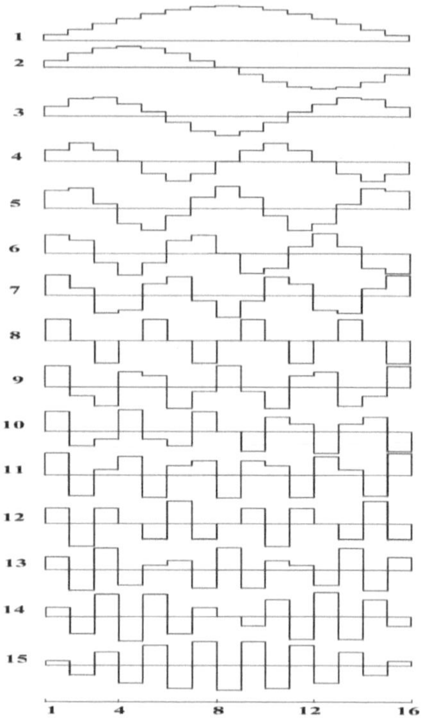

Figure 3.2: DST-I basis functions for $N = 15$.

moved to either the forward or the inverse transform. The basis functions of DST-I for $N = 15$ are shown in Fig. 3.2.

It is clear that the shift properties for the DSTs similar to those for the DCTs can be developed (see Problem 3.5.). We conclude this section by noting that although the shift property is derived for a unit time shift, transforms of more general time shifts can be obtained by the repeated application of the appropriate formulae.

THE DIFFERENCE PROPERTY

Consider a sequence of data points, consisting of differences between adjacent samples, i.e.,

$$d_n = x_{n+1} - x_n \qquad 0, 1, \ldots, N, \tag{3.54a}$$

$$\mathbf{d} = \mathbf{x}_+ - \mathbf{x}. \tag{3.54b}$$

Equation (3.54b) is expressed in terms of the data vectors. It is clear that the DCT of d_n can be obtained by applying the shift property derived in the last section.

(1). DCT-I:

$$\mathbf{D}^{C(1)} = \mathbf{X}_{+}^{C(1)} - \mathbf{X}^{C(1)}, \tag{3.55}$$

where $\mathbf{D}^{C(1)}$ denotes the DCT-I of the difference sequence (3.54b). Other DCTs are similarly obtained,

(2). DCT-II:

$$\mathbf{D}^{C(2)} = \mathbf{X}_{+}^{C(2)} - \mathbf{X}^{C(2)}, \tag{3.56}$$

(3). DCT-III:

$$\mathbf{D}^{C(3)} = \mathbf{X}_{+}^{C(3)} - \mathbf{X}^{C(3)}, \tag{3.57}$$

(4). DCT-IV:

$$\mathbf{D}^{C(4)} = \mathbf{X}_{+}^{C(4)} - \mathbf{X}^{C(4)}. \tag{3.58}$$

THE CONVOLUTION PROPERTY

For discrete finite sequences x_n and y_n, two types of convolutions are defined (see Elliott and Rao [122]). The circular convolution a_i is defined for sequences x_n and y_n which are both periodic with a period of N. a_i is given by

$$a_i = x_n \circledast y_n = \frac{1}{\sqrt{N}} \sum_{m=0}^{N-1} x_m y_{i-m} \qquad i = 0, 1, \ldots, N-1, \tag{3.59}$$

where $\circledast$ denotes circular convolution. The linear convolution b_i for two nonperiodic sequences x_n and y_n, of lengths L and M, respectively, is defined by

$$b_i = x_n * y_n = \frac{1}{\sqrt{N}} \sum_{n=0}^{N-1} x_n y_{i-n} \qquad i = 0, 1, \ldots, N-1, \tag{3.60}$$

where $N \geq L + M - 1$. By augmenting x_n and y_n with zeros to equal lengths (N), b_i can be obtained as a portion of a circular convolution of the augmented sequences. In this section, we derive the convolution property for DCT-II only. We shall see that, because DCT-II can be easily related to DFT, the convolution property of DFT can be borrowed to obtain a comparable result for DCT-II. We shall also see that instead of the conventional convolution as expressed by (3.59) and (3.60), the convolution of even extension of given sequences must be considered.

Consider the sequence $f_n,\ n = 0,1,\ldots,N-1$. Construct a $2N$ sequence defined as

$$\begin{aligned} \hat{f}_n &= f_n & n &= 0,1,\ldots,N-1 \\ &= f_{2N-n-1} & &= N,N+1,\ldots,2N-1. \end{aligned} \tag{3.61}$$

Thus, $\hat{f}_n$ is an even extension of f_n. Now, the $2N$-point DFT of this sequence is given by

$$\hat{F}_{F,k} = \frac{1}{\sqrt{2N}} \sum_{n=0}^{2N-1} \hat{f}_n \exp\left(-\frac{j2\pi kn}{2N}\right). \tag{3.62}$$

Using the symmetry in (3.61), we can now rewrite (3.62) as

$$\hat{F}_{F,k} = \frac{1}{\sqrt{2N}} \sum_{n=0}^{N-1} f_n \left\{ \exp\left(-\frac{j2\pi kn}{2N}\right) + \exp\left(\frac{j2\pi k(n+1)}{2N}\right) \right\}, \tag{3.63}$$

Let us now introduce a phase shift and a weight factor into (3.63),

$$k_k \hat{F}_{F,k} \exp\left(-\frac{j\pi k}{2N}\right) = \sqrt{\frac{2}{N}} \sum_{n=0}^{N-1} k_k f_n \cos\left\{\frac{k(n+0.5)\pi}{N}\right\}, \tag{3.64}$$

where

$$k_m = \begin{cases} \frac{1}{\sqrt{2}} & \text{for } m = 0, \\ 1 & \text{for } m = 1,2,\ldots,N-1\ , \\ 0 & \text{otherwise}\ . \end{cases}$$

Using $F_k^{C(2)}$ to denote the DCT-II for the sequence we have

$$F_k^{C(2)} = k_k \exp\left(-\frac{j\pi k}{2N}\right) \hat{F}_{F,k} \qquad k = 0,1,\ldots,N-1. \tag{3.65}$$

Equation (3.65) provides the relation between the DCT-II of the unappended sequence f_n and the Fourier transform of the appended sequence $\hat{f}_n$. It is not difficult to show that the sequence f_n can be recovered from the transformed sequence $F_k^{C(2)}$, as a result we have

$$f_n = \sqrt{\frac{2}{N}} Re\left\{ \sum_{k=0}^{2N-1} k_k F_k^{C(2)} \exp\left(\frac{j\pi k}{2N}\right) \exp\left(\frac{2j\pi kn}{2N}\right) \right\},$$

$$n = 0,1,\ldots,N-1. \tag{3.66}$$

This equation states that the inverse N-point DCT-II of the sequence $F_k^{C(2)}$ is given by twice the real part of a $(2N-1)$-point DFT inverse of the weighted and phase-shifted $F_k^{C(2)}$.

Next, consider the product of two DCT-II transforms $F_k^{C(2)}$ and $G_k^{C(2)}$ of the sequences f_n and g_n, respectively,

$$W_k^{C(2)} = F_k^{C(2)} G_k^{C(2)} \qquad k = 0, 1, \ldots, N-1. \tag{3.67}$$

Using (3.64), we obtain

$$W_k^{C(2)} = k_k^2 \exp\left(-\frac{j\pi k}{N}\right) \hat{F}_{F,k}^{C(2)} \hat{G}_{F,k}^{C(2)}, \tag{3.68}$$

where $\hat{F}_{F,k}^{C(2)}$ and $\hat{G}_{F,k}^{C(2)}$ are as defined in (3.63). To obtain the DCT-II inverse of (3.68), we apply the result to (3.66)

$$w_n = 2Re\left\{ F_F^{-1}\left[k_k^3 \exp\left(-\frac{j\pi kn}{2N}\right) \hat{F}_{F,k} \hat{G}_{F,k} \right] \right\}. \tag{3.69}$$

Here F_F^{-1} denotes the inverse DFT operator. It is evident that the convolution theorem for the Fourier transform can now be applied, resulting in the following:

$$w_n = 2Re\left\{ F_F^{-1}\left[k_k^3 \exp\left(-\frac{j\pi kn}{2N}\right) \circledast \hat{f}_n \circledast \hat{g}_n \right] \right\}. \tag{3.70}$$

The remaining inverse discrete Fourier transform (IDFT) can be explicitly evaluated (see Chen and Fralick [72]) to give

$$\begin{aligned} F_F^{-1}\left[k_k^3 \exp\left(-\frac{j\pi kn}{2N}\right) \right] &= \frac{\hat{h}_n}{2} \\ &= \left\{ \left(\frac{1}{2\sqrt{2}} - 1\right) + \exp\left(\frac{j(2n-1)(N-1)\pi}{4N}\right) \right. \\ &\quad \left. \times \frac{\sin\left[\frac{(2n-1)\pi}{4}\right]}{\sin\left[\frac{(2n-1)\pi}{4N}\right]} \right\} / \sqrt{(2N)}. \end{aligned} \tag{3.71}$$

Since $\hat{f}_n$ and $\hat{g}_n$ are real, we get

$$w_n = \hat{h}_n \circledast \hat{f}_n \circledast \hat{g}_n. \tag{3.72}$$

With these results, we are now finally ready to state the convolution theorem for the DCT-II:

$$F^{C(2)}\left[\hat{h}_n \circledast \hat{f}_n \circledast \hat{g}_n \right] = F_k^{C(2)} G_k^{C(2)}, \qquad k = 0, 1, \ldots, N-1, \tag{3.73}$$

where $\hat{h}_n$ is given by (3.71). The result in (3.73) indicates that the product of the DCT-II transforms of two sequences is the DCT-II of the convolution of these two sequences and a third function given by $\hat{h}_n$. In addition, the convolutions are circular convolutions of the evenly extended sequences (3.61). Similar convolution properties may be derived for the other members of the DCT family (see Problem 3.11.). Such properties are found to be useful for transform domain filtering (see Ngan and Clarke [349]). Another convolution property that is simpler than (3.73) has been developed for DCT-II by Chitprasert and Rao [82].

RELATION TO THE KLT

In Chapter 2 we studied the bases for the KLT expansion and its properties which are a direct consequence of the diagonalization problem in (2.10) . The set of basis vectors obtained will diagonalize the auto-covariance matrix $[R]$. It is also immediately clear that the basis functions are dependent on the auto-covariance matrix and, therefore, cannot be predetermined.

There are only a few cases in which analytical solutions are available. Such is the case when the signal statistics produce an auto-covariance matrix of the form in (2.23) for $0 < \rho < 1$, where ρ is the adjacent correlation coefficient. Such a signal is said to be a stationary Markov-1 signal. Davenport and Root reported an example in the continuous domain as a solution to an integral equation (see Chapter 3 of Davenport and Root [104]). In the discrete case, Ray and Driver [393] provided the solution as follows:

$$\begin{aligned} [\phi]_{nm} &= \Phi_{nm} \\ &= \sqrt{\frac{2}{(N+\lambda_m)}} \sin\left\{ \omega_m \left[(n+1) - \frac{N+1}{2} \right] + (m+1)\frac{\pi}{2} \right\} \\ m,n &= 0,1,\ldots,N-1, \end{aligned} \tag{3.74}$$

where $[\phi]_{nm}$ is the mn-element of matrix $[\phi]$ and Φ_{nm} is the nth component of the mth eigenvector. The eigenvalues are calculated as $\lambda_m = (1-\rho^2)/[1-2\cos\omega_m+\rho^2]$ and $\omega_m s$ are real positive roots of the transcendental equation,

$$\tan(N\omega) = -\frac{(1-\rho^2)\sin(\omega)}{[\cos(\omega) - 2\rho + \rho^2\cos(\omega)]}. \tag{3.75}$$

However, practical implementation of KLT involves the estimation of the auto-covariance matrix of the data sequence, its diagonalization and the construction of the basis vectors. The inability to predetermine the basis vectors in the transform domain has made KLT an ideal but impractical tool. A natural question is whether there are predetermined basis vectors that are good approximations to the KLT. To answer this question attempts were made to examine the diagonalization of matrices that are asymptotically equivalent to (2.23). Stationary zero-mean Markov-1 signals are deemed to be sufficiently general in signal processing to warrant detailed studies. Such studies have led to the construction of other discrete transforms. Although the KLT provides no easy solutions to the problem of actual decorrelation, it does provide a benchmark against which other discrete transforms may be judged. It may be noted that techniques for estimating the basis functions and the associated eigenvalues and their convergence to the desired values have been developed [56].

Notice that (2.23) is a symmetric Toeplitz matrix. In fact, a Toeplitz matrix may be treated as asymptotically equivalent (i.e., as N, the size of the matrix, increases) to a circulant matrix (see, for example, Davis [105]), whose eigenvectors are the basis vectors for the discrete Fourier transform (DFT). From this point of view, DFT and KLT are asymptotically equivalent.

One can also examine the asymptotic behavior of the auto-covariance matrix as $\rho \to 0$ or as $\rho \to 1$. Thus, when one speaks of the asymptotic equivalence of the matrix R, it should be clearly understood whether it applies to N or ρ. Yet another approach may be taken. Instead of examining the full covariance matrix, it is possible to examine an approximation to R, where only a few diagonals are retained and the diagonalization is considered for this approximate matrix.

The above approaches have been taken by Ahmed and Flickner [6], Jain [220], Kitajima [255], and Yemini and Pearl [556]. From these studies have come not only the DCTs and DSTs, but also other discrete unitary transforms. We shall present the derivations for the DCT-I and DCT-II in the following discussions.

DCT-I

In Section 3, we obtained the symmetric cosine transform or DCT-I by simply discretizing the Fourier cosine transform. This particular approach of obtaining the DCT-I bears no relation to the optimal KLT that we have discussed in the previous section. The original derivation by Kitajima [255] was motivated by an attempt to approximate the KLT basis functions by simpler basis functions. The derivation is based on a recursive relation of the Tchebyshev polynomial of the first kind, the diagonalization of a tridiagonal matrix, and its asymptotic equivalence (as N tends to infinity) to the inverse of the Markov-1 auto-covariance matrix. The derivation follows.

First, we recall a simple matrix property. If a non-singular matrix R is diagonalized by a similarity transformation S, i.e.,

$$S^{-1}RS = [\cap] = \operatorname{diag}[\lambda_0, \lambda_1, \ldots, \lambda_{N-1}], \tag{3.76}$$

then its inverse is also diagonalized by S so that

$$S^{-1}R^{-1}S = [\cap]^{-1} = \operatorname{diag}[\lambda_0^{-1}, \lambda_1^{-1}, \ldots, \lambda_{N-1}^{-1}]. \tag{3.77}$$

We note also that the symmetric Toeplitz matrix R for the auto-covariance of the Markov-1 signal has an inverse in tridiagonal form (see Theory and Applications of Toeplitz Matrices, Yip and Agrawal [559]) given by

$$R^{-1} = (1-\rho^2)^{-1}\begin{bmatrix} 1 & -\rho & 0 & 0 & \cdots & \cdots & \cdots \\ -\rho & 1+\rho^2 & -\rho & 0 & \cdots & \cdots & \cdots \\ 0 & -\rho & 1+\rho^2 & -\rho & 0 & \cdots & \cdots \\ \cdots & 0 & -\rho & 1+\rho^2 & -\rho & 0 & \cdots \\ \cdots & \cdots & \cdots & \cdots & \cdots & 1+\rho^2 & -\rho \\ \cdots & \cdots & \cdots & \cdots & \cdots & -\rho & 1 \end{bmatrix}. \quad (3.78)$$

It is therefore evident that the transformation that diagonalizes (3.78) will also diagonalize the matrix R. Instead of considering the diagonalization of (3.78), which will lead to the KLT basis functions, a matrix that is asymptotically equivalent (as $N \to \infty$) to R^{-1} is considered.

Let R^{-1} in (3.78) be decomposed into two parts, $R^{-1} = B + A$ such that

$$B = (1-\rho^2)^{-1}\begin{bmatrix} 1+\rho^2 & -\sqrt{2}\rho & 0 & \cdots & \cdots \\ -\sqrt{2}\rho & 1+\rho^2 & -\rho & \cdots & \cdots \\ 0 & -\rho & 1+\rho^2 & -\rho & \cdots \\ \cdots & \cdots & \cdots & \cdots & \cdots \\ \cdots & \cdots & \cdots & -\sqrt{2}\rho & 1+\rho^2 \end{bmatrix}, \quad (3.79)$$

and

$$A = (1-\rho^2)^{-1}\begin{bmatrix} -\rho^2 & (\sqrt{2}-1)\rho & 0 & \cdots & \cdots \\ (\sqrt{2}-1)\rho & 0 & 0 & \cdots & \cdots \\ \cdots & \cdots & \cdots & 0 & (\sqrt{2}-1)\rho \\ \cdots & \cdots & \cdots & (\sqrt{2}-1)\rho & -\rho^2 \end{bmatrix}. \quad (3.80)$$

We note that A is very close to being a null matrix. The matrix B can be further decomposed into a linear combination of a matrix B_1 and an identity matrix as follows,

$$B = \left\{\frac{-2\rho}{1-\rho^2}\right\}B_1 + \left\{\frac{1+\rho^2}{1-\rho^2}\right\}I_N,$$

with

$$B_1 = \begin{bmatrix} 0 & \frac{1}{\sqrt{2}} & 0 & 0 & \cdots & \cdots & \cdots \\ \frac{1}{\sqrt{2}} & 0 & \frac{1}{2} & 0 & \cdots & \cdots & \cdots \\ 0 & \frac{1}{2} & 0 & \frac{1}{2} & \cdots & \cdots & \cdots \\ \cdots & \cdots & \cdots & \cdots & \frac{1}{2} & 0 & \frac{1}{\sqrt{2}} \\ \cdots & \cdots & \cdots & \cdots & 0 & \frac{1}{\sqrt{2}} & 0 \end{bmatrix}. \quad (3.81)$$

The transformation that diagonalizes B_1 will also diagonalize B. The diagonalization of B_1 depends on a recurrence relation for Tchebyshev polynomials of the first kind, which are given by

$$T_n(x) = \cos(n\theta), \qquad \cos(\theta) = x, \qquad n = 0, 1, \cdots \quad (3.82)$$

The relevant three-term recurrence relation is (see *Special Functions for Physics and Chemistry*, Sneddon [443]) :

$$\frac{1}{2}T_{n-1}(x)+\frac{1}{2}T_{n+1}(x)=xT_n(x). \tag{3.83}$$

If the first N polynomials are considered at the discrete values of x given by

$$x_k=\cos\left\{\frac{k\pi}{N-1}\right\}, \qquad k=0,1,\cdots N-1, \tag{3.84}$$

then

$$\frac{1}{2}T_{n-1}(x_k)+\frac{1}{2}T_{n+1}(x_k)=x_kT_n(x_k), \qquad n=1,2,\cdots,N-2,$$
$$T_1(x_k)=x_kT_0(x_k),$$

and

$$T_{N-2}(x_k)=x_kT_{N-1}(x_k). \tag{3.85}$$

[Note: (3.85) can easily be verified by the compound angle expansion of cosine functions.] Equation (3.85) can be represented in matrix form if we define an N-dimensional vector given by,

$$v_k=\left\{\frac{T_0(x_x)}{\sqrt{2}},T_1(x_x),\cdots,\frac{T_{N-1}(x_x)}{\sqrt{2}}\right\}^T. \tag{3.86}$$

Then

$$B_1v_k=x_kv_k, \qquad k=0,1,\cdots,N-1. \tag{3.87}$$

Equation (3.87) states that x_k and v_k are, respectively, the eigenvalue and eigenvector of the matrix B_1. For the matrix B we have,

$$Bv_k=w_kv_k, \qquad k=0,1,\cdots,N-1, \tag{3.88}$$

where the eigenvalues $w_k s$ are given by

$$w_k=(1-\rho^2)^{-1}\left\{1+\rho^2-2\rho\cos\left(\frac{k\pi}{N-1}\right)\right\}. \tag{3.89}$$

In (3.79) we note that as $N\to\infty$, B should asymptotically approach A^{-1}. This is valid, at least in the transform domain, as we show in the following:

Let

$$V=[\hat{v}_0,\hat{v}_1,\cdots,\hat{v}_{N-1}]^T, \tag{3.90}$$

where $\hat{v}_k$ is the normalized eigenvector in (3.88). Since $\hat{v}_k$ is real, the matrix in (3.90) is real and unitary. Thus $VV^T=I_N$. Applying the similarity transform to (3.79) gives

$$V^TR^{-1}V=V^TBV+V^TAV=\text{diag }(w_0,w_1,\cdots,w_{N-1})+V^TAV. \tag{3.91}$$

By direct matrix multiplication, it can be shown that the $nm-$element of V^TAV is given by

$$(V^TAV)_{nm} = \begin{cases} 0 & \text{if } n+m = \text{odd} \\ k_n k_m 2\rho \frac{\left\{-\rho+(2-\sqrt{2})\left[\cos\left(\frac{m\pi}{N-1}\right)+\cos\left(\frac{n\pi}{N-1}\right)\right]\right\}}{(N-1)(1-\rho^2)} & \text{if } n+m = \text{even,} \end{cases} \tag{3.92}$$

where

$$k_m = \begin{cases} \frac{1}{\sqrt{2}} & \text{if } m = 0 \text{ or } N-1, \\ 1 & \text{otherwise.} \end{cases}$$

For $\rho \neq 1$, (3.92) vanishes as $N \to \infty$. Therefore, using matrix property (3.76) and (3.77), we obtain asymptotically as $N \to \infty$ that

$$V^TAV \approx \text{diag}\,[\lambda_0, \lambda_1, \cdots, \lambda_{N-1}],$$

where

$$\lambda_k = \frac{1}{w_k}$$

and

$$V_{nm} = k_n k_m \sqrt{\frac{2}{N-1}} \cos\left(\frac{nm\pi}{N-1}\right), \qquad n,m = 0,1,\cdots,N-1. \tag{3.93}$$

Equation (3.93) is identical to the definition of DCT-I given in (3.20) if $N-1$ is replaced by N. This is an example of asymptotic equivalence to the KLT as $N \to \infty$. We note that the equivalence is valid for $\rho \neq 1$. In fact, as $\rho \to 1$ $(V^TAV)_{nm}$ in (3.92) becomes indeterminate even as $N \to \infty$. This predicts that the decorrelation power of DCT-I decreases as ρ increases to 1, even when N is very large.

FAST ALGORITHMS

The development of efficient algorithms for the computation of DCT (more specifically DCT-II) began soon after Ahmed, Natarajan and Rao [7] reported their work on DCT. It was natural for initial attempts to focus on the computation of DCT by using the fast Fourier transform (FFT) algorithms. Although DCT-II was developed not as a discretized version of the FCT, its relations to the DFT were exploited in the initial developments of its computational algorithms. Haralick [174] reported the computation of an N-point DCT using two N-point FFTs. Other work along the same line soon followed. Narashima and Peterson [343], Tseng and Miller [483], Makhoul [300], Duhamel and Hollmann [119], and Vetterli and Nussbaumer [499] all contributed to the advances in this direction.

Fast computational algorithms can also be obtained by considering the factorization of the DCT matrix. When the component factors of this factorization are sparse, the decomposition represents a fast algorithm. Chen, Smith, and Fralick [73] reported such a development. Since matrix factorization is not unique, there exist other forms of fast algorithms. Corrington [96], Wang [521], Lee [268], [269], and Suehiro and Hatori [459], among others, contributed to this development. Some of the factorization schemes fall into the decimation in time (DIT) category. Others fall into the decimation in frequency (DIF) category (see Rao, Kim and Hwang [388], [561], [562], [563]). Fast DCT algorithms can also be obtained through the computation of other discrete transforms (see Hein and Ahmed [178]; Jones, Hein, and Knauer [234]; Venkataraman et al. [495]; Malvar [305], [306], [307]; Nagesha [338]), through prime factor decomposition (see Yang and Narashima [554]), through recursive computation (see Hou [183] and Lee [268], [269]), and by planar rotations [283], [284], [289], [290]. Other algorithms include split radix [464].

DCT VIA FFT

In Section 3 the properties of the Fourier cosine transform are recalled, (3.2) shows that the Fourier transform of an even function results in the definition for the Fourier cosine transform [388]. This very simple property can be exploited for the computation of the DCT. Let $\{x_n\}$, $n = 0, 1, \cdots, N-1$ be a given sequence. Then an extended sequence $\{y_n\}$ symmetric about the $(2N-1)/2$ point can be constructed so that,

$$y_n = \begin{cases} x_n & n = 0, 1, \cdots, N-1, \\ x_{2N-n-1} & n = N, N+1, \cdots, 2N-1. \end{cases} \tag{3.94}$$

Example 3.4. *Suppose N = 4, show* $\{x_n\}$ *and* $\{y_n\}$,

$$\{x_n\} = \{x_0, x_1, x_2, x_3\}.$$
$$\{y_n\} = \{x_0, x_1, x_2, x_3, x_3, x_2, x_1, x_0\}.$$

Let W_{2N} to denote $\exp(-j2\pi/2N)$; it can be seen that the DFT of y_n given by

$$Y_m = \sum_{n=0}^{2N-1} y_n W_{2N}^{nm}, \tag{3.95}$$

is easily reduced to

$$\begin{aligned} Y_m &= \sum_{n=0}^{N-1} x_n W_{2N}^{nm} + \sum_{n=N}^{2N-1} y_n W_{2N}^{nm} \\ &= \sum_{n=0}^{N-1} x_n W_{2N}^{nm} + \sum_{n=N}^{2N-1} x_{2N-n-1} W_{2N}^{nm} \\ &= \sum_{n=0}^{N-1} x_n W_{2N}^{nm} + \sum_{n=0}^{N-1} x_n W_{2N}^{(2N-n-1)m} \\ &= \sum_{n=0}^{N-1} x_n [W_{2N}^{nm} + W_{2N}^{-(n+1)m}], \qquad m = 0, 1, \cdots, 2N-1. \end{aligned} \tag{3.96}$$

If we now multiply both sides of (3.96) by a factor of $\frac{1}{2}W_{2N}^{m/2}$, we obtain,

$$\frac{1}{2}W_{2N}^{m/2}Y_m = \sum_{n=0}^{N-1} x_n \cos\left[(2n+1)\frac{m\pi}{2N}\right], \qquad m = 0, 1, \cdots, N-1. \tag{3.97}$$

It is easy to see that (3.97) is the DCT-II of (3.20) of the N-point sequence x_n except for the required scale factors. In (3.95), $\{Y_m\}$ is the $2N$-point DFT of $\{y_n\}$ and (3.97) shows that for $m = 0, 1, \cdots, N-1$, the transformed sequence $\{Y_m\}$, properly scaled, becomes the DCT-II of the N-point sequence $\{x_n\}$. Then, (3.97) provides a means to compute the DCT-II of $\{x_n\}$ via a $2N$-point DFT of the extended sequence $\{y_n\}$. This concept was utilized in building a feasibility model based on high-speed pipeline FFT processing techniques [317].

When $\{x_n\}$ is real and $\{y_n\}$ is real and symmetric, $\{Y_m\}$ can be obtained via two N-point FFTs rather than by a single $2N$-point FFT [174]. Since an N-point FFT requires $N\log_2 N$ complex operations in general, such an approach represents a savings of $2N$ complex operations. Tseng and Miller [483], however, pointed out that when the input sequence $\{x_n\}$ is real, the radix-2 DFT of the real, even, $2N$-point sequence $\{y_n\}$ requires only $(N/2)\log_2 N + N/2$ complex operations. The $N/2$ term arises from the scaling required as in (3.97). This represents a significant reduction in the complexity from that of computing two N-point DFTs.

A different rearrangement of the input sequence was proposed by Narasimha and Peterson [343]. This provides an even more efficient algorithm. Consider an N-point sequence $\{x_n\}$ and its DCT (specifically DCT-II). If N is even, the sequence $\{x_n\}$ can be relabelled as

$$y_n = x_{2n}$$

and

$$y_{N-1-n} = x_{2n+1}, \qquad n = 0, 1, \cdots, \frac{N}{2} - 1. \tag{3.98}$$

The DCT-II of $\{x_n\}$ can now be reduced to

$$\begin{aligned} X_m^{C(2)} &= \sum_{n=0}^{N/2-1} y_n \cos\left[\frac{(4n+1)m\pi}{2N}\right] \\ &+ \sum_{n=0}^{N/2-1} y_{N-n-1} \cos\left[\frac{(4n+3)m\pi}{2N}\right] \qquad m = 0, 1, \cdots, N-1. \end{aligned} \tag{3.99}$$

If the sum over n is replaced by a sum over $N-n-1$ in the second term, the two summations in (3.99) can be combined into one, resulting in

$$X_m^{C(2)} = \sum_{n=0}^{N-1} y_n \cos\left[\frac{(4n+1)m\pi}{2N}\right], \qquad m = 0, 1, \cdots, N-1. \tag{3.100a}$$

The right-hand side of this equation is the real part of a scaled N-point DFT of the sequence $\{y_n\}$. Thus,

$$X_m^{C(2)} = \mathrm{Re}[Z_m],$$

where

$$Z_m = W_{4N}^m Y_m = W_{4N}^m \sum_{n=0}^{N-1} y_n W_N^{nm}, \qquad m = 0, 1, \cdots, N-1. \tag{3.100b}$$

In (3.100b), $\{Y_m\}$ is the N-point DFT of $\{y_n\}$. Also, it is easy to verify that

$$Z_{N-m} = -jZ_m^*, \tag{3.101}$$

where * denotes complex conjugate, provided $\{y_n\}$ is a real sequence. Therefore,

$$X_m^{C(2)} = \mathrm{Re}[Z_m],$$

and

$$X_{N-m}^{C(2)} = -\mathrm{Im}[Z_m], \qquad m = 0, 1, \cdots, N/2. \tag{3.102}$$

Equation (3.102) summarizes the procedure. The DCT-II of the N-point real sequence $\{x_n\}$ is obtained by evaluating the real and imaginary parts of the first $(N/2+1)$ components of the N-point DFT of the rearranged sequence $\{y_n\}$. Using a radix-2 algorithm for the DFT, the resulting number of complex multiplications is $(N\log_2 N - N + 1)/4$, which is about twice as fast as the algorithm in (3.97), [483]. Although we have compared the number of complex operations based on radix-2 algorithms for the DFT, the formulations are general. For example, the Winograd algorithm (see Silverman [435, 436]) may be used to facilitate the DFT computation when N is not a power of 2. The expansions are valid regardless of what algorithm is being used in the computations of the DFTs.

In the rearrangement of the data sequence, (3.98) indicates that N should be even. This restriction can in fact be removed, as pointed out by Makhoul [300]. As a result, (3.98) groups all the even-numbered points of x_n in increasing order followed by the odd-numbered points in reversed order. In Figures 3.3 and 3.4 the block diagrams for DCT-II representing the two algorithms discussed above are shown.

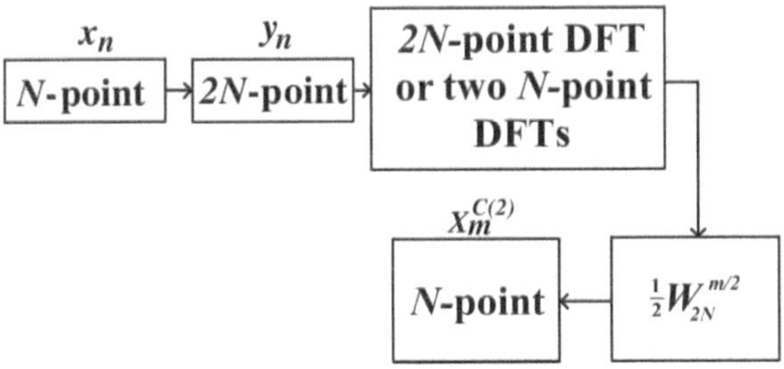

Figure 3.3: N-point DCT via $2N$-point DFT.

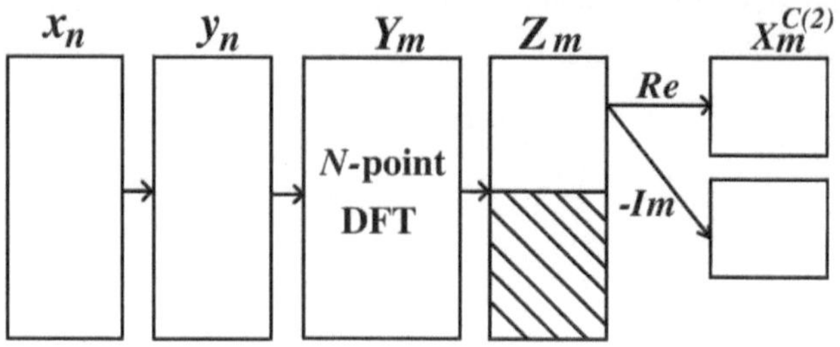

Figure 3.4: N-point DCT via N-point DFT.

The relation between the DCT and DFT can be exploited in a different way, as demonstrated by Vetterli and Nussbaumer [499]. The approach taken is a recursive one, whereby the dimension of the DCT is repeatedly reduced until only trivial ($N =$ 2 or 4) transforms. In particular, when the input sequence $\{x_n\}$ is real, the real and imaginary parts of its DFT, $\{X_m\}$, can be written as

$$\begin{aligned}
\mathrm{Re}[X_m] &= \sum_{n=0}^{N/2-1} x_{2n}\cos\left[\frac{2nm\pi}{(N/2)}\right] \\
&+ \sum_{n=0}^{N/4-1} \{x_{2n+1}+x_{N-2n-1}\}\cos\left[\frac{2(2n+1)m\pi}{N}\right], \\
\mathrm{Im}[X_m] &= \sum_{n=0}^{N/2-1} x_{2n}\sin\left[\frac{2nm\pi}{(N/2)}\right] \\
&+ \sum_{n=0}^{N/4-1} \{x_{2n+1}-x_{N-2n-1}\}\sin\left[\frac{2(2n+1)m\pi}{N}\right], \\
m &= 0,1,\cdots,N-1. \qquad (3.103)
\end{aligned}$$

In the real and imaginary parts of (3.103), the first terms on the right-hand side are essentially the real and imaginary parts of an $N/2$-point of the even-numbered points in x_n. The second term for the real part is a DCT-II of $N/4$-points, and the second term of the imaginary part can be further reduced to a $N/4$-point DCT-II by using the identity,

$$\sin\left[\frac{2(2n+1)m\pi}{N}\right] = (-1)^n\cos\left[\frac{(2n+1)(N-4m)\pi}{2N}\right]. \qquad (3.104)$$

It can also be verified that

$$X_m^{C(2)} = \cos\left[\frac{m\pi}{2N}\right]\mathrm{Re}\left[Y_m\right] - \sin\left[\frac{m\pi}{2N}\right]\mathrm{Im}\left[Y_m\right], \qquad (3.105)$$

where Y_m is the N-point DFT of the sequence y_n defined in (3.98). The computation of the DCT-II in (3.105) is summarized by the block diagram of Figure 3.5 (also see Problem 3.22.).

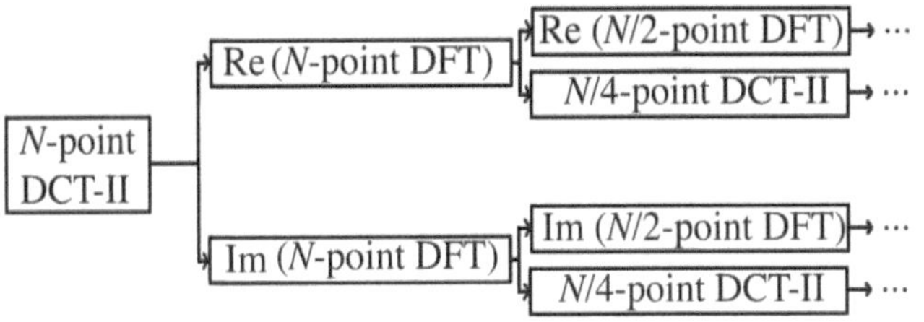

Figure 3.5: Recursive computation of DCT-II [499].

Figure 3.5 shows the initial steps of recursive decomposition, excluding the actual scale factors. The rearrangement and recombinations of the input sequences are also seen. A continuation of this procedure will result in trivial transforms with $N = 2$ or 4. The resulting number of real multiplications required for an N-point DCT (radix-2) is given by

$$\left(\frac{N}{2}\right) \log_2 N. \tag{3.106}$$

The number of real multiplications described by (3.106) compares favorably with the number of complex multiplications required for the other algorithms discussed earlier in this section. It should be pointed out that the reduced number of multiplications is obtained at the cost of increased complexity in the permutation and combination of the input data samples and of the output transformed samples. This increased complexity in the topology of the overall transform is its major drawback when compared with the other algorithms.

Example 3.5. *Use the block diagram of Figure 3.3 to calculate the DCT of the sequence* $\{x_n\} = \{10, 7, 12, 15\}$. *Compute the scaling factors to obtain coefficients of the DCT-II. Write a MATLAB code to implement the DCT-II using Figure 3.3.*

$N = 4$, $M = 8$

$\{y_n\} = \{10, 7, 12, 15, 15, 12, 7, 10\}$

$\{Y_m\} = \{88, -12.071 + 5j, 6 - 6j, 2.071 - 5j, 0, 2.071 + 5j, 6 + 6j, -12.071 - 5j\}$

$\frac{1}{2} W_{2N}^{m/2} Y_m = \{44, -6.5328, 4.2426, 2.7059, 0, -2.7059, -4.2426, 6.5328\}$

Take the first N coefficients and multiply them by the scale factors as indicated in (3.36) *for DCT-II*

Scale factors = $\sqrt{\frac{2}{N}}\{\frac{1}{\sqrt{2}}, 1, 1, 1\}$

$X^{C(2)} = \{22, -4.6193, 3, 1.9134\}$

MATLAB code:

```
% N-point DCT-II via 2N-point DFT

clear all;
% N - point
xn=[10, 7, 12, 15];

% 2N - point
yn= [xn, xn(end:-1:1)];

% FFT of yn
Ym = fft(yn);
M = length(Ym);

W = 0.5*(cos(pi*(0:M-1)/M) - j*sin(pi*(0:M-1)/M));
DCT2N = real(Ym.*W);

N = length(xn);
ScaleFactors = sqrt(2/N).*[1/sqrt(2), ones(1,N-1)];
DCT_II = ScaleFactors.*DCT2N(1:N);
```

Example 3.6. *Use the block diagram of Figure 3.4 to calculate the DCT of the same sequence as in Example 3.5. Compute the scaling factors to obtain coefficients of the DCT-II, compare the results with those obtained in Example 3.5. Write the MATLAB code to implement the DCT-II using Figure 3.4.*

$\{x_n\} = \{10, 7, 12, 15\}$

$N = 4$

$\{y1_n\} = x_{2n} = \{10, 12\}$

$\{y2_n\} = \{y_{N-n-1}\} = x_{2n+1} = \{15, 7\}$

$\{y_n\} = \{y1_n,\ y2_n\} = \{x_0, x_2, x_3, x_1\} = \{10, 12, 15, 7\}$

$\{Y_m\} = \{44, -5-5j, 6, -5, 5j\} \rightarrow$ *4-point DFT*

$W_{4N}^{m} = \{1, 0.9239 - 0.3827j, 0.7071 - 0.7071j, 0.3827 - 0.9239j\}$

DCT coefficients

$Real\{W_{4N}^{m} Y_m\} = \{44, -6.5328, 4.2426, 2.7060\}$

Scale factors = $\sqrt{\frac{2}{N}}\{\frac{1}{\sqrt{2}}, 1, 1, 1\}$

DCT-II coefficients

$X^{C(2)} = \{22, -4.6193, 3, 1.9134\}$

MATLAB code:

```
% N-point DCT-II via N-point DFT

clear all;

% N - point
xn=[10, 7, 12, 15];

N = length(xn);

y1=xn(1:2:end);
y2 = xn(2:2:end);
yn= [y1, y2(end:-1:1)];

Ym = fft(yn)

W =  cos((2*pi.*(0:(N-1)))/(4*N))-j.*sin((2*pi.*(0:(N-1))/(4*N)));
DCT = real(W.*Ym);

ScaleFactors = sqrt(2/N).*[1/sqrt(2), ones(1,N-1)];
DCT_II = ScaleFactors.*DCT;
```

DIRECT COMPUTATION BY SPARSE MATRIX FACTORIZATIONS

Consider the computation of the DCT-II of an input sequence $\{x_n\}$ and let this sequence be represented by the column vector x. The transformed sequence can be expressed in vector notations as follows:

$$X^{C(2)} = \sqrt{\frac{2}{N}} A_N X. \tag{3.107}$$

Here A_N is an $(N \times N)$ matrix and is equal to $[C_N^{II}]$, defined in (3.20) except for the normalization factor. When the matrix A_N (or o $[C_N^{II}]$) is decomposed into sparse matrices, a fast algorithm will result. For example, if we consider

$$A_4 = \begin{bmatrix} \frac{1}{\sqrt{2}} & \frac{1}{\sqrt{2}} & \frac{1}{\sqrt{2}} & \frac{1}{\sqrt{2}} \\ C_8^1 & C_8^3 & C_8^5 & C_8^7 \\ C_8^2 & C_8^6 & C_8^6 & C_8^2 \\ C_8^3 & C_8^7 & C_8^1 & C_8^5 \end{bmatrix}, \tag{3.108}$$

where $C_k^m = \cos(m\pi/k)$. We can decompose A_4 as follows:

$$\begin{aligned} A_4 &= \begin{bmatrix} 1 & 0 & 0 & 0 \\ 0 & 0 & 0 & 1 \\ 0 & 1 & 0 & 0 \\ 0 & 0 & 1 & 0 \end{bmatrix} \begin{bmatrix} \frac{1}{\sqrt{2}} & \frac{1}{\sqrt{2}} & \frac{1}{\sqrt{2}} & \frac{1}{\sqrt{2}} \\ C_4^1 & C_4^3 & C_4^3 & C_4^1 \\ C_8^3 & -C_8^1 & C_8^1 & -C_8^3 \\ C_8^1 & C_8^3 & -C_8^3 & -C_8^1 \end{bmatrix} \\ &= \begin{bmatrix} 1 & 0 & 0 & 0 \\ 0 & 0 & 0 & 1 \\ 0 & 1 & 0 & 0 \\ 0 & 0 & 1 & 0 \end{bmatrix} \begin{bmatrix} \frac{1}{\sqrt{2}} & \frac{1}{\sqrt{2}} & 0 & 0 \\ C_4^1 & C_4^3 & 0 & 0 \\ 0 & 0 & -C_8^1 & C_8^3 \\ 0 & 0 & C_8^3 & C_8^1 \end{bmatrix} \begin{bmatrix} 1 & 0 & 0 & 1 \\ 0 & 1 & 1 & 0 \\ 0 & 1 & -1 & 0 \\ 1 & 0 & 0 & -1 \end{bmatrix}. \end{aligned} \tag{3.109}$$

Here, (3.109) can be expressed in more compact form as

$$A_4 = \bar{P}_4 \begin{bmatrix} A_2 & 0 \\ 0 & \bar{R}_2 \end{bmatrix} B_4, \tag{3.110}$$

where $\bar{P}_4$ is a permutation matrix, B_4 is a butterfly matrix, and $\bar{R}_2$ is the remaining (2×2) block in the second factor matrix. The key is that A_4 is reduced in terms of A_2. In general, for N being a power of 2, (3.110) is in the form of

$$A_N = \bar{P}_N \begin{bmatrix} A_{N/2} & 0 \\ 0 & \bar{R}_{N/2} \end{bmatrix} B_N, \tag{3.111}$$

the matrix $\bar{P}_N$ permutes the even rows in increasing order in the upper half and the odd rows in decreasing order in the lower half. The butterfly matrix B_N can be expressed in terms of the identity matrix $I_{N/2}$ and the opposite identity matrix $\tilde{I}_{N/2}$ as follows:

$$B_N = \begin{bmatrix} I_{N/2} & \tilde{I}_{N/2} \\ \tilde{I}_{N/2} & -I_{N/2} \end{bmatrix}. \tag{3.112}$$

The matrix $\bar{R}_N$ is obtained by reversing the orders of both the rows and columns of the matrix R_N whose $ik-$element is given by

$$R_{N_{ik}} = \cos\left[\frac{(2i+1)(2k+1)\pi}{4N}\right], \qquad i,k = 0,1,\cdots,N-1. \tag{3.113}$$

Therefore, (3.113) essentially defines the transform matrix for DCT-IV.

Example 3.7. *Implement step by step a signal flowgraph of* (3.109)*, assuming that the input vector is* $x = [x_0, x_1, x_2, x_3]^T$.

We will use the following identities:

$$\begin{aligned} C_4^1 &= \cos(\pi/4) = 1/\sqrt{2}, \\ C_4^3 &= \cos(3\pi/4) = -C_4^1 = -1\sqrt{2}, \\ C_8^1 &= \cos(\pi/8) = \sin(3\pi/8) = S_8^3, \\ C_8^3 &= \cos(3\pi/8) = \sin(\pi/8) = S_8^1. \end{aligned} \tag{3.114}$$

Equation (3.109) *can be expressed in a more general form*

$$A_4 x = \begin{bmatrix} 1 & 0 & 0 & 0 \\ 0 & 0 & 0 & 1 \\ 0 & 1 & 0 & 0 \\ 0 & 0 & 1 & 0 \end{bmatrix} \begin{bmatrix} C_4^1 & C_4^1 & 0 & 0 \\ C_4^1 & -C_4^1 & 0 & 0 \\ 0 & 0 & -C_8^1 & C_8^3 \\ 0 & 0 & C_8^3 & C_8^1 \end{bmatrix} \begin{bmatrix} 1 & 0 & 0 & 1 \\ 0 & 1 & 1 & 0 \\ 0 & 1 & -1 & 0 \\ 1 & 0 & 0 & -1 \end{bmatrix} \begin{bmatrix} x_0 \\ x_1 \\ x_2 \\ x_3 \end{bmatrix}.$$

The first product is

$$B_4 x = \begin{bmatrix} 1 & 0 & 0 & 1 \\ 0 & 1 & 1 & 0 \\ 0 & 1 & -1 & 0 \\ 1 & 0 & 0 & -1 \end{bmatrix} \begin{bmatrix} x_0 \\ x_1 \\ x_2 \\ x_3 \end{bmatrix} = \begin{bmatrix} x_0 + x_3 \\ x_1 + x_2 \\ x_1 - x_2 \\ x_0 - x_3 \end{bmatrix} = \begin{bmatrix} a \\ b \\ c \\ d \end{bmatrix}.$$

Figure 3.6 shows the butterfly signal flowgraph for the 4 points.

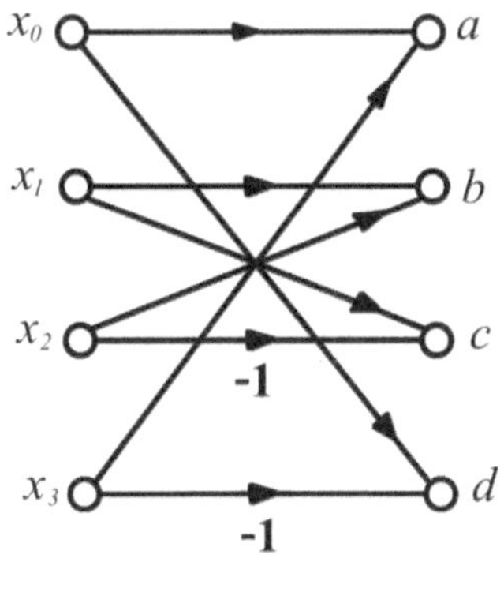

Figure 3.6: Butterfly signal flowgraph of the product $B_4 x$.

The next product is

$$\begin{bmatrix} C_4^1 & C_4^1 & 0 & 0 \\ C_4^1 & -C_4^1 & 0 & 0 \\ 0 & 0 & -C_8^1 & C_8^3 \\ 0 & 0 & C_8^3 & C_8^1 \end{bmatrix} \begin{bmatrix} a \\ b \\ c \\ d \end{bmatrix} = \begin{bmatrix} C_4^1 a + C_4^1 b \\ C_4^1 a - C_4^1 b \\ -C_8^1 c + C_8^3 d \\ C_8^3 c + C_8^1 d \end{bmatrix} = \begin{bmatrix} e \\ f \\ g \\ h \end{bmatrix}.$$

The permutation matrix $\bar{P}_4$ moves the last element of the vector $[e, f, g, h]^T$ to the second row, the second element to the third row and the third element to the last row as follows $[e, h, f, g]^T$. The new positions of each element can be expressed in two binary bits (00, 01, 10, 11). To find the final positions we apply the bit reversal order of the elements as (00, 10, 01, 11) in other words, the final positions of the output elements are $[e, f, h, g]^T$. Also, we can use the identities defined in (3.114) *at the output stage as shown in Figure 3.7.*

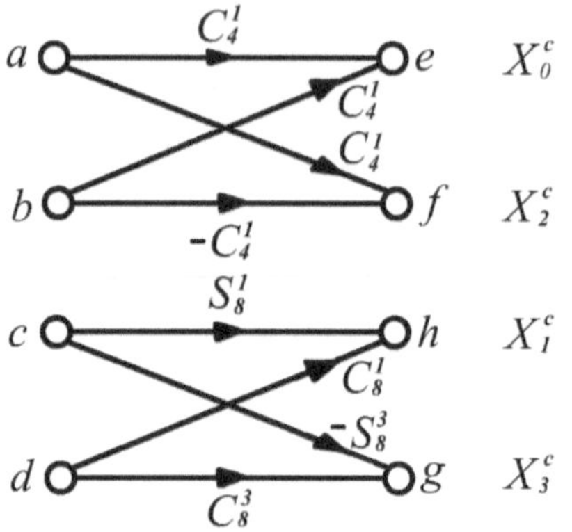

Figure 3.7: Output stage of the 4-point DCT.

Figure 3.8 shows the complete signal flowgraph.

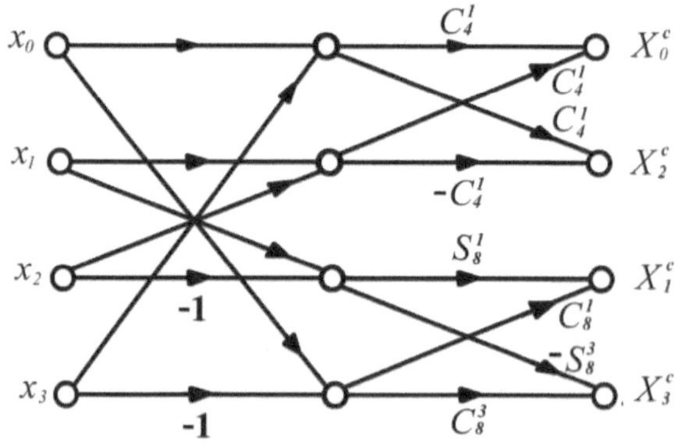

Figure 3.8: Signal flowgraph of the 4-point DCT.

As it can be seen in (3.111), the factorization is only partly recursive, since the matrix R_N cannot be recursively factored. There is, however, regularity in its factorization, and it is found that it can be decomposed into five types of matrix factors, all of which have no more than two non-zero elements in each row. The interested reader is referred to the work by Wang [521] and Chen, Smith, and Fralick [73] for more detailed descriptions of the actual matrix factors for R_N. This algorithm requires

$$\begin{aligned}&\frac{3N}{2}(\log_2 N - 1) + 2 \quad \text{real additions, and}\\ &N\log_2 N - \frac{3N}{2} + 4 \quad \text{real multiplications.}\end{aligned} \tag{3.115}$$

DECIMATION-IN-TIME (DIT) AND DECIMATION-IN-FREQUENCY (DIF) ALGORITHMS

Of all the algorithms described so far, only the scheme proposed by Vetterli and Nussbaumer [499] is strictly recursive. The factorization presented in Section 3 is only partly recursive, since the factor cannot be so factored. Although its factorization is quite regular, it still lacks the desired property. The scheme proposed by Vetterli and Nussbaumer [499], on the other hand, involves both the DFT and DCT of reduced dimensions, and the topology of the index mapping is known to be complex.

The rearrangement of $\{x_n\}$ in (3.98) makes possible the computation of DCT via an N-point DFT and is, in fact, typical of some of the fast algorithms for computing the DCT. Lee [268, 269] studied just such an algorithm, where the key concept is to reduce an N-point DCT to an $N/2$-point DCT by permutation of the input sample points. The resulting algorithm contains the desirable recursive modularity of a fast decomposition. This arrangement can be accomplished either in the time domain or in the frequency domain. In the former case, it is generally referred to as decimation in time (DIT) and in the latter case, as decimation in frequency (DIF). In this section, we describe in detail the DIT and DIF algorithms reported by Yip and Rao [561, 563] for DCT-II.

The DIT Algorithm

In (3.36) we shall leave out the scale factor for convenience. Then, the transformed sequence $\{X_m^{C(2)}\}$ is given by

$$X_m^{C(2)} = \sum_{n=0}^{N-1} x_n C_{2N}^{(2n+1)m}, \qquad m = 0, 1, \cdots, N-1, \tag{3.116}$$

where $C_k^m = \cos(m\pi/k)$ as previously defined.

The first stage in the DIT algorithm consists of rearrangement of the input sample points. The second stage reduces the N-point transform to $N/2$-point transforms to establish the recursive aspect of the algorithm. These stages are described by the following equations:

Let

$$G_m = \sum_{n=0}^{N/2} [x_{2n} + x_{2n-1}] C_{N/2}^{nm},$$

and

$$H_m = \sum_{n=0}^{N/2-1} [x_{2n} + x_{2n+1}] C_N^{(2n+1)m}, \qquad m = 0, 1, \cdots, N/2-1, \tag{3.117}$$

with $x_{-1} = x_N = 0$. It is not difficult to show, using properties of the cosine functions, that (3.117) can be used in (3.116), resulting in

$$X_m^{C(2)} = \frac{1}{2} \frac{[G_m + H_m]}{C_N^m},$$

and

$$X_{N-m-1}^{C(2)} = \frac{1}{2} \frac{[G_{m+1} - H_{m+1}]}{C_{2N}^{N-m-1}}, \qquad m = 0, 1, \cdots, N/2-1. \tag{3.118}$$

In (3.117), the sequence $\{H_m\}$ is obtained as the DCT-II of $N/2$ sample points. However, $\{G_m\}$ is obtained as a DCT-I of $(N/2+1)$ sample (see Section 3). This part of the transform can be further reduced (see Yip and Rao [561]), leading to the desired recursive structure Figure 3.9.

In (3.117) the sequence $\{H_m\}$ is obtained as the DCT-II of $N/2$ sample. However, $\{G_m\}$ is obtained as a DCT-I of $(N/2+1)$ sample points (see Section 3). This part of the transform can be further reduced (see Yip and Rao [561]), leading to the desired recursive structure. Figure 3.9 shows the signal flowgraph for the DIT computation of an $N = 8$ DCT-II of the sequence $\{x_n\}$. We note here that in [561], DCT-II has been incorrectly labeled as DCT-III.

Example 3.8. *Using the DIT algorithm, calculate the DCT-II of the 8-point vector*

$$x = \begin{bmatrix} 10 & 12 & 16 & 8 & 6 & 10 & 14 & 16 \end{bmatrix}.$$

The input to the $N/2+1$-point DCT-I is

$$\begin{aligned} g_n =& x_{2n} + x_{2n-1} \\ =& \begin{bmatrix} 10 & 28 & 14 & 24 & 16 \end{bmatrix}. \end{aligned}$$

The input to the $N/2$-point DCT-II is

$$\begin{aligned} h_n =& x_{2n} + x_{2n+1} \\ =& \begin{bmatrix} 22 & 24 & 16 & 30 \end{bmatrix}. \end{aligned}$$

According to (3.117) *the DCT-I and DCT-II of the sequences g_n and h_n respectively are*

$$\begin{aligned} G_m &= \begin{bmatrix} 92 & -3.1716 & 12 & -8.8284 & -12 \end{bmatrix}, \\ H_m &= \begin{bmatrix} 92 & -4.3296 & 8.4853 & -10.4525 \end{bmatrix}. \end{aligned}$$

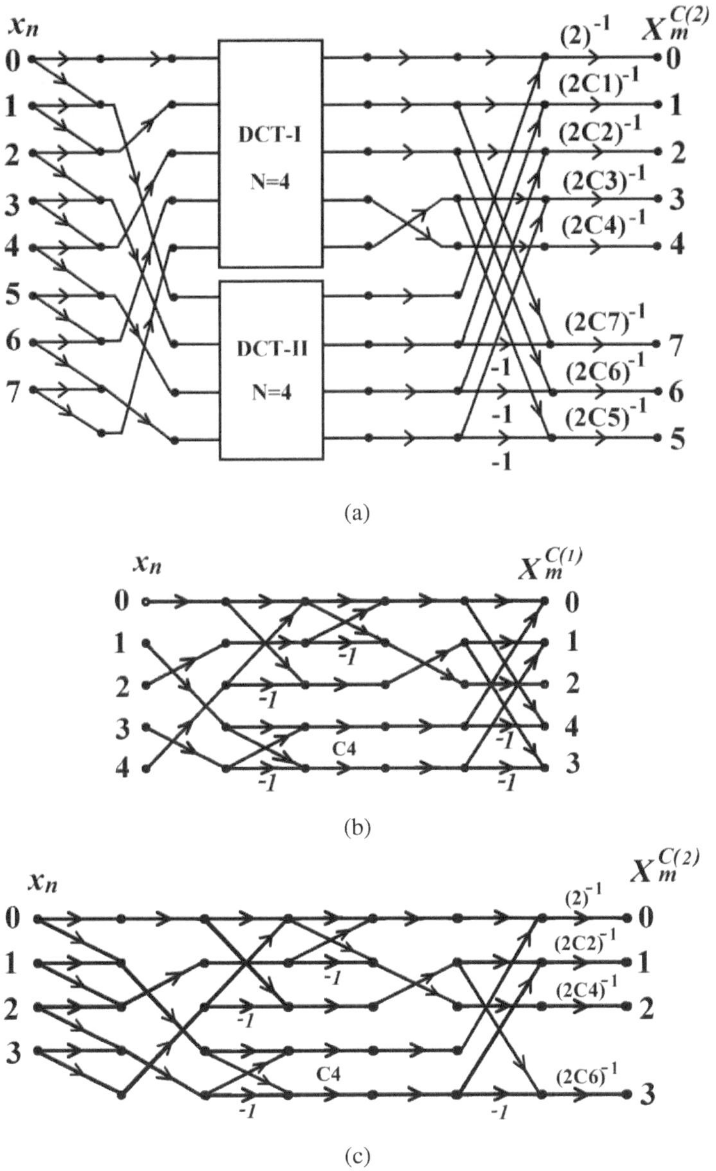

Figure 3.9: Signal flowgraph for DIT DCT-II for $N = 8$. $Ck = \cos(k\pi/16)$. (a) DIT DCT-II for $N = 8$. (b) DIT DCT-I for $N = 4$. (c) DIT DCT-II for $N = 4$ [390].

Then,

$$X_m^{C(2)} = \begin{bmatrix} 92 & -3.8240 & 11.0866 & -11.5945 & -8.4853 \end{bmatrix}$$
$$X_{N-m-1}^{C(2)} = \begin{bmatrix} 2.9678 & 4.5922 & 1.4616 \end{bmatrix}$$

The output coefficients must be rearranged as

$$X^{C(2)} = \begin{bmatrix} 92 & -3.824 & 11.086 & -11.594 & -8.485 & 1.461 & 4.5922 & 2.967 \end{bmatrix}.$$

Multiplying $X^{C(2)}$ by the scaling factor $\sqrt{\frac{2}{N}}$ and also the first coefficient by the factor $\frac{1}{\sqrt{2}}$ we have

$$X^{C(2)} = \begin{bmatrix} 32.526 & -1.912 & 5.543 & -5.797 & -4.242 & 0.730 & 2.296 & 1.483 \end{bmatrix},$$

which is the DCT-II of the vector x using the DIT strategy.

The MATLAB code is shown following, compare this result using the inbuilt MATLAB function dct( ):

```
% DCT DIT
clear;
x = [10 12 16 8 6 10 14 16]; %This is the input
N = length(x);

x2n= [x(1:2:end) 0];
x2n_minus1= [0 x(2:2:end)];
gn = x2n + x2n_minus1

CI = zeros(N/2+1,N/2+1);
for m = 0:N/2
    for n = 0:N/2
        CI(m+1, n+1) = cos(m*n*pi/(N/2));%%%
    end
end
Gm = CI*gn';

x2n= [x(1:2:end)];
x2n_plus1= [x(2:2:end)];
hn = x2n + x2n_plus1;

CII = zeros(N/2,N/2);
for m = 0:N/2-1
    for n = 0:N/2-1
        CII(m+1, n+1) = cos(m*(2*n+1)*pi/(N));%%%
    end
end
Hm = CII*hn';
```

```
% Sum
k=0:N/2;
ScaleSum = cos(k.*pi./16)';
XC2sum= 0.5*(Gm + [Hm;0])./ScaleSum;

% Dif
m = 0:(N/2)-2;
k = N-m-1;
ScaleDif = cos(k*pi/16);
XC2dif = 0.5*(Gm(m+2)-Hm(m+2))./ScaleDif';

% DCT-II
XC2=[XC2sum; fliplr(XC2dif')'].*sqrt(2/N);
XC2(1) = XC2(1)/sqrt(2);
```

The matrix factorization corresponding to the procedure expressed by (3.117) and (3.118) is given by

$$\left[C_N^{II}\right] = \text{diag}\left\{\left[M_{N/2+1}^{I}\right], \left[M_{N/2-1}^{IV}\right]\right\}\left[A_N^{IV}\right]\text{diag}\left\{\left[I_{N/4+1}\right], \left[\tilde{I}_{N/4}\right], \left[I_{N/2}\right]\right\}$$
$$\times\,\text{diag}\left\{\left[C_{N/2+1}^{I}\right], \left[C_{N/2}^{II}\right]\right\}\left[P_{N+1}\right]\left[B_{N+1}^{I}\right], \tag{3.119}$$

where $\left[C_{N+1}^{I}\right]$ is the DCT-I transform matrix given by

$$\left[C_{N+1}^{I}\right] = \left[A_{N+1}^{I}\right]\text{diag}\left\{\left[I_{N/4+1}\right], \left[\tilde{I}_{N/4+1}\right], \left[M_{N/4+1}^{I}\right], \left[M_{N/4-1}^{II}\right]\right\}$$
$$\times\,\text{diag}\left\{\left[C_{N/2+1}^{I}\right], \left[A_{N/2}^{II}\right]\left[C_{N/2+1}^{I}\right]\left[B_{N/2+1}^{I}\right]\right\}\left[P_{N+1}\right]. \tag{3.120}$$

The salient feature of these factors is the presence of the DCT-I and DCT-II matrices of half the original size. The other factors, although complicated in appearance, are in fact quite simple. They are either butterfly matrices with +1 or -1 as elements, or are permutation matrices of some kind. The $[M]$ matrices are either diagonal or opposite identity (diagonal), serving as scaling factor matrices. For completeness, we describe the other factor matrices in what follows:

$$[P_N] = \begin{bmatrix} 1 & 0 & 0 & 0 & \cdots & \cdots & \cdots \\ 0 & 0 & 1 & 0 & \cdots & \cdots & \cdots \\ \cdots & \cdots & \cdots & \cdots & \cdots & \cdots & \cdots \\ 0 & 1 & 0 & 0 & \cdots & \cdots & \cdots \\ 0 & 0 & 0 & 1 & \cdots & \cdots & \cdots \\ \cdots & \cdots & \cdots & \cdots & \cdots & \cdots & \cdots \end{bmatrix},$$

i.e., an $N \times N$ even-odd permutation matrix;

$$\left[B^{I}_{N+1}\right] = \begin{bmatrix} 1 & 0 & 0 & 0 & \cdots & \cdots & \cdots \\ 1 & 1 & 0 & 0 & \cdots & \cdots & \cdots \\ 0 & 1 & 1 & 0 & \cdots & \cdots & \cdots \\ \cdots & \cdots & \cdots & \cdots & \cdots & \cdots & \cdots \\ \cdots & \cdots & \cdots & \cdots & \cdots & 1 & 1 \\ \cdots & \cdots & \cdots & \cdots & \cdots & 0 & 1 \end{bmatrix},$$

i.e., an $(N+1) \times N$ matrix made up of an $N \times N$ lower bi-diagonal matrix appended by a unit Nth row vector;

$$\left[A^{I}_{N+1}\right] = \begin{bmatrix} \left[I_{N/2}\right] & \vdots & \left[I_{N/2}\right] \\ \cdots & 1 & \cdots \\ \left[I_{N/2}\right] & \vdots & \left[I_{N/2}\right] \end{bmatrix}$$

i.e., an $(N+1) \times (N+1)$ matrix with unit elements;

$$\left[A^{II}_{N}\right] = \begin{bmatrix} & \left[I_{N-1}\right] & & 0 & 0 \\ & \vdots & & \vdots & \vdots \\ 0 & 0 & \cdots & 0 & 1 \end{bmatrix},$$

i.e., an $N \times (N+1)$ matrix with unit elements;

$$\left[A^{IV}_{N}\right] = \begin{bmatrix} & \left[I_{N/2}\right] & \vdots & & \left[I_{N/2}\right] \\ & \cdots & 1 & & \cdots \\ \vdots & \left[I_{N/2-1}\right] & \vdots & \vdots & -\left[I_{N/2-1}\right] \end{bmatrix},$$

i.e., an $N \times (N+1)$ matrix with unit elements;

$$\left[M^{I}_{N+1}\right] = \text{diag}\{1, 1/C^{1}_{4N}, \cdots, 1/C^{N}_{4N}\};$$

$$\left[M^{II}_{N-1}\right] = \text{opposite diag}\{1/C^{N+1}_{4N}, \cdots, 1/C^{2N-1}_{4N}\};$$

and

$$\left[M^{IV}_{N-1}\right] = \text{diag}\{1/C^{2N-1}_{4N}, 1/C^{2N-2}_{4N}, \cdots, 1/C^{N+1}_{4N}\}. \tag{3.121}$$

The sparseness of these factor matrices is quite apparent, and simple difference equations can be established to delineate the computational complexity in terms of the numbers of additions and multiplications. Excluding scaling and normalization, it is found that for an N-point (radix-2) sequence $\{x_n\}$, this DIT algorithm for DCT-II requires

$$\left(\frac{3N}{2}-1\right)\log_2 N+\frac{N}{4}+1 \qquad \text{real additions}$$

and

$$\left(\frac{N}{2}\right)\log_2 N+\frac{N}{4} \qquad \text{real multiplications.}$$

These numbers compare favorably with those of other algorithms.

The DIF Algorithm

When the rearrangement of the sample points results in the transformed sequence being grouped into even- and odd-frequency indexed parts, the decomposition is said to constitute a DIF algorithm. In the case of DCT-II, we shall see a very obvious recursive structure. We proceed by rearranging the transformed sequence $\{X_m^{C(2)}\}$ into even and odd portions so that,

$$X_m^{C(2)} = G_m,$$

and

$$X_{2m+1}^{C(2)} = H_m + H_{m+1}, \qquad m = 0, 1, \cdots, \frac{N}{2}-1, \tag{3.122}$$

where the sequences $\{G_m\}$ and $\{H_m\}$ are DCT-IIs of the two rearranged and recombined sequences, as shown by the following equations

$$G_m = \sum_{n=0}^{N/2-1} [x_n + x_{N-n-1}]\, C_N^{(2n+1)m},$$

and

$$H_m = \sum_{n=0}^{N/2-1} [x_n - x_{N-n-1}]\, \frac{C_N^{(2n+1)m}}{C_N^{(2n+1)}}. \tag{3.123}$$

Compared with (3.116), it is easy to see that both $\{G_m\}$ and $\{H_m\}$ are DCT-IIs of $N/2$ points. Figure 3.10 shows the signal flowgraph for $N = 16$.

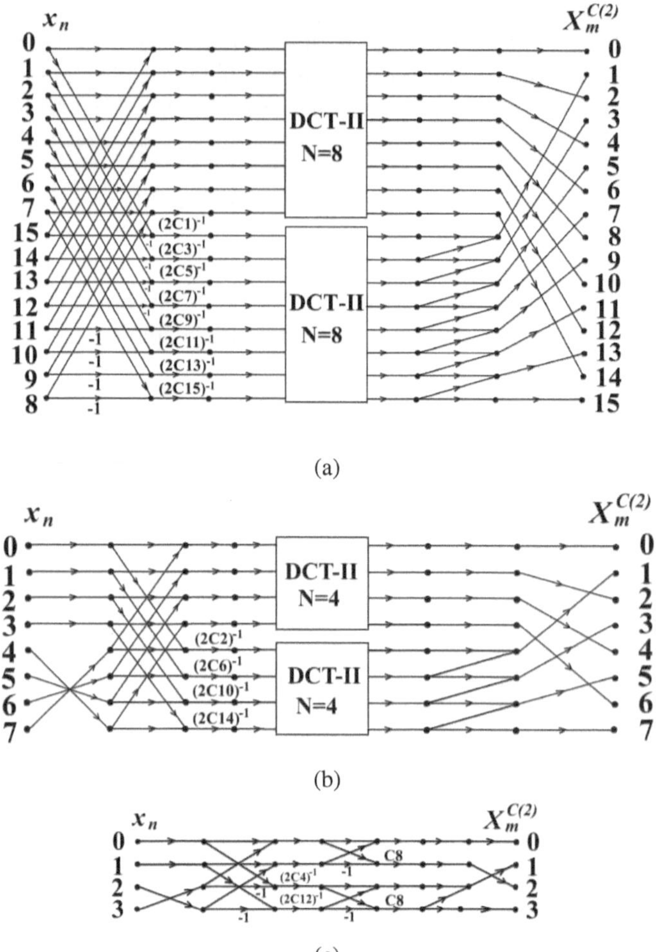

Figure 3.10: Flowgraph for DIF computation of 16-point DCT-II. $Ck = \cos[k\pi/32]$. (a) DIF DCT-II for $N = 16$. (b) DIF DCT-II for $N = 8$. (c) DIF DCT-II for $N = 4$ [390].

The matrix factorization corresponding to the above procedure is given by

$$\left[C_N^{II}\right] = [P_N]^T \operatorname{diag}\left\{\left[I_{N/2}\right], \left[B_{N/2}^{I}\right]^T\right\} \operatorname{diag}\left\{\left[C_{N/2}^{II}\right], \left[C_{N/2}^{II}\right]\right\}$$
$$\times \operatorname{diag}\left\{\left[I_{N/2}\right], \left[M_{N/2}^{III}\right]\right\} \left[A_N^{III}\right] \operatorname{diag}\left\{\left[I_{N/2}\right], \left[\tilde{I}_{N/2}\right]\right\}. \qquad (3.124)$$

One popular DIF algorithm to compute the DCT-II (3.36) is that developed by B. G. Lee [268], [269]. By doing $\hat{X}_m^{C(2)} = k_m X_m^{C(2)}$ and $C_{2N}^{(2k+1)n} = \cos[n(2k+1)\pi/2N]$

the N-point IDCT becomes,

$$x_n = \sum_{m=0}^{N-1} \hat{X}_m^{C(2)} C_{2N}^{(2k+1)n}, \qquad n = 0, 1, \cdots, N-1; \tag{3.125}$$

Then, x_n can be split into even and odd indices of m and assuming N even. Without loss of generality (3.125) can be rewritten as

$$x_n = g_n + h_n^{'} \tag{3.126a}$$

$$x_{N-n-1} = g_n - h_n^{'} \qquad n = 0, 1, \cdots, N/2-1 \tag{3.126b}$$

where

$$\begin{aligned} g_n &= \sum_{m=0}^{N/2-1} \hat{X}_{2m} C_{2N}^{(2n+1)2m} \\ &= \sum_{m=0}^{N/2-1} \hat{X}_{2m} C_{2(N/2)}^{(2n+1)m} \end{aligned} \tag{3.127a}$$

$$h_n^{'} = \sum_{m=0}^{N/2-1} \hat{X}_{2m+1} C_{2N}^{(2n+1)(2m+1)}$$

$$2C_{2N}^{2n+1} h_n^{'} = \sum_{m=0}^{N/2-1} [\hat{X}_{2m+1} + \hat{X}_{2m-1}] C_{2(N/2)}^{(2n+1)m}, \tag{3.127b}$$

which is expressed as

$$h_n^{'} = \frac{1}{2C_{2N}^{2n+1}} h_n$$

where

$$h_n = \sum_{m=0}^{N/2-1} [\hat{X}_{2m+1} + \hat{X}_{2m-1}] C_{2(N/2)}^{(2n+1)m}, \qquad n = 0, 1, \cdots, N/2-1. \tag{3.128}$$

Finally,

$$x_n = g_n + \frac{1}{2C_{2N}^{2n+1}} h_n \tag{3.129a}$$

$$x_{N-n-1} = g_n - \frac{1}{2C_{2N}^{2n+1}} h_n. \tag{3.129b}$$

The algorithm splits an N-point inverse discrete cosine transform (IDCT) into the sum of two $N/2$-point IDCTs. The process can be carried out recursively. The DCT can be decomposed in a similar manner or it can be obtained by simply reversing the directions of the arrows in a IDCT (transposing) since the DCT is an orthogonal transform. Figure 3.11 shows the signal flowgraph for this 8-point forward DCT.

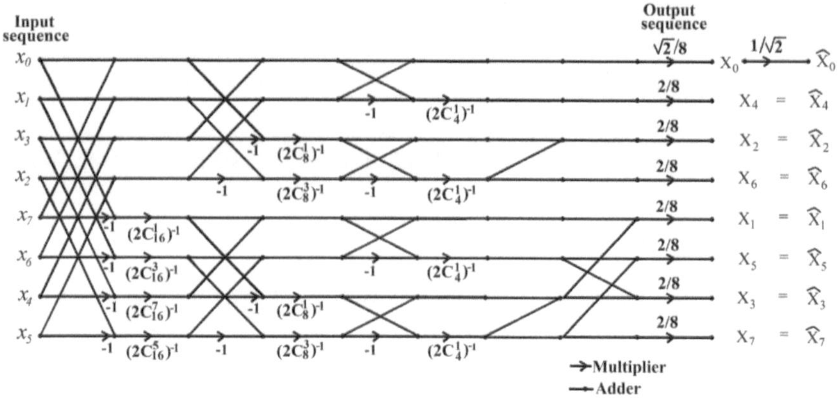

Figure 3.11: B.G. Lee signal flowgraph for the $N = 8$-point forward DCT. $C_i^k = \cos(k\pi/i)$ [268], [269].

Example 3.9. *Figure 3.12 shows the signal flowgraph for the inverse DCT (IDCT) defined by B. G. Lee [268], [269]. Write down the sparse matrix factors (SMFs) of each stage (S_k, $k = 1, \cdots, 8$) of the flowgraph and the final result.*

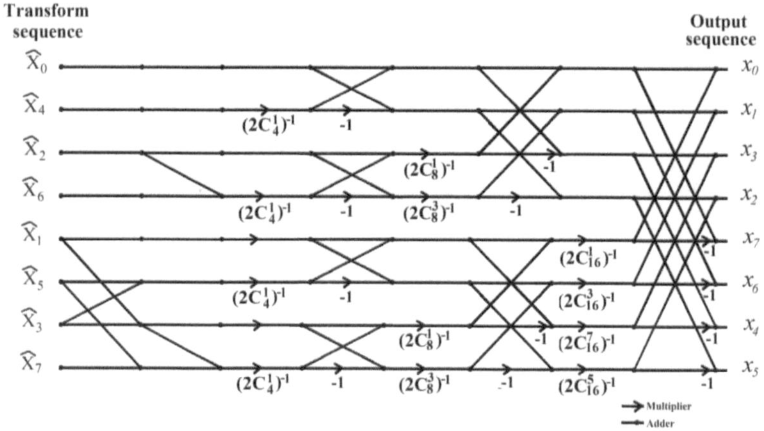

Figure 3.12: B.G. Lee signal flowgraph for the $N = 8$-point inverse DCT. $C_i^k = \cos(k\pi/i)$ [268], [269].

Clearly, from stages S_1 to S_7, the diagram can be divided into an upper and a lower part. The upper part can be defined by the upper coefficients $\hat{X}_0, \hat{X}_4, \hat{X}_2$ and $\hat{X}_6$ and the lower part by the lower coefficients $\hat{X}_1, \hat{X}_5, \hat{X}_3$ and $\hat{X}_7$. Observe the symmetry of the upper and lower parts from S_2 to S_6. Therefore, similar matrices define these two parts. The last stage is an 8×8 matrix. Also notice that $C_k^i = \cos(k\pi/i)$

- *Stage 1*
  Upper part

$$\begin{bmatrix} 1 & 0 & 0 & 0 \\ 0 & 1 & 0 & 0 \\ 0 & 0 & 1 & 0 \\ 0 & 0 & 0 & 1 \end{bmatrix} = I_4$$

  Lower part

$$\begin{bmatrix} 1 & 0 & 0 & 0 \\ 0 & 1 & 1 & 0 \\ 1 & 0 & 1 & 0 \\ 0 & 1 & 0 & 1 \end{bmatrix} = \left[\begin{array}{c|cc} I_2 & 0 & 0 \\ & 1 & 0 \\ \hline I_2 & I_2 & \end{array}\right]$$

- *Stage 2*
  Upper part

$$\left[\begin{array}{cc|cc} 1 & 0 & 0 & 0 \\ 0 & 1 & 0 & 0 \\ \hline 0 & 0 & 1 & 0 \\ 0 & 0 & 1 & 1 \end{array}\right] = diag\left\{I_2, \begin{bmatrix} 1 & 0 \\ 1 & 1 \end{bmatrix}\right\}$$

  Lower part

$$\left[\begin{array}{cc|cc} 1 & 0 & 0 & 0 \\ 0 & 1 & 0 & 0 \\ \hline 0 & 0 & 1 & 0 \\ 0 & 0 & 1 & 1 \end{array}\right] = diag\left\{I_2, \begin{bmatrix} 1 & 0 \\ 1 & 1 \end{bmatrix}\right\}$$

- *Stage 3*
  Upper part

$$\begin{bmatrix} 1 & 0 & 0 & 0 \\ 0 & (2C_4^1)^{-1} & 0 & 0 \\ 0 & 0 & 1 & 0 \\ 0 & 0 & 0 & (2C_4^1)^{-1} \end{bmatrix} = diag\left\{1, (2C_4^1)^{-1}, 1, (2C_4^1)^{-1}\right\}$$

  Lower part

$$\begin{bmatrix} 1 & 0 & 0 & 0 \\ 0 & (2C_4^1)^{-1} & 0 & 0 \\ 0 & 0 & 1 & 0 \\ 0 & 0 & 0 & (2C_4^1)^{-1} \end{bmatrix} = diag\left\{1, (2C_4^1)^{-1}, 1, (2C_4^1)^{-1}\right\}$$

- *Stage 4*
  Upper part

$$\begin{bmatrix} 1 & 1 & 0 & 0 \\ 1 & -1 & 0 & 0 \\ 0 & 0 & 1 & 1 \\ 0 & 0 & 1 & -1 \end{bmatrix} = diag\left\{\begin{bmatrix} 1 & 1 \\ 1 & -1 \end{bmatrix}, \begin{bmatrix} 1 & 1 \\ 1 & -1 \end{bmatrix}\right\}$$

Lower part

$$\begin{bmatrix} 1 & 1 & 0 & 0 \\ 1 & -1 & 0 & 0 \\ 0 & 0 & 1 & 1 \\ 0 & 0 & 1 & -1 \end{bmatrix} = diag\left\{\begin{bmatrix} 1 & 1 \\ 1 & -1 \end{bmatrix}, \begin{bmatrix} 1 & 1 \\ 1 & -1 \end{bmatrix}\right\}$$

- *Stage 5*
  Upper part

$$\begin{bmatrix} 1 & 0 & 0 & 0 \\ 0 & 1 & 0 & 0 \\ 0 & 0 & (2C_8^1)^{-1} & 0 \\ 0 & 0 & 0 & (2C_8^3)^{-1} \end{bmatrix} = diag\left\{I_2, (2C_8^1)^{-1}, (2C_8^3)^{-1}\right\}$$

Lower part

$$\begin{bmatrix} 1 & 0 & 0 & 0 \\ 0 & 1 & 0 & 0 \\ 0 & 0 & (2C_8^1)^{-1} & 0 \\ 0 & 0 & 0 & (2C_8^3)^{-1} \end{bmatrix} = diag\left\{I_2, (2C_8^1)^{-1}, (2C_8^3)^{-1}\right\}$$

- *Stage 6*
  Upper part

$$\left[\begin{array}{cc|cc} 1 & 0 & 1 & 0 \\ 0 & 1 & 0 & 1 \\ \hline 1 & 0 & -1 & 0 \\ 0 & 1 & 0 & -1 \end{array}\right] = diag\left\{\left[\begin{array}{c|c} I_2 & I_2 \\ \hline I_2 & -I_2 \end{array}\right]\right\}$$

Lower part

$$\left[\begin{array}{cc|cc} 1 & 0 & 1 & 0 \\ 0 & 1 & 0 & 1 \\ \hline 1 & 0 & -1 & 0 \\ 0 & 1 & 0 & -1 \end{array}\right] = diag\left\{\left[\begin{array}{c|c} I_2 & I_2 \\ \hline I_2 & -I_2 \end{array}\right]\right\}$$

- *Stage 7*
  Upper part

$$\begin{bmatrix} 1 & 0 & 0 & 0 \\ 0 & 1 & 0 & 0 \\ 0 & 0 & 1 & 0 \\ 0 & 0 & 0 & 1 \end{bmatrix} = I_4$$

Lower part

$$diag\left\{(2C_{16}^1)^{-1}, (2C_{16}^3)^{-1}, (2C_{16}^7)^{-1}, (2C_{16}^5)^{-1}\right\}$$

- *Stage 8*

$$\left[\begin{array}{cccc|cccc} 1 & 0 & 0 & 0 & 1 & 0 & 0 & 0 \\ 0 & 1 & 0 & 0 & 0 & 1 & 0 & 0 \\ 0 & 0 & 1 & 0 & 0 & 0 & 1 & 0 \\ 0 & 0 & 0 & 1 & 0 & 0 & 0 & 1 \\ \hline 1 & 0 & 0 & 0 & -1 & 0 & 0 & 0 \\ 0 & 1 & 0 & 0 & 0 & -1 & 0 & 0 \\ 0 & 0 & 1 & 0 & 0 & 0 & -1 & 0 \\ 0 & 0 & 0 & 1 & 0 & 0 & 0 & -1 \end{array}\right] = \left[\begin{array}{c|c} I_4 & I_4 \\ \hline I_4 & -I_4 \end{array}\right]$$

The final result is,

$$\begin{aligned} [x_0, x_1, x_3, x_2, x_7, x_6, x_4, x_5]^T &= \left[\begin{array}{c|c} I_4 & I_4 \\ \hline I_4 & -I_4 \end{array}\right] \\ &\times diag\left\{I_4, (2C_{16}^1)^{-1}, (2C_{16}^3)^{-1}, (2C_{16}^7)^{-1}, (2C_{16}^5)^{-1}\right\} \\ &\times diag\left\{\left[\begin{array}{c|c} I_2 & I_2 \\ \hline I_2 & -I_2 \end{array}\right], \left[\begin{array}{c|c} I_2 & I_2 \\ \hline I_2 & -I_2 \end{array}\right]\right\} \\ &\times diag\left\{I_2, (2C_8^1)^{-1}, (2C_8^3)^{-1}, I_2, (2C_8^1)^{-1}, (2C_8^3)^{-1}\right\} \\ &\times diag\left\{\begin{bmatrix} 1 & 1 \\ 1 & -1 \end{bmatrix}, \begin{bmatrix} 1 & 1 \\ 1 & -1 \end{bmatrix}, \begin{bmatrix} 1 & 1 \\ 1 & -1 \end{bmatrix}, \begin{bmatrix} 1 & 1 \\ 1 & -1 \end{bmatrix}\right\} \\ &\times diag\left\{1, (2C_4^1)^{-1}, 1, (2C_4^1)^{-1}, 1, (2C_4^1)^{-1}, 1, (2C_4^1)^{-1}\right\} \\ &\times diag\left\{I_2, \begin{bmatrix} 1 & 0 \\ 1 & 1 \end{bmatrix}, I_2, \begin{bmatrix} 1 & 0 \\ 1 & 1 \end{bmatrix}\right\} diag\left\{I_4, \left[\begin{array}{c|cc} I_2 & 0 & 0 \\ & 1 & 0 \\ \hline I_2 & I_2 & \end{array}\right]\right\} \\ &\times [\hat{X}_0^{C(2)}, \hat{X}_4^{C(2)}, \hat{X}_2^{C(2)}, \hat{X}_6^{C(2)}, \hat{X}_1^{C(2)}, \hat{X}_5^{C(2)}, \hat{X}_3^{C(2)}, \hat{X}_7^{C(2)}]^T \end{aligned}$$

THE DCT VIA THE WALSH–HADAMARD TRANSFORM (WHT)

The Walsh–Hadamard transform (WHT) [178, 234, 495] is known to be fast since the computation involves no multiplications. Thus, an algorithm for DCT via WHT may well utilize this advantage. The WHT is a suboptimal, real, orthogonal transform that projects a signal into rectangular waveforms called Walsh functions. The functions have only two values, +1 or -1.

Let x be an $(N \times 1)$ column vector representing the input data and $X^{C(2)}$ and X_W be $(N \times 1)$ column vectors representing the DCT-II and the WHT of x, respectively. Then,

$$X^{C(2)} = \left[C_N^{II}\right] x \qquad \text{and} \qquad X_W = [H_W] x, \tag{3.130}$$

i.e., for $N = 2$, $N = 4$ and $N = 8$, the natural-ordered matrix $[H_W]_N$ is

$$[H_W]_2 = \begin{bmatrix} 1 & 1 \\ 1 & -1 \end{bmatrix}.$$

$$[H_W]_4 = \begin{bmatrix} 1 & 1 & 1 & 1 \\ 1 & -1 & 1 & -1 \\ 1 & 1 & -1 & -1 \\ 1 & -1 & -1 & 1 \end{bmatrix}.$$

$$[H_W]_8 = \begin{bmatrix} 1 & 1 & 1 & 1 & 1 & 1 & 1 & 1 \\ 1 & -1 & 1 & -1 & 1 & -1 & 1 & -1 \\ 1 & 1 & -1 & -1 & 1 & 1 & -1 & -1 \\ 1 & -1 & -1 & 1 & 1 & -1 & -1 & 1 \\ 1 & 1 & 1 & 1 & -1 & -1 & -1 & -1 \\ 1 & -1 & 1 & -1 & -1 & 1 & -1 & 1 \\ 1 & 1 & -1 & -1 & -1 & -1 & 1 & 1 \\ 1 & -1 & -1 & 1 & -1 & 1 & 1 & -1 \end{bmatrix}.$$

In general, we can construct $[H_W]_N$ recursively by doing

$$[H_W]_N = \begin{bmatrix} [H_W]_{N/2} & [H_W]_{N/2} \\ [H_W]_{N/2} & -[H_W]_{N/2} \end{bmatrix}.$$

Notice that: $[H_W]_N * [H_W]_N^T = NI_N$ where I_N is the $N \times N$ identity matrix.

$\left[C_N^{II}\right]$ and $[H_W]$ are the $(N \times N)$ transform matrices for the DCT and WHT (Walsh ordered) [8], [122], [386], respectively. Since $[H_W]$ is real and orthonormal, we see that

$$\begin{aligned} X^{C(2)} &= \left[C_N^{II}\right] [H_W]^T [H_W] x \\ &= [T_N] X_W, \end{aligned} \tag{3.131a}$$

where $[T_N] = \left[C_N^{II}\right] [H_W]^T$ is the conversion matrix that takes the Walsh domain vector X_W and converts it to the DCT domain vector $X^{C(2)}$. This conversion matrix has a block diagonal structure if the transformed vectors X_W and $X^{C(2)}$ are arranged in bit-reversed order (BRO) (see Hein and Ahmed [178]). Using the subscript BRO to denote such bit-reversed ordering, (3.131a) can be rewritten as

$$\begin{aligned} X_{BRO}^{C(2)} &= \left[C_N^{II}\right]_{BRO} [H_W]_{BRO}^T [H_W]_{BRO} x, \\ &= [T_N]_{BRO} [H_W]_{BRO} x. \end{aligned} \tag{3.131b}$$

For example, when $N = 8$, the conversion matrix is given explicitly as

$$[T_8]_{BRO} = \begin{bmatrix} [1.0] & & & \\ & [1.0] & & 0 \\ & & \begin{bmatrix} 0.923 & 0.383 \\ -0.383 & 0.923 \end{bmatrix} & \\ & 0 & & \begin{bmatrix} 0.907 & -0.075 & 0.375 & 0.180 \\ 0.214 & 0.768 & -0.513 & 0.318 \\ -0.318 & 0.513 & 0.768 & 0.214 \\ -0.180 & -0.375 & -0.075 & 0.907 \end{bmatrix} \end{bmatrix}. \quad (3.132)$$

Since the WHT requires no multiplications, (3.132) accounts for all the multiplications required to compute $X_{BRO}^{C(2)}$. For $N = 8$, (3.132) indicates 20 real multiplications, as compared with 24 real multiplications needed for its computation using FFT (see Section 3). This computational advantage quickly disappears as N increases beyond 16. However, when many small transforms have to be computed, such as in the case of two-dimensional image processing, the accumulated savings may be substantial. In addition, the BRO as well as the block diagonal structure of the conversion matrix completely separates the even-frequency components from the odd-frequency components in the transformed vector, providing a possible means for data compression in the frequency domain. It may be noted that [495] has extended the conversion matrix up to $N = 32$ (see Problems 3.24. and 3.25.).

THE DCT VIA THE DISCRETE HARTLEY TRANSFORM (DHT)

The discrete Hartley transform (DHT) is very similar to DFT and has some computational advantages if only real input sequences are to be processed.

Recall that the kernel for the DFT of N points is given by

$$W_N^{mn} = \exp\left(-\frac{j2mn\pi}{N}\right) = \cos\left[\frac{2mn\pi}{N}\right] - j\sin\left[\frac{2mn\pi}{N}\right],$$
$$m,n = 0,1,\cdots,N-1 \quad (3.133)$$

The kernel of an N-point DHT is given by

$$H_N^{mn} \equiv \text{cas}\left(\frac{2mn\pi}{N}\right) = \cos\left[\frac{2mn\pi}{N}\right] + \sin\left[\frac{2mn\pi}{N}\right],$$
$$m,n = 0,1,\cdots,N-1 \quad (3.134)$$

where $\text{cas}(\theta) = \cos(\theta) + \sin(\theta)$. If X_F and X_H are used to denote the DFT and the DHT of x, respectively, then it is not difficult to establish the relation

$$X_H = \text{Re}\{X_F\} - \text{Im}\{X_F\}, \quad (3.135)$$

provided x is real. Equation (3.135) indicates that for any fast DFT algorithm, there should be a corresponding fast algorithm for DHT, as was indeed demonstrated by Sorensen et al. [447]. Therefore, there exists a relation between DCT and DHT analogous to the relation between DCT and DFT discussed in Section 3.

The following discussion (see [184], [185], [305], [306], [307], [338]) delineates this relation and provides an algorithm to compute DCT via DHT. A similar relationship has also been developed by Hou [184], [185]. The recursive structure of the fast Hartley transform (FHT) is exploited.

Let the DCT-II transformed sequence $\{X_m^{C(2)}\}$ be given as in (3.116) by

$$X_m^{C(2)} = \sum_{n=0}^{N-1} x_n C_{2N}^{(2n+1)m}, \qquad m = 0, 1, \cdots, N-1, \tag{3.116}$$

We consider the DHT of a reordered sequence of $\{x_n\}$ and its relation to $\{X_m^{C(2)}\}$ in (3.116). First, we construct the sequence $\{y_n\}$ from $\{x_n\}$ by letting

$$\begin{aligned} y_n &= x_{2n}, & n &= 0, 1, \cdots, N/2-1, \\ &= x_{2N-2n-1}, & n &= N/2, \cdots, N-1. \end{aligned} \tag{3.136}$$

The DHT of $\{y_n\}$ is given by

$$Y_H^m = \sum_{n=0}^{N-1} y_n \text{cas}\left(\frac{2mn\pi}{N}\right) = \sum_{n=0}^{N-1} y_n H_N^{mn}. \tag{3.137}$$

Now, using (3.136) we obtain,

$$Y_H^m = \sum_{\textit{even } n} x_n \text{cas}\left[\frac{mn\pi}{N}\right] + \sum_{\textit{odd } n} x_n \text{cas}\left[-\frac{m(n+1)\pi}{N}\right]. \tag{3.138}$$

Notice that cas(α)cas(β)=cas($\alpha - \beta$), therefore,

$$Y_m^{C(2)} = \frac{1}{2}\left\{Y_H^m \text{cas}\left(-\frac{m\pi}{2N}\right) + Y_H^{N-m} \text{cas}\left(\frac{m\pi}{2N}\right)\right\}. \tag{3.139}$$

Using H_N^{mn} to denote cas($2mn\pi/N$), (3.139) can be written more concisely,

$$\begin{bmatrix} X_m^{C(2)} \\ X_{N-m}^{C(2)} \end{bmatrix} = \frac{1}{2} \begin{bmatrix} H_{4N}^{-m} & H_{4N}^{m} \\ H_{4N}^{N-m} & -H_{4N}^{-(N-m)} \end{bmatrix} \begin{bmatrix} Y_H^m \\ Y_H^{N-m} \end{bmatrix}. \tag{3.140a}$$

For $m = 0$ and $N/2$, the transformed points are determined by

$$X_0^{C(2)} = Y_H^0 \quad \text{and} \quad X_{N/2}^{C(2)} = \frac{Y_H^{N/2}}{\sqrt{2}} \tag{3.140b}$$

The amount of computation necessary for an N-point sequence compares favorably with that of other algorithms. Figure 3.13 shows the signal flowgraph for the computation of an $N = 16$ DCT-II via the DHT.

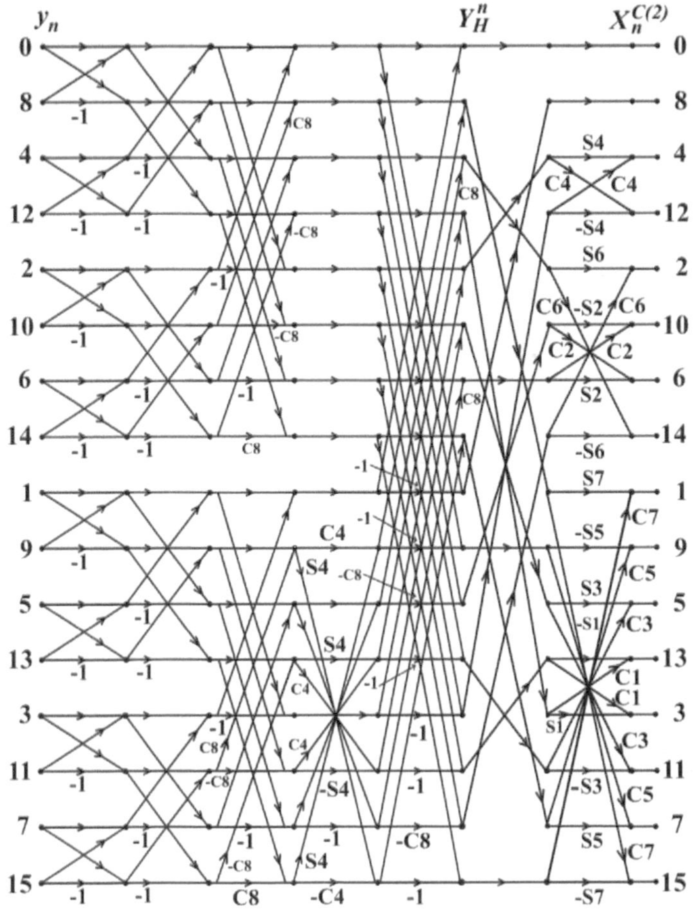

Figure 3.13: Signal flowgraph for computing DCT via DHT. $Ck = \cos(k\pi/32)$ and $Sk = \sin(k\pi/32)$, $N = 16$ [390].

OTHER FAST ALGORITHMS

In addition to the algorithms described in the previous sections, there are many more. Some are developed for specific applications. Others are developed for specific numbers of input sample points. Radix-2 algorithms, described in previous sections, may not be the most appropriate. A case in point is the implementation of transmultiplexers for the conversion between time-division multiplexing (TDM) and frequency-division multiplexing (FDM) in voice transmission over a telecommunication channel [341, 344, 342]. The required sampling rates dictate that the DCT-II used to implement the transmultiplexers have lengths of 14 and 72 points. Another situation arises in the consideration of systolic arrays implementation, where fast computation of DCT-II may be accomplished through the use of the Winograd Fourier transform algorithm (WFTA) [435], [436], [482]. In this case, radix-2 factorization is a special case. In Section 3, the prime factor algorithm (PFA) [554] is discussed as representative of the factorizations not based on radix-2.

Many of the algorithms discussed have been considered for very-large-scale integration (VLSI) implementation [283, 416, 272, 162, 187, 263, 166, 5, 227, 366, 431, 583, 322, 426, 491, 369, 238, 103, 18, 17, 239, 440, 460]. In such an implementation, the orderly architecture and the numerical stability are important considerations, in addition to the recursive and modular nature of the algorithm. In Section 3 an algorithm that has these important ingredients is discussed. Unlike some of the others, this algorithm produces a 2^m-point DCT from two 2^{m-1}-point DCTs in much the same way as the Cooley–Tukey algorithm for the DFT [95]. Another aspect to be considered in the physical realization of the algorithm is the number of different basic processors required. In this regard, Ligtenberg and O'Neil [283] and Loeffler et al. [289, 290] proposed a procedure to compute the DCT using only planar rotations.

Prime Factor Algorithm (PFA)

The prime factor algorithm (PFA) for DCT-II was developed (see [554]) based on the PFA for DFT [435], [436], [482]. Since DCT-II can be directly related to the DFT, as shown in Section 3, we describe first the PFA for DFT.

Consider an input sequence $\{y_n\}$ of $N = N_1 \times N_2$ points. If N_1 and N_2 are relatively prime, the one-dimensional sequence $\{y_n\}$ can be mapped to a two-dimensional array $\{y_{n_1,n_2}\}$ where the indices are related by the equation,

$$n = n_1 r_1 N_2 + n_2 r_2 N_1 \text{ modulo } N. \tag{3.141}$$

Here r_1 and r_2 are the inverses of N_2 modulo N_1 and N_1 modulo N_2 respectively, i.e.,

$$r_1 N_2 = 1 \text{ modulo } N_1 \text{ and } r_2 N_1 = 1 \text{ modulo } N_2$$

or

$$r_1 N_2 = 1 \text{ and } r_2 N_1 = 1$$

For example, when $N_1 = 3$ and $N_2 = 4$, $r_1 = 1$ and $r_2 = 3$. The input index mapping for n is given by Table 3.1.

Table 3.1
Input index mapping for n

n_2 / n_1	0	1	2	3
		n		
0	0	9	6	3
1	4	1	10	7
2	8	5	2	11

$\{Y_m\}$, the DFT of $\{y_n\}$, can be expressed as a cascaded transform of the two-dimensional array $\{y_{n_1,n_2}\}$ given by

$$\begin{aligned} Y_m &= Y_{m_1,m_2} \\ &= \sum_{n_1=0}^{N_1-1}\left[\sum_{n_2=0}^{N_2-1} y_{n_1,n_2} W_{N_2}^{m_2 n_2}\right] W_{N_1}^{m_1 n_1}, \end{aligned} \tag{3.142a}$$

where the output indices are mapped by the relation

$$m = m_1 N_2 + m_2 N_1 \text{ modulo } N. \tag{3.142b}$$

Figure 3.14 illustrates the essential features of this decomposition for $N_1 = 3$ and $N_2 = 4$.

Since the DCT-II is related to the DFT, we consider a scaled output sequence of $\{Y_m\}$ given by

$$Z_m = W_{4N}^m Y_m = W_{4N}^m \sum_{n=0}^{N-1} y_n W_N^{mn}. \tag{3.143}$$

Then, as is similarly shown in Section 3, the DCT-II of a sequence $\{x_n\}$ is given by the real part of $\{Z_m\}$, provided it is related to the sequence $\{y_n\}$ by

$$y_n = y_{2n} \text{ and } y_{N-n-1} = x_{2n+1}, \qquad n = 0, 1, \cdots, N/2 - 1. \tag{3.144a}$$

In short, if $\{X_m^{C2}\}$ is the DCT-II of $\{x_n\}$, then

$$\{X_m^{C2}\} = \text{Re}[Z_m], \qquad m = 0, 1, \cdots, N-1 \tag{3.144b}$$

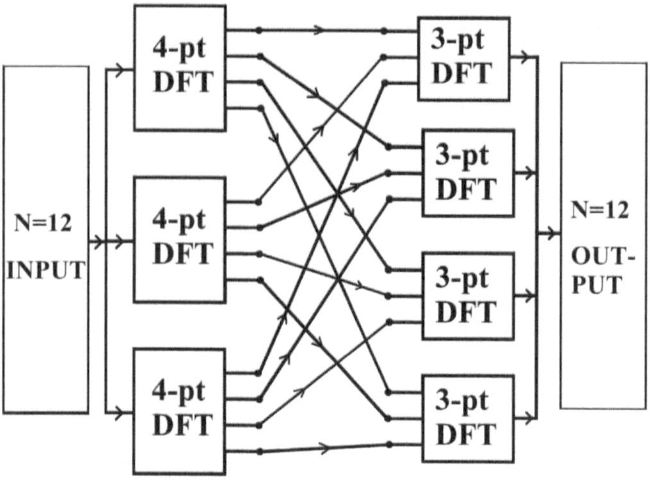

Figure 3.14: Signal block diagram for the PFA of a 12-point DFT. pt = point [390].

Whereas (3.144a) is valid only for even N, a generalization is possible whereby the even-numbered input sample points are arranged in increasing order and the odd-numbered ones are arranged in decreasing order. Thus

$$\begin{aligned} y_n = y_{2n} \quad & n = 0, 1, \cdots, (N-1)/2, && \text{odd } N, \\ & \quad = 0, 1, \cdots, N/2 - 1, && \text{even } N; \\ = x_{2N-2n-1} \quad & n = N-1, \cdots, (N+1)/2, && \text{odd } N, \\ & \quad = N-1, \cdots, N/2, && \text{even } N. \end{aligned} \tag{3.145}$$

It is easy to see that the DCT-II of $\{x_n\}$ can be indirectly obtained by the PFA of DFT (or more correctly, the "phasor-modified" DFT). The drawback, as shown in (3.143), is that $\{Z_m\}$ is a complex sequence, even though $\{y_n\}$ (and in turn, $\{x_n\}$) is a real sequence. Such a drawback may be overcome by directly examining the DCT-II of the two-dimensional sequence $\{y_{n_1,n_2}\}$ whose indices are related to n, as indicated in (3.141). For a real input sequence, (3.102) ensures that

$$Z_0 = X_0^{C(2)},$$

and

$$Z_m = X_m^{C(2)} - jX_{N-m}^{C(2)}, \quad m = 1, 2, \cdots, N-1. \tag{3.146}$$

Since $\{Z_m\}$ is the phasor-modified DFT of $\{y_n\}$ it is clear that $\{X_m^{C(2)}\}$ is given by the DCT-II of $\{y_{n_1,n_2}\}$. The idea of cascading the transform is now applied directly to compute the DCT-II of $\{y_{n_1,n_2}\}$. First we identify a two-dimensional array $\{Z_{m_1,m_2}\}$ with $\{Z_m\}$ using the mapping in (3.142b). Then, we combine (3.142a), (3.142b) and (3.143) to get

$$\{Z_{m_1,m_2}\} = W_{4N}^{m_1N_2 + m_2N_1 - hN} Y_{m_1,m_2}, \tag{3.147a}$$

where

$$h = \begin{cases} 0 & \text{if } m_1N_2 + m_2N_1 < N, \\ 1 & \text{otherwise,} \end{cases}$$

since the index is evaluated modulo N in (3.142b). The cascaded form of (3.147a) is now readily obtained,

$$\{Z_{m_1,m_2}\} = W_4^{-h} Y_{1,m_1,m_2}, \tag{3.147b}$$

where

$$Y_{1,m_1,m_2} = W_{4N_1}^{m_1} \left[\sum_{n_1=0}^{N_1-1} Y_{2,n_1,m_2} W_{N_1}^{m_1n_1} \right],$$

and

$$Y_{2,n_1,m_2} = W_{4N_2}^{m_2} \left[\sum_{n_2=0}^{N_2-1} y_{n_1,n_2} W_{N_2}^{m_2n_2} \right].$$

We note that Y_{2,n_1,m_2} and Y_{1,m_1,m_2} are phasor-modified DFTs of y_{n_1,n_2} on n_2 and Y_{2,n_1,m_2} on n_1 respectively, and are in exactly the same form as (3.143). Therefore, for $\{Y_{2,n_1,m_2}\}$ we have (since $\{y_{n_1,n_2}\}$ is real)

$$Y_{2,n_1,m_2} = Y_{2,n_1,m_2}^{C(2)} - jY_{2,n_1,N_2-m_2}^{C(2)}, \tag{3.148}$$

where $\{Y_{2,n_1,m_2}\}$ is the DCT-II of a sequence $\{y_{2,n_1,n_2}\}$ which maps to the sequence y_{n_1,n_2} on n_2, as $\{x_n\}$ maps to $\{y_n\}$ in (3.145) has a DCT-II denoted by

$$Y_{1,m_1,m_2} = Y_{1,m_1,m_2}^{C(2)} - jY_{1,N_1-m_1,m_2}^{C(2)} - jY_{1,m_1,N_2-m_2}^{C(2)} - Y_{1,N_1-m_1,N_2-m_2}^{C(2)}. \tag{3.149}$$

Here $j = \sqrt{-1}$. The result in (3.149), when combined with (3.147b) and (3.146), gives the relation between $\{X_m^{C(2)}\}$ and $\{Y_{1,m_1,m_2}^{C(2)}\}$. Thus

$$\begin{aligned} X_m^{C(2)} &= Y_{1,m_1,m_2}^{C(2)} \qquad m_1m_2 = 0, \\ &= -Y_{1,N_1-m_1,m_2}^{C(2)} - Y_{1,m_1,N_2-m_2}^{C(2)} \qquad \text{otherwise,} \\ m &= m_1N_2 + m_2N_1 \text{ modulo } N. \end{aligned} \tag{3.150}$$

It is useful to briefly summarize the procedure for the PFA of DCT-II, given an input sequence $\{x(n)\}$ of $N = N_1N_2$ points:

(1). Map $\{x_n\}$ to $\{y_n\}$ using (3.145).
(2). Map $\{y_n\}$ to $\{y_{n_1,n_2}\}$ using (3.141).
(3). Compute the N_2-point DCT-II of $\{y_{n_1,n_2}\}$ on the index n_2 to obtain $\{Y_{2,n_1,m_2}^{C(2)}\}$.
(4). Compute the N_1-point DCT-II of $\{Y_{2,n_1,m_2}\}$ on n_1 to obtain $\{Y_{1,m_1,m_2}\}$.

(5). Obtain $X_m^{C(2)}$ from (3.150).

Because the indices are rearranged each time a DCT-II is computed, the index mappings are not simple when one also includes the mappings required for cascading the transforms. Yang and Narasimha [554] have worked out the mappings, and we will state them here without proof.

A. *Input index mapping:* $n \longleftrightarrow (n_1, n_2)$, for $N = N_1N_2$,

$$n = \begin{cases} \frac{\bar{n}-1}{2} & \text{if } \bar{n} < 2N, \\ \frac{4N-\bar{n}-1}{2} & \text{if } \bar{n} > 2N \end{cases} \tag{3.151}$$

where

$$\bar{n} = \left|(2n_1+1)r_1N_2 - (2n_2+1)(N_2-r_2)N_1\right| \bmod 4N, \quad n_1+n_2 \text{ even,} \tag{3.152}$$

$$= \left|(2n_1+1)r_1N_2 + (2n_2+1)(N_2-r_2)N_1\right| \bmod 4N, \quad n_1+n_2 \text{ odd,} \tag{3.153}$$

mod is the abbreviation for modulo.
Here, r_1 and r_2 are inverses of N_2 and N_1 modulo N_1 and N_2 respectively, [see (3.141)].

B. *Output index mapping:* $m \longleftrightarrow (m_1, m_2)$, for $N = N_1N_2$,

$$m = m_1N_2 + m_2N_1 \qquad \text{if } m_1m_2 = 0. \tag{3.154}$$

For $m_1m_2 \neq 0$, the butterfly shown in Figure 3.15 is required to determine $X_m^{C(2)}$ from $Y_{1,m_1,m_2}^{C(2)}$, as seen in (3.150).

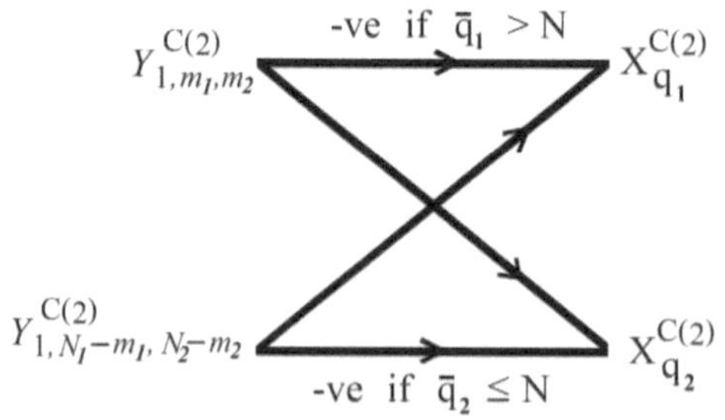

Figure 3.15: Butterfly diagram for the last stage of PFA.

Here,

$$q_1 = |m_1 N_2 - m_2 N_1|, \tag{3.155}$$

$$q_2 = \begin{cases} |\bar{q}_2| & \text{if } \bar{q}_2 \le N, \\ |\bar{q}_2 - 2N| & \text{if } \bar{q}_2 > N, \end{cases}$$

where

$$\bar{q}_2 = (m_1 N_2 + m_2 N_1). \tag{3.156}$$

Figure 3.16 shows the detailed signal flowgraph of $N = 3 \cdot 4$ DCT-II using the PFA. The complete 12-point DCT-II requires 23 multiplications and 46 additions (see Problem 3.27.). Lee [270] has provided tables for mapping the indices for various sizes in the PFA of DCT-II.

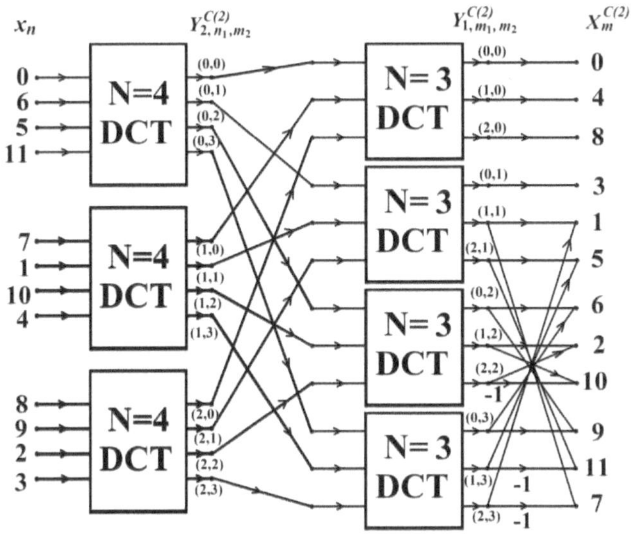

Figure 3.16: Signal flowgraph for the PFA of DCT-II. $N = 3 \cdot 4$ [390].

A Fast Recursive Algorithm

Hou [183] proposed a fast recursive algorithm for computing the DCT-II. The index permutation involved follows closely that of the Cooley-Tukey algorithm for the DFT [95]. It is therefore not surprising that it finds its beginning at the same place as the Narasimha and Peterson algorithm [343] discussed in Section 3. Recall that DCT-II can be obtained as the real part of a scaled DFT if the input sequence undergoes a reordering first, given by

$$y_n = x_{2n},$$

and

$$y_{N-1-n} = x_{2n+1}, \qquad n = 0, 1, \cdots, \frac{N}{2} - 1. \tag{3.98}$$

The transform sequence $\{X_m^{C(2)}\}$

$$X_m^{C(2)} = \sum_{n=0}^{N-1} y_n \cos\left[\frac{(4n+1)m\pi}{2N}\right], \qquad m = 0, 1, \cdots, N-1. \tag{3.100a}$$

Instead of viewing (3.100a) as the real part of a scaled JV-point DFT, Hou delves into the symmetry of the kernel in (3.100a) and partitions the transform matrix to obtain the recursive structure. The concept is best understood by examining the structures of the $N = 4$ and $N = 8$ transforms. For the sake of simplicity, we shall drop the $C(2)$ superscript in the transformed sequence in what follows.

For $N = 4$, we have

$$\begin{bmatrix} X_0 \\ X_2 \\ X_1 \\ X_3 \end{bmatrix} = \begin{bmatrix} 1 & 1 & 1 & 1 \\ \alpha & -\alpha & \alpha & -\alpha \\ \beta & -\delta & -\beta & \delta \\ \delta & \beta & -\delta & -\beta \end{bmatrix} \begin{bmatrix} x_0 \\ x_2 \\ x_3 \\ x_1 \end{bmatrix}. \tag{3.157}$$

For $N = 8$, we have

$$\begin{bmatrix} X_0 \\ X_4 \\ X_2 \\ X_6 \\ X_1 \\ X_5 \\ X_3 \\ X_7 \end{bmatrix} = \begin{bmatrix} 1 & 1 & 1 & 1 & 1 & 1 & 1 & 1 \\ \alpha & -\alpha & \alpha & -\alpha & \alpha & -\alpha & \alpha & -\alpha \\ \beta & -\delta & -\beta & \delta & \beta & -\delta & -\beta & \delta \\ \delta & \beta & -\delta & -\beta & \delta & \beta & -\delta & -\beta \\ \sigma & \mu & -\tau & \varepsilon & -\sigma & -\mu & \tau & -\varepsilon \\ \mu & \tau & -\varepsilon & \sigma & -\mu & -\tau & \varepsilon & -\sigma \\ \varepsilon & -\sigma & \mu & \tau & -\varepsilon & \sigma & -\mu & -\tau \\ \tau & \varepsilon & \sigma & \mu & -\tau & -\varepsilon & -\sigma & -\mu \end{bmatrix} \begin{bmatrix} x_0 \\ x_2 \\ x_4 \\ x_6 \\ x_7 \\ x_5 \\ x_3 \\ x_1 \end{bmatrix}, \tag{3.158}$$

where $\alpha = \frac{1}{\sqrt{2}}$,$\beta = C_8^1$, $\delta = S_8^1$, $\sigma = C_{16}^1$, $\varepsilon = C_{16}^3$, $\mu = S_{16}^3$, $\tau = S_{16}^1$ with $S_k^i = \sin\left(\frac{i\pi}{k}\right)$ and $C_k^i = \cos\left(\frac{i\pi}{k}\right)$. In these input and output permutations, (3.157) and (3.158) suggest that if $[\hat{T}_N]$ is used to represent the permuted transform matrix for the DCT-II, then

$$\begin{bmatrix} [\hat{T}_{N/2}] & [\hat{T}_{N/2}] \\ [\hat{D}_{N/2}] & -[\hat{D}_{N/2}] \end{bmatrix}. \tag{3.159}$$

If $[\hat{D}_{N/2}]$ can be further expressed in terms of $[\hat{T}_{N/2}]$ in a simple manner, the required recursive structure will have been established. We can demonstrate this relation by using the submatrices in (3.158) defined as

$$[\hat{T}_4] = \begin{bmatrix} 1 & 1 & 1 & 1 \\ \alpha & -\alpha & \alpha & -\alpha \\ \beta & -\delta & -\beta & \delta \\ \delta & \beta & -\delta & -\beta \end{bmatrix}$$

$$[\hat{D}_4] = \begin{bmatrix} \sigma & \mu & -\tau & \varepsilon \\ \mu & \tau & -\varepsilon & \sigma \\ \varepsilon & -\sigma & \mu & \tau \\ \tau & \varepsilon & \sigma & \mu \end{bmatrix},$$

where the regularity of these matrices can be further examined if we note that by defining $\phi = (4n+1)\pi/2N$ for $N = 4$ we have

$$[\hat{T}_4] = \begin{bmatrix} 1 & 1 & 1 & 1 \\ \cos(4\phi_0) & \cos(4\phi_1) & \cos(4\phi_2) & \cos(4\phi_3) \\ \cos(2\phi_0) & \cos(2\phi_1) & \cos(2\phi_2) & \cos(2\phi_3) \\ \cos(6\phi_0) & \cos(6\phi_1) & \cos(6\phi_2) & \cos(6\phi_3) \end{bmatrix}$$

$$[\hat{D}_4] = \begin{bmatrix} \cos(\phi_0) & \cos(\phi_1) & \cos(\phi_2) & \cos(\phi_3) \\ \cos(5\phi_0) & \cos(5\phi_1) & \cos(5\phi_2) & \cos(5\phi_3) \\ \cos(3\phi_0) & \cos(3\phi_1) & \cos(3\phi_2) & \cos(3\phi_3) \\ \cos(7\phi_0) & \cos(7\phi_1) & \cos(7\phi_2) & \cos(7\phi_3) \end{bmatrix}. \tag{3.160a}$$

Since $\cos[(2k+1)\phi] = 2[\cos(2k\pi)][\cos(\phi)] - \cos[(2k-1)\phi]$, the elements in the matrix $[\hat{D}_4]$ can be rewritten as

$$[\hat{D}_4] = \begin{bmatrix} \cos(\phi_0), & \cdots \\ \cos(\phi_0)\{1+2\cos(4\phi_0)-2\cos(2\phi_0)\}, & \cdots \\ \cos(\phi_0)\{-1+2\cos(2\phi_0)\}, & \cdots \\ \cos(\phi_0)\{-1-2\cos(4\phi_0)+2\cos(2\phi_0)+2\cos(6\phi_0)\}, & \cdots \end{bmatrix}, \tag{3.161}$$

where we have shown elements in the first column only. The remaining columns contain elements of the same form with ϕ_0 replaced by ϕ_1, ϕ_2 and ϕ_3 respectively. With this form, it is not difficult to verify the factorization for (3.161):

$$[\hat{D}_4] = [P_4'][L_4][P_4'][\hat{T}_4][Q_4], \tag{3.162}$$

where

$$[P_4'] = \begin{bmatrix} 1 & 0 & 0 & 0 \\ 0 & 0 & 1 & 0 \\ 0 & 1 & 0 & 0 \\ 0 & 0 & 0 & 1 \end{bmatrix},$$

is a 4×4 binary bit-reversal matrix,

$$[L_4] = \begin{bmatrix} 1 & 0 & 0 & 0 \\ -1 & 2 & 0 & 0 \\ 1 & -2 & 2 & 0 \\ -1 & 2 & -2 & 2 \end{bmatrix},$$

is a lower triangular matrix of the coefficients of the cosines in (3.161), and

$$[Q_4] = \text{diag}\{\cos(\phi_0), \cos(\phi_1), \cos(\phi_2), \cos(\phi_3)\}.$$

Equation (3.162), which delineates the relation of $[\hat{D}]$ to $[\hat{T}]$ in the 4×4 case, is easily extended to the general $(N \times N)$ case, thus completing the recursive structure. The formal structure is made clear by the following equation:

$$\begin{bmatrix} (X_e)_{BRO} \\ (X_o)_{BRO} \end{bmatrix} = \begin{bmatrix} [\hat{T}_{N/2}] & [\hat{T}_{N/2}] \\ [K_{N/2}][\hat{T}_{N/2}][Q_{N/2}] & -[K_{N/2}][\hat{T}_{N/2}][Q_{N/2}] \end{bmatrix} \begin{bmatrix} (x_e) \\ (x_o)_{RO} \end{bmatrix}, \quad (3.163)$$

where (X_e) and (X_o) are the even- and odd-numbered transformed components, and (x_e) and (x_e) are the even and odd input points. BRO and RO stand for bit-reversed order and reversed order, respectively. The matrix $[K]$ is used to denote the product $[P'][L][P']$ as shown in (3.162). Thus,

$$[K_{N/2}] = [P'_{N/2}][L_{N/2}][P'_{N/2}],$$

and

$$[Q_{N/2}] = \text{diag}\{\cos(\phi_0), \cdots, \phi_{N/2-1}\}, \quad (3.164a)$$

here, $[L_{N/2}]$ is the lower $(N/2) \times (N/2)$ triangular matrix given by

$$[L_{N/2}] = \begin{bmatrix} 1 & 0 & 0 & 0 & \cdots & 0 \\ -1 & 2 & 0 & 0 & \cdots & 0 \\ 1 & -2 & 2 & 0 & \cdots & 0 \\ -1 & 2 & -2 & 2 & \cdots & 0 \\ \vdots & \vdots & \vdots & \vdots & \cdots & 0 \\ -1 & 2 & -2 & 2 & \cdots & 2 \end{bmatrix}. \quad (3.164b)$$

It is clear from (3.163) that the factorization equation for the transform matrix is

$$[\hat{T}_N] = \begin{bmatrix} [\hat{T}_{N/2}] & [\hat{T}_{N/2}] \\ [K_{N/2}][\hat{T}_{N/2}][Q_{N/2}] & -[K_{N/2}][\hat{T}_{N/2}][Q_{N/2}] \end{bmatrix}. \quad (3.165)$$

It should be noted that in (3.163) the even- and odd-numbered components of the input and transformed vectors are separated, reversed, or bit reversed. It is therefore not strictly possible to classify this algorithm as being either decimation in time (DIT) or decimation in frequency (DIF).

The computational complexity of the algorithm can be delineated by considering the block matrix factorization of (3.165). This gives

$$[\hat{T}_N] = \begin{bmatrix} [I_{N/2}] & 0 \\ 0 & [K_{N/2}] \end{bmatrix} \begin{bmatrix} [\hat{T}_{N/2}] & 0 \\ 0 & [\hat{T}_{N/2}] \end{bmatrix} \times \begin{bmatrix} [I_{N/2}] & 0 \\ 0 & [Q_{N/2}] \end{bmatrix} \begin{bmatrix} [I_{N/2}] & [I_{N/2}] \\ [I_{N/2}] & -[I_{N/2}] \end{bmatrix}. \quad (3.166)$$

Now, $[Q_{N/2}]$ requires $(N/2)$ multiplications and $[K_{N/2}]$ can be implemented using shift-and-adds, which can be done using $N(N-2)/4$ additions. The last matrix factor can be implemented using N additions. Thus, the recurrence equations for the number of multiplications, μ_N, and for the number of additions, α_N, are given by

$$\mu_N = 2\mu_{N/2} + (N/2),$$

and

$$\alpha_N = 2\alpha_{N/2} + [N(N-2)/4] + N, \qquad N \geq 4. \quad (3.167a)$$

These equations can be solved by using μ_2=1, α_2=1, giving

$$\mu_N = \left(\frac{N}{2}\right) \log_2 N,$$

and

$$\alpha_N = \left(\frac{3N}{2}\right) \log_2 N + \frac{N^2}{2} - N, \qquad N \geq 4. \quad (3.167b)$$

It is noted in [183] that the second block matrix factor in (3.166) contains identical blocks and that the computation can be implemented using multiplexing techniques. This reduces the number of multipliers and adders required in the computation, making this algorithm quite competitive with other algorithms discussed in this chapter.

DCT-II Realization via Planar Rotations

One of the important considerations in physically realizing a DCT fast algorithm is the variety of processing units required. In this regard Ligtenberg and O'Neal [283] and Loeffler, Ligtenberg and Moschytz [289], [290] proposed an algorithm to compute the DCT that requires only rotations. This single type of processing unit replaces the adders, multipliers, and rotators required in the implementation of other algorithms. To help us understand the basis of this algorithm, we briefly outline the theory. It is well known that any matrix $[A]$ can be factored so that

$$[A] = [Q][R], \tag{3.168}$$

where $[Q]$ is an orthogonal matrix and $[R]$ is an upper triangular matrix. If, in addition, $[A]$ is an orthogonal matrix, then $[R]$ will also be orthogonal. Since an orthogonal upper triangular matrix must be diagonal, we have

$$[R]^T[R] = [I]. \tag{3.169}$$

The diagonal elements of $[R]$ can only be +1 or -1. From (3.168) this implies that the columns of $[A]$ and $[Q]$ are identical, except along the ith column, wherever $[R]_{ii} = -1$ in which case they differ by a minus sign. The factorization in (3.168) is called a QR–decomposition and can be accomplished by using the so-called Givens rotations. The essential feature of this procedure is to null the lower off-diagonal elements of $[A]$ by means of a rotation of the proper angle. It can be best illustrated by an example in the two-dimensional case.

Consider a two-dimensional vector $[x_0, x_1]^T$. A (2×2) rotation matrix $[T(\theta)]$ can be constructed based on the values of x_0 and x_1 so that

$$[T(\theta)]\begin{bmatrix} x_0 \\ x_1 \end{bmatrix} = \begin{bmatrix} \hat{x}_0 \\ 0 \end{bmatrix}. \tag{3.170}$$

It is easy to see that if

$$[T(\theta)] = \begin{bmatrix} \cos(\theta) & \sin(\theta) \\ -\sin(\theta) & \cos(\theta) \end{bmatrix} \tag{3.171}$$

then (3.170) is obtained when

$$\cos(\theta) = \frac{x_0}{\sqrt{x_0^2 + x_1^2}}, \quad \text{and} \quad \sin(\theta) = \frac{x_1}{\sqrt{x_0^2 + x_1^2}}. \tag{3.172}$$

The nulling of the element x_1 can be applied, in general, to null the ik–element of the matrix $[A]$. The required matrix is seen to be $[T_{ik}(\theta)]$, where the angle of rotation θ is determined by the elements $[A]_{ik}$ and $[A]_{i-1,k}$ such that

$$\cos(\theta) = \frac{[A]_{i-1,k}}{D}, \quad \text{and} \quad \sin(\theta) = \frac{[A]_{ik}}{D}, \tag{3.173}$$

where $D = \sqrt{[A]^2_{i-1,k} + [A]^2_{ik}}$.

If $[A]$ is an $(N\times N)$ orthogonal matrix, $[T_{ik}(\theta)]$ is seen to be an $(N\times N)$ identity matrix, except at the $(i-1, i-1)$, $(i-1, i)$, $(i, i-1)$ and (i, i) positions, where $\cos(\theta)$,

$\sin(\theta)$, $-\sin(\theta)$, and $\cos(\theta)$ are located. We also note that in nulling the $ik-$element of $[A]$ the angle of rotation can be determined using $[A]_{ik}$ and any other element located above it in the same column. We have chosen to follow the more conventional procedure of using $[A]_{i-1,k}$. A systematic application of this type of rotations will eventually reduce the matrix $[A]$ to an upper triangular form and, in particular, when $[A]$ is orthogonal to begin with, a diagonal matrix with +1 or -1 as diagonal elements. This sequence of rotations on the matrix $[A]$ can be represented by

$$\left\{\prod_{k=1}^{N}\prod_{i=k+1}^{N}[T_{ik}(\theta)]\right\}[A]=[R]. \tag{3.174}$$

Thus

$$\begin{aligned}[A] &= \left\{\prod_{k=1}^{N}\prod_{i=k+1}^{N}[T_{ik}(\theta)]\right\}^{T}[R], \\ &= \left\{\prod_{i=1}^{N}\prod_{k=i+1}^{N}[T_{ik}(\theta)]\right\}[R]. \end{aligned} \tag{3.175}$$

From (3.168), it is evident that the orthogonal matrix $[Q]$ is given by

$$[Q]=\prod_{i=1}^{N}\prod_{k=i+1}^{N}[T_{ik}(\theta)]. \tag{3.176}$$

Since $[A]$ and $[Q]$ have the same columns (up to a minus sign), (3.176) also represents a decomposition of $[A]$ into factor matrices, each of which is a planar rotation. Now, the DCT-II matrix is a real orthogonal matrix. It is therefore clear that the DCT-II matrix, $[C^{II}]$, can be factored into a sequence of planar rotations also. For an $(N\times N)$ matrix, there are $N(N-1)/2$ elements to be nulled, and the number of rotations required is thus of order N^2.

Example 3.10. *Using the DCT-II matrix for $N=4$ compute the sequence of rotation matrices $[T_{ik}(\theta)]$ and the matrices $[Q]$ and $[R]$.*

The matrix for the DCT-II is given by

$$[C^{II}]_4=\begin{bmatrix}0.50 & 0.50 & 0.50 & 0.50\\ 0.65 & 0.27 & -0.27 & -0.65\\ 0.50 & -0.50 & -0.50 & 0.50\\ 0.27 & -0.65 & 0.65 & -0.27\end{bmatrix} \tag{3.177}$$

For $N=4$ six rotation matrices are required

$$[T_{41}]=\begin{bmatrix}1 & 0 & 0 & 0\\ 0 & 1 & 0 & 0\\ 0 & 0 & 0.88 & 0.48\\ 0 & 0 & -0.48 & 0.88\end{bmatrix};\quad [T_{31}]=\begin{bmatrix}1 & 0 & 0 & 0\\ 0 & 0.75 & 0.66 & 0\\ 0 & -0.66 & 0.75 & 0\\ 0 & 0 & 0 & 1\end{bmatrix};$$

$$[T_{21}] = \begin{bmatrix} 0.50 & 0.87 & 0 & 0 \\ -0.87 & 0.50 & 0 & 0 \\ 0 & 0 & 1 & 0 \\ 0 & 0 & 0 & 1 \end{bmatrix}; \quad [T_{42}] = \begin{bmatrix} 1 & 0 & 0 & 0 \\ 0 & 1 & 0 & 0 \\ 0 & 0 & 0.91 & 0.41 \\ 0 & 0 & -0.41 & 0.91 \end{bmatrix};$$

$$[T_{32}] = \begin{bmatrix} 1 & 0 & 0 & 0 \\ 0 & 0.58 & 0.82 & 0 \\ 0 & -0.82 & 0.58 & 0 \\ 0 & 0 & 0 & 1 \end{bmatrix}; \quad [T_{43}] = \begin{bmatrix} 1 & 0 & 0 & 0 \\ 0 & 1 & 0 & 0 \\ 0 & 0 & 0.71 & 0.71 \\ 0 & 0 & -0.71 & 0.71 \end{bmatrix}. \tag{3.178}$$

The product of these matrices gives the orthogonal matrix $[Q]$ in the QR–decomposition of the matrix $[C^{II}]$, therefore

$$[Q] = \begin{bmatrix} 0.50 & -0.50 & 0.50 & -0.50 \\ 0.65 & -0.27 & -0.27 & 0.65 \\ 0.50 & 0.50 & -0.50 & -0.50 \\ 0.27 & 0.65 & 0.65 & 0.27 \end{bmatrix} \tag{3.179}$$

The resulting $[R]$ as expected, is diagonal and is given by

$$[R] = \text{diag}\{1, -1, 1, -1\}. \tag{3.180}$$

In general, when N is an integer power of 2, it is possible to first convert the transform matrix to a block diagonal one before applying the rotations. This approach will reduce the number of rotations to the order of $N \log_2 N$. Each rotation can be implemented by either four multiplications and two additions, or three multiplications and three additions, depending on the architecture one desires.

It is also possible to implement the QR–decomposition by another type of matrix operation. For the two-dimensional case, the operation is given by

$$[T'(\theta)] \begin{bmatrix} x_0 \\ x_1 \end{bmatrix} = \begin{bmatrix} \hat{x}'_0 \\ 0 \end{bmatrix}, \tag{3.181}$$

where

$$[T'(\theta)] = \begin{bmatrix} \sin(\theta) & \cos(\theta) \\ \cos(\theta) & -\sin(\theta) \end{bmatrix}$$

and the angle θ is determined by (3.172) as before. We note here that $[T'(\theta)]$ is a symmetric matrix and its determinant is -1. Therefore $[T'(\theta)]$ is really a reflection of some sort. Vetterli and Ligtenberg [498] have combined this type of operation with the recursive algorithm of Vetterli and Nussbaumer [499] to propose a layout for an actual transform chip (see also Section 3). For example, the computation of an $N = 8$ DCT-II is accomplished via two $N = 2$ DCT and one $N = 4$ DFT, which are in turn implemented using $[T'(\theta)]$ operations. We note that in their work, they have referred to these operations as rotations also. Figure 3.17 shows the block diagram

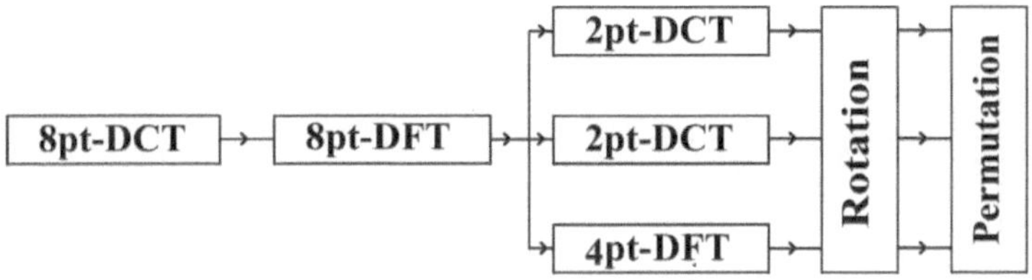

Figure 3.17: Stages of $N = 8$ DCT-based on [499] and rotations, pt = point [390].

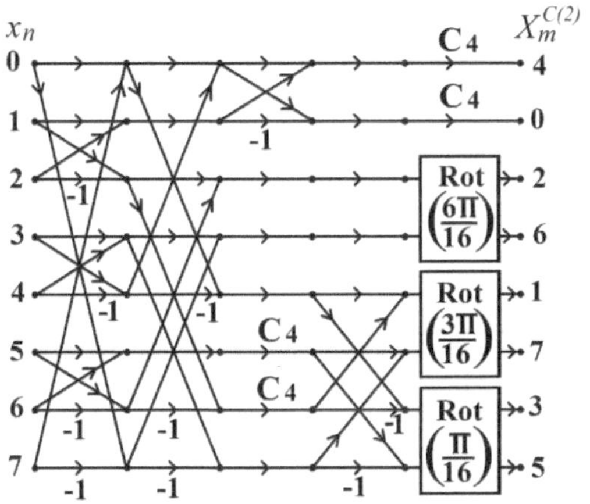

Figure 3.18: Signal flowgraph for $N = 8$ DCT-II. The rotator here is defined in (3.181). $Ck = \cos(k\pi/16)$.

of the stages of the $N = 8$ DCT. The signal flowgraph for the forward DCT-II, $N = 8$ is shown in Figure 3.18

Figure 3.19 shows the signal flowgraph for DCT-II, $N = 8$ using rotators of the type in (3.171) only.

The input to the rotator is the two-dimensional vector x and the output is the two-dimensional vector y. Therefore, the rotator here is given by,

$$[y] = [\text{Rot}(\theta)][x],$$
$$\begin{bmatrix} y_i \\ y_k \end{bmatrix} = \begin{bmatrix} \cos(\theta) & \sin(\theta) \\ -\sin(\theta) & \cos(\theta) \end{bmatrix} \begin{bmatrix} x_i \\ x_k \end{bmatrix}.$$

Note that the rotation matrix here is not exactly a Given's rotation that will null the second component of the vector x. It may be considered as a combination of several Given's rotations. For $N = 8$, we see that only 13 rotations are required.

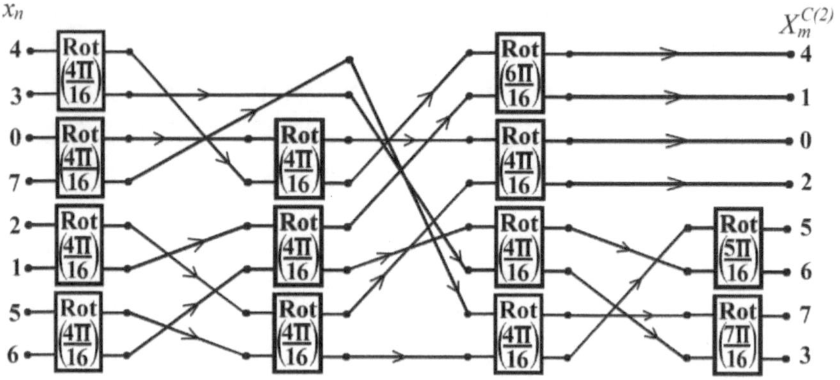

Figure 3.19: Signal flowgraph of DCT-II, $N = 8$ using only rotations. The rotator here is as defined in (3.171).

APPLICATIONS

Since the introduction of the DCT, there has been an increasing interest in using this transform for different applications. For example, in data compression, the DCT provides an energy packing close to that of KLT (see Chapter 2) [70], [463], [591]. In pattern recognition, it allows a classification scheme to be implemented with fewer features. Other applications include: Progressive image transmission PIT [490], interpolation [4], [525], pre- and post-filtering for block DCT-based systems [102], reversible data hiding [285], [191], mirroring and rotation [144], focus/blurriness measure [582], face detection and recognition [587], feature transform for image retrieval and reconstruction [514], regularized extrema visual recovery [516], signal and noise-level estimation [553], echo and feedback cancellation [157], signal decorrelation for voice-activity detection [152], surveillance (i.e., high resolution image reconstruction, hallucinating face, background modelling) [266], [276], [365], [472], [579], video analytics [46], multi-view image and video compression [275], image in-painting [281], variable block size integer DCTs for next generation video codecs [75], DCT pyramid for error concealment [572], 3D high efficiency video coding HEVC [478], (DCTQM) DCT quality degradation metric (QM quality metric) [29], [412], [413], spatio-temporal and foveated masking effects [30], corner detection [271], edge detection [382], directional block-wise image representation [337], watermarking [98], compression of medical images [541].

Free (GPL) C libraries for fast DCTs/DCTs (types I–IV) computation in one or more dimensions, of arbitrary size, can be found in [142], [249], [362].

SUMMARY

The DCT separates the image into parts (or spectral subbands) of differing importance and performs close to KLT for a first-order Markov source with large correlation coefficient. The DCT has a vast field of applications because it is a real (avoids redundancy) and orthogonal transform. Also, it is derived from DFT, all the desirable properties of DFT (such as the fast algorithm) are preserved. The satisfactory performance of DCT in terms of energy compaction makes it suitable for applications in image and video compression and it has become the heart of many international standards such as JPEG, H.26x and MPEG families. There are several types of DCT transforms. However, the one labeled as DCT-II is the most used. Many fast algorithms have been developed and some of them take advantage of the relationships between DCT and other fast algorithms, including fast Fourier transform, Walsh-Hadamard transform and the discrete Hartley transform, and some of them are recursive and based on sparse factorizations of the DCT matrix as presented in this chapter. Also, in this chapter, the DCT is defined and its relevant properties are defined. The discussion of the properties of the Fourier cosine transform forms the basis upon which corresponding properties for the DCT have been built. It is observed that both the shift property and the convolution property for the DCT are more complex than those for the Fourier cosine transform. However, there are situations under which these properties may be utilized. In the updating of an instantaneous DCT of a continuously incoming data stream, the shift property is considered to be useful. In applications that involve transform domain filtering, the convolution property is essential [349].

PROBLEMS

3.1. In (3.9), the Fourier cosine transform of the derivative of a function is developed. Extend this to higher order derivatives and obtain a general formula for

$$F_c\left[\frac{d^n}{dt^n}x(t)\right].$$

3.2. In Section 3 the properties of Fourier cosine transform are developed. Obtain similar properties for the Fourier sine transform.

3.3. In Section 3 the unitarity of DCT-I is shown in detail [see (3.22) through (3.32)]. Prove this property for the remaining DCTs.

3.4. In Section 3 the shift property of DCT-I is developed. In (3.49), (3.50) and (3.51), this property is listed for the other DCTs. Derive the same in detail.

3.5. Develop shift properties for the DSTs defined in (3.52).

3.6. Sketch the waveforms of DCTs for $N = 4$.

3.7. Repeat Problem 3.6. for DSTs.

3.8. In Section 3 the difference property for DCT-I is shown. Write down the expressions for the difference properties for all the DCTs.

3.9. Derive (3.63).

3.10. Derive (3.71).

3.11. The convolution property for DCT-II is derived in Section 3. Derive, if any, the same property for the other DCTs defined in 3.20.

3.12. Investigate if the convolution property exists for the family of DSTs defined in (3.53). If so, derive the same.

3.13. Derive (3.11).
3.14. Derive (3.24).
3.15. Derive (3.25).
3.16. Using (3.20), develop the matrices for all the DCTs for $N = 4$. Show that these matrices are orthogonal. (Hint: Compute the matrix multiplication of each DCT matrix by its transpose.)
3.17. Repeat Problem 3.16. for the DSTs using (3.53). Are any of the DSTs interrelated? If so, how?
3.18. A simple convolution property for DCT-II has been developed by Chitprasert and Rao [82]. The only restriction is that one of the functions to be convolved must be real and even. This property has been utilized in human visual weighted progressive image transmission [83]. Investigate if similar properties exist for the remaining DCTs defined in (3.20).
3.19. Repeat Problem 3.18. for the family of DSTs defined in (3.53).
3.20. Kou and Fjallbrant [257] have developed a fast algorithm for computing the DCT-II coefficients of a signal block composed of adjacent blocks. The composite block consists of the second half of the data of one block and the first half of the data of the following block. Generalize this technique to other cases. For example, a composite block is formed from the last quarter (or last third) of one block and the first three quarters (or first two thirds) of the following block. Compare the advantages of this technique with the traditional method.
3.21. Wang [520] has developed a technique for implementing DST-II via DCT-II. Develop a similar technique for implementing DCT-II via DST-II. Show all steps in detail.
3.22. The recursive computation of an N-point DCT-III (radix-2) has been developed by Vetterli and Nussbaumer [499] (see Fig. 3.5). Verify this algorithm. For $N = 8$ and 16, develop flowgraphs and sparse matrix factors for DCT-II and its inverse. Write computer programs based on these flowgraphs and verify the same using random data. (Hint: DCT-II of a sequence followed by its inverse recovers the original sequence subject to computational errors-round-off, truncation, etc.).
3.23. In Section 3, the DIT and DIF algorithms for DCT-II are presented [561, 562, 563]. Repeat Problem 3.22. for these algorithms.
3.24. Derive 3.132. Sketch the flowgraph for implementing the DCT-II via WHT for $N = 8$ [178]. Extend this to $N = 16$ [495].
3.25. Sketch the flowgraphs for implementing the IDCT via WHT for $N = 8$ and 16 (see Prob. 3.24.).
3.26. Similar to radix-2 FFT [122], Lee [268, 269] has developed an algorithm for IDCT and has illustrated the algorithm for $N = 8$. Develop a similar algorithm for the DCT. Verify the validity of this algorithm by a flowgraph for $N = 8$. (Hint: Use the orthogonality property of the DCT matrix).
3.27. In Section 3 the PFA for DCT-II [554] is discussed. This algorithm is illustrated by a flowgraph/block diagram format for $N = 12$ in Fig. 3.14. Develop the flowgraph in detail and obtain the sparse matrix factors. Using the

orthogonality property of the DCT-II, obtain the sparse matrix factors and then the flowgraph for inverse DCT (IDCT), $N = 12$.

3.28. Duhamel [118] has developed the split-radix FFT algorithm. As DIT and DIF radix-2 algorithms [561, 562, 563] for the DCT have been developed, investigate whether split-radix DCT-II algorithms (both DIT and DIF) can be developed (see [464]). Compare the efficiency of these algorithms with the radix-2 algorithms.

3.29. In designing transmultiplexers, Narasimha and Peterson [344] have developed a fast 14-point DCT-II algorithm. The flowgraph for the 14-point IDCT is shown in Figure 10 of [344]. Obtain the matrix factors based on this flowgraph and obtain the number of multiplications and additions required for this algorithm. See also [341].

3.30. From Problem 3.29., transpose the matrix factors and draw the flowgraph based on the transposition. Show that this flowgraph yields the 14-point DCT-II as defined by (28) of [344]. Show that this flowgraph is a mirror image of the flowgraph shown in Figure 10 of [344].

3.31. In realizing a 60-channel transmultiplexer, Narasimha, Yang and Lee [342] utilized a 72-point DCT-II, which was implemented by the PFA [554]. Obtain the flowgraph for this DCT-II and comment on the computational complexity of this algorithm.

3.32. Repeat Problem 3.31. for a 72-point IDCT-II.

PROJECTS

P.3.1. Watermarking proposes to solve the problem of protecting the intellectual property. Recently, in [245] a robust watermarking approach based on DCT that combines quick response (QR) code and chaotic system was proposed. Inspect the possibility to adapt this watermarking system in the transform blocks (TBs) of the high efficiency video coding (HEVC) scheme [387].

P.3.2. Currently, 3D video is becoming widely available. The high correlation between views allows to carry out processing operations very efficiently. Develop a Bayesian denoising framework in the DCT domain and test it with the new multiview HEVC (MV-HEVC) [387].

P.3.3. Develop a denoising Wiener filter [475], [555] in the DCT domain and test it with the new MV-HEVC [387].

P.3.4. Fong, Han and Cham [135] have developed recursive integer cosine transform (RICT). This has orthogonal basis vectors and also has recursive property. Based on this, a flowgraph for order 32 (RICT) with flow graphs for orders 4, 8 and 16 embedded is shown in Figure 1. They also have implemented the RICT in HEVC HM13.0 and have demonstrated that the RICT has similar coding efficiency as the core transform in HEVC (see Tables VI thru XI). Using Figure 1, write down the sparse matrix factors (SMFs) for orders 4, 8, 16 and 32 RICTs.

P.3.5. See P.3.4. Using Table I and the SMFs for orders 4, 8, 16 and 32 RICTs write down the corresponding transform matrices. Verify the transform matrix for order 32 with that shown in Equation 16 of [135].

P.3.6. See P.3.4. Forward and inverse architecture of order 4 RICT is shown in Fig.4. This structure is symmetric (hardware of order 4 inverse RICT can be shared by the order 4 forward RICT). Develop similar symmetric hardware structures for orders 8, 16 and 32.

P.3.7. The authors in [135] state that higher order RICTs such as orders 64 and 128 can be derived using their fast algorithms. Derive these.

P.3.8. See P.3.7. Draw the flow graph for order 64 RICT with the flow graphs for orders 4, 8, 16 and 32 embedded similar to that shown in Figure 1.

P.3.9. Repeat P.3.8. for order 128 RICT.

P.3.10. See P.3.9. Extend the suggested values for $b_{N,i}$ for orders 64 and 128 RICTs (see Table I in [135]).

P.3.11. See P.3.6. Develop similar symmetric hardware structures for orders 64 and 128 RICT.

P.3.12. See Table V in [135]. Extend the comparison to orders 64 and 128. This requires extending Loeffler's method to orders 64 and 128.

P.3.13. Jridi and Meher [239] have developed approximate DCT architectures for efficient HEVC compliant video coding. Full-parallel and area constrained architectures for the proposed approximate DCT are described. Show that the equations given by (6a) and (6b) yield the 4-point DCT kernel given by (4). Write down the SMFs based on Figure 1 for the 4-point integer 1D DCT.

P.3.14. See P.3.13. Show that the equations given by (7)-(9) yield Eq. (5) for the 8-point DCT kernel.

P.3.15. See P.3.14. Write down the SMFs for the 8-point integer 1D DCT-based on Figure 2. Show that these SMFs yield Eq. (5) for the 8-point DCT kernel.

P.3.16. See [239]. Parallel architecture for approximate 8-point 1D DCT is shown in Figure 2. Develop similar architecture for approximate 16-point 1D DCT.

P.3.17. Based on P.3.16. write down the corresponding SMFs and the resulting 16x16 1D DCT matrix.

P.3.18. See P.3.16. Extend this to approximate 32-point 1D DCT.

P.3.19. See P.3.17. Extend this to 32-point 1D DCT.

P.3.20. Chen, Han and Cham [78] have developed a set of low-complexity integer cosine transforms (LCICTs) and used them in HEVC HM13. These have fully factorizable structures and high circuit reusability. They conclude that order-64 integer cosine transforms (ICTs) can significantly improve the coding performance under high QP (low bitrate) which makes them beneficial in low bitrate applications such as video conferencing and video surveillance. Using equations 5-10, write down the LCICT matrices for order 8, 16, 32 and 64.

P.3.21. See P.3.20. Using Figure 3 (flow graph) write down the SMFs for LCICTs of order 8, 16, and 32.

P.3.22. See P.3.20. Extend the flow graph for LCICT of order 64.

P.3.23. See P.3.20. Write down the SMFs for LCICT of order 64.

P.3.24. See [78]. In Table IV the BD bitrates for LCICT are compared with those for another order 64 ICT (see reference 14 at the end) using various test sequences under AI, RA, and LDB. Confirm these results.

P.3.25. See [78]. The flow graph for order-32 LCICT (flow graphs for order-8 and 16 LCICT are embedded) is shown in Figure 1. Draw the flow graph for the inverse LCICTs.

P.3.26. See P.3.25. Write down the corresponding SMFs for these inverse LCICTs.

P.3.27. See P.3.25. and P.3.26. Evaluate the orthogonality of these LCICTs. (Hint: matrix multiply the forward LCICT with the matrix for corresponding inverse LCICT). Are these matrices truly orthogonal?

P.3.28. See [533] Section 7.3.3. Determination of the interpolation filter (IF) coefficients pages 194–196 for the theory behind how these IF coefficients are derived using DCT. Derive these IF coefficients independently and confirm that they match with the coefficients listed in Tables 5.4 and 5.5 in this paper. Plot the corresponding transfer functions (see Figures 7.16 and 7.1, pages 201–202 of the paper).

P.3.29. Sjövall, Viitamki and Vanne [440] have developed a high level synthesis implementation of HEVC 2D DCT/DST (8/16/32- point DCT units and 4-point DCT/DST unit) on FPGA. Develop this architecture and verify the results shown in Tables 1 and 2.

P.3.30. See P.3.29. Extend the high level synthesis to 2D 64-point DCT.

P.3.31. See [360]. The authors have developed low-complexity integer approximation of discrete Tchebichef transforms (DTT) for sizes 4 and 8. (see T4* and T8*). Extend this to T16*.

P.3.32. See P.3.31. Develop the sparse matrix factors (SMF) for T16* similar to those for T8*. Draw the signal flow graph for T16* similar to that shown in Figure 1 of [360]. Compute the number of additions and bit shifting operations based on these SMF. See Table II [360]. What is the diagonal matrix before T16* that yields T16* hat?

P.3.33. See [360]. Extend the discussion under Section B (orthogonality and invertibility for T8*) to T16*.

P.3.34. In HEVC transform sizes can go up to 32×32 (integer approximations of DCT). Develop T32* [360].

P.3.35. See P.3.34. Derive the SMF for $T32^*$ and draw the signal flow graph similar to Figure 1 in [360]. Compute the number of additions and bit shifting operations based on these SMF. See Table II [360]. What is the diagonal matrix before $T32*$ that yields $\hat{T}32^*$?

P.3.36. See [360]. The authors have compared the performance of original JPEG with those where the 8×8 DCT is replaced by the DTT (See Section 5 and Figures 1 and 2 for graphs and Figure 3 for Lena image). Verify these results. The software for using the DTT can be requested from the authors. Spectral residual-based similarity (SR-SIM) is described in reference [64] cited in [360].

P.3.37. The performance of T8* in video coding (CIF resolution) is shown in Figure 4 and the first frame of the decoded Foreman sequence is shown in Figure 5 [360]. Confirm these results.

P.3.38. Wang, Xu and You [516] have applied multi-scale DCT in visual recovery that is superior to the state-of-the-art methods. They also state that the proposed DCT regularizer can be independently embedded into most image processing tasks, e.g., image in painting and image denoising. Implement this DCT regularizer and confirm the results shown in this paper.

P.3.39. See [238]. The authors have developed recursive approximation of DCT that satisfies the three features (see pages 449–450). Using Figure 1 write down the SMF for $\hat{C}_8$. Compute the approximated DCT for $N = 8$. Using Equation (10), compute D_8 the diagonal matrix. The recursive relationship for the approximated DCT is shown in Equation (9).

P.3.40. See P.3.39. Using Figure 2 write down the SMF for $\hat{C}_{16}$. Compute the approximated DCT for $N = 16$. Using Equation (10), compute D_{16} the diagonal matrix.

P.3.41. See [238]. The proposed reconfigurable architecture for approximate DCT of lengths $N = 8$ and 16 is shown in Figure 3. Verify this. (Hint: Use Figure 2).

P.3.42. See P.3.40. Sketch the block diagram of the proposed SCT for $N = 32$, similar to that shown in Figure 2 for $N = 16$. Compute D_{32}, the diagonal matrix.

P.3.43. See P.3.42. Verify the proposed reconfigurable architecture for approximate DCT of lengths $N = 8$, 16 and 32 shown in Figure 4.

P.3.44. See [238]. The performance of the approximated DCT in HEVC (main profile all intra) is compared with other DCT approximations in HEVC (see Figures 5–7). Verify this comparison. This requires extensive simulations.

P.3.45. See [238]. As future work the authors state "the coding loss could be reduced by an appropriate modification of context models used in CABAC coder." Explore this.

P.3.46. See [30]. In the conclusion the authors state "as future work we plan to apply our DCT-based pvc where perceptual redundancy is effectively removed thus improving the compression efficiency of HEVC-based pvc encoders." Explore this.

4 The Discrete Sine Transform

INTRODUCTION

Similar to the (DCT), the discrete sine transform (DST) is mathematically related to the discrete Fourier transform (DFT). Both, the DCTs and the DSTs are members of the class of sinusoidal unitary transforms over a set of cosine and sine basis functions [101]. Nevertheless, the DST has no zero frequency coefficient (DC); the basis vectors are sine functions; is an orthogonal and separable transform; the DST matrix can be sparse matrix factored; it has a fast algorithm and it is a recursive transform. For many years, the DST escaped the attention of researchers in the image and video coding field because, unlike the DCT, the DST does not decorrelate highly correlated signals. In other words, the energy compaction of DST distributes into the non-zero frequency coefficients (AC) since the transform has no DC component. This characteristic makes the DST suitable for low correlated signals.

Recently, the new HEVC standard specifies an alternate 4-point integer transform based on the discrete sine transform (AC) for coding 4×4 luma intra prediction residual blocks [296], [387], [419], [462] because the DST basis functions better fit the statistical property that the residual amplitudes tend to increase as the distance from the boundary samples that are used for prediction becomes larger [462].

This chapter is devoted to the basic properties of the DST, some fast algorithms and several applications.

THE FOURIER SINE TRANSFORM

Definitions : Equations (3.1a) and (3.1b) define the Fourier transform (FT) pair. Consider the $y(t)$ the odd extension of $x(t)$,

$$\begin{aligned} y(t) &= x(t) \quad t \geq 0, \\ &= -x(-t) \quad t \leq 0. \end{aligned}$$

Then the FT of $y(t)$ is given by

$$\begin{aligned} F[y(t)] &= \sqrt{\frac{1}{2\pi}}\left[\int_0^\infty x(t)e^{-j\omega t}dt - \int_{-\infty}^0 x(-t)e^{-j\omega t}dt\right] \\ &= \sqrt{\frac{1}{2\pi}}\int_0^\infty x(t)[e^{-j\omega t} - e^{j\omega t}]dt \\ &= -j\sqrt{\frac{2}{\pi}}\int_0^\infty x(t)\sin(\omega t)dt. \end{aligned} \tag{4.1}$$

Taking the negative imaginary part of $F[y(t)]$ yields the FST

$$X_s(\omega) \equiv F_s[x(t)] = -jF[y(t)] = \sqrt{\frac{2}{\pi}} \int_0^{\infty} x(t) \sin(\omega t) dt, \tag{4.2}$$

where, $j = \sqrt{-1}$, $\omega = 2\pi f$; ω is the radian frequency and f the frequency in Hertz.
By applying (3.1b) to (4.2) $y(t)$ is recovered

$$y(t) = x(t) \equiv F_s^{-1}[X_s(\omega)] = \sqrt{\frac{2}{\pi}} \int_0^{\infty} X_s(\omega) \sin(\omega t) d\omega \quad (t \geq 0). \tag{4.3}$$

Equations (4.2) and (4.3) define the FST pair. Some of the properties are as follows:

PROPERTIES OF THE FST

(a). *Inversion:*

$$F_s \equiv F_s^{-1}. \tag{4.4}$$

It is clear from (4.2) and 4.3 that

$$F_s^{-1}\{F_s[x(t)]\} = x(t) \quad t \geq 0.$$

(b). *Linearity:*

$$F_s[\alpha x(t) + \beta y(t)] = \alpha X_s(\omega) + \beta Y_s(\omega), \tag{4.5}$$

where α and β are constants. F_s is clearly a linear operator.

(c). *Scaling in time:* Scaling in time is obtained in a fashion similar to that in (3.7). Therefore,

$$F_s[x(\alpha t)] = \frac{1}{\alpha} X_s\left(\frac{\omega}{\alpha}\right) \quad \text{for real } \alpha \geq 0. \tag{4.6}$$

(d). *Shift in time:*

$$F_s[x(t-\alpha)] = F_s[x(t)] \cos(\alpha\omega) + F_c[x(t)] \sin(\alpha\omega) \tag{4.7}$$

Using the compound angle expansion and defining the odd extension x_o of the function $x(t)$ as

$$x_o(t) = \frac{t}{|t|} x(|t|),$$

we can get

$$F_s[x_o(t+\alpha) + x_o(t-\alpha)] = 2X_s(\omega) \cos(\alpha\omega). \tag{4.8}$$

(e). *Differentiation in time:*

$$F_s\left[\frac{d}{dt}x(t)\right] = \sqrt{\frac{2}{\pi}}\int_0^\infty \left[\frac{d}{dt}x(t)\right]\sin(\omega t)dt.$$

Note that even orders involve sine transforms only and odd orders involve cosine transforms only. Integrating by parts and using the fact that $\sin(\omega t) = 0$ for $t = 0$ and assuming $x(t)$ vanishes as $t \to \infty$ we obtain

$$F_s\left[\frac{d}{dt}x(t)\right] = -\omega F_c[x(t)]. \tag{4.9}$$

(f). *The convolution property:* Let the odd extended functions of $x(t)$ and $y(t)$ be $x_o(t)$ and $y_o(t)$ respectively defined by

$$x_o(t) = \frac{t}{|t|}x(|t|) \quad \text{and} \quad y_o(t) = \frac{t}{|t|}y(|t|),$$

when $x_o(t)$ and $y_o(t)$ are convolved, the FCT reduces to the product of the FSTs of $x(t)$ and $y(t)$ with the integral on the right-hand side representing the convolution of the odd extended functions.

$$2\pi X_s(\omega)Y_s(\omega) = F_c\left\{\int_0^\infty x(\tau)[y(t+\tau) + y_o(t-\tau)]d\tau\right\} \tag{4.10}$$

Observe that the integral on the right-hand side is an even function resulting in an FCT. This property is obtained by applying the Fourier transform to the convolution and by using the convolution property. Also, notice that the convolution of one of the functions with odd extension with the even extended function of the other function yields the following results,

$$2\pi X_s(\omega)Y_c(\omega) = F_s\left\{\int_0^\infty x(\tau)[y(|t-\tau|) - y(t+\tau)]d\tau\right\}, \tag{4.11a}$$

and

$$2\pi X_s(\omega)Y_c(\omega) = F_s\left\{\int_0^\infty y(\tau)[x(t+\tau) + x_o(t-\tau)]d\tau\right\}. \tag{4.11b}$$

Example 4.1. *The unit rectangular pulse (Heaviside unit step function) is defined by,*

$$x(t) = U(t) - U(t-1), \text{ where } U(t) = \begin{cases} 0 & \text{for } t < 0, \\ 1 & \text{for } t > 0. \end{cases}$$

The FST is given by,

$$X_s(\omega) = \sqrt{\frac{2}{\pi}}\int_0^1 \sin\omega t dt = \sqrt{\frac{2}{\pi}}\left(\frac{1-\cos\omega}{\omega}\right). \tag{4.12}$$

Example 4.2. *The exponential function,*

$$x(t) = e^{-j\alpha t}, \quad \alpha > 0$$

Then, the FST is,

$$X_s(\omega) = \sqrt{\frac{2}{\pi}} \int_0^\infty e^{-j\alpha t} \sin(\omega t) dt.$$

Using Euler's formula

$$\begin{aligned}
X_s(\omega) &= \frac{1}{2j}\sqrt{\frac{2}{\pi}} \int_0^\infty e^{-j\alpha t}[e^{j\omega t} - e^{-j\omega t}]dt, \\
&= \frac{1}{2j}\sqrt{\frac{2}{\pi}} \left[\int_0^\infty e^{-j(-\omega+\alpha)t} dt - \int_0^\infty e^{-j(\omega+\alpha)t} dt\right], \\
&= -\frac{1}{2}\sqrt{\frac{2}{\pi}} \left[\frac{1}{(-\omega+\alpha)} - \frac{1}{(\omega+\alpha)}\right], \\
&= \sqrt{\frac{2}{\pi}} \left[\frac{\omega}{(\omega^2+\alpha^2)}\right].
\end{aligned} \tag{4.13}$$

Example 4.3. *The sinc function,*

$$x(t) = \frac{\sin(\alpha t)}{t}, \quad \alpha > 0.$$

The FST is,

$$\begin{aligned}
X_s(\omega) &= \sqrt{\frac{2}{\pi}} \int_{-\infty}^\infty \frac{1}{t} \sin(\alpha t)\sin(\omega t) dt. \\
&= \sqrt{\frac{2}{\pi}} \ln\left|\frac{\omega+\alpha}{\omega-\alpha}\right|.
\end{aligned} \tag{4.14}$$

Example 4.4. *The Bessel function of the first kind ,*

$$x(t) = J_0(\alpha t), \quad \alpha > 0.$$

The result is similar to the FCT (3.14)

$$X_s(\omega) = \sqrt{\frac{2}{\pi}} \int_0^\infty J_0(\alpha t) \sin \omega t dt = \begin{cases} \sqrt{\frac{2}{\pi}}\left(\frac{1}{\sqrt{\omega^2-\alpha^2}}\right) & \text{for } \omega > \alpha, \\ \infty & \text{for } \omega = \alpha, \\ 0 & \text{for } 0 < \omega < \alpha. \end{cases} \tag{4.15}$$

THE DISCRETE SINE TRANSFORM

Like the DCT, the discrete sine transform DST belongs to the class of sinusoidal unitary transforms with a kernel defined by a set of complete, orthogonal or orthonormal discrete cosine or sine basis functions [52], [386], [453]. Each transform is of type I, II, III or IV. The importance of transformations lies in their implementation. This requires fast algorithms and numerical stability. One of the advantages of DCT and DST is their excellent numerical stability.

The set of equations (3.52) can be expressed as [52],

$$\text{DST-I:} \quad \left[S_{N-1}^{I}\right]_{mn} = \sqrt{\frac{2}{N}}\left\{\sin\left[\frac{(m+1)(n+1)\pi}{N}\right]\right\}, \quad m,n = 0,1,\cdots,N-2,$$

$$\text{DST-II:} \quad \left[S_{N}^{II}\right]_{mn} = \sqrt{\frac{2}{N}}\left\{k_m \sin\left[\frac{(m+1)(2n+1)\pi}{2N}\right]\right\}, \quad m,n = 0,1,\cdots,N-1,$$

$$\text{DST-III:} \quad \left[S_{N}^{III}\right]_{mn} = \sqrt{\frac{2}{N}}\left\{k_n \sin\left[\frac{(2m+1)(n+1)\pi}{2N}\right]\right\}, \quad m,n = 0,1,\cdots,N-1,$$

$$\text{DST-IV:} \quad \left[S_{N}^{IV}\right]_{mn} = \sqrt{\frac{2}{N}}\left\{\sin\left[\frac{(2m+1)(2n+1)\pi}{4N}\right]\right\}, \quad m,n = 0,1,\cdots,N-1, \tag{4.16}$$

where

$$k_p = \begin{cases} \frac{1}{\sqrt{2}} & p = N-1, \\ 1 & \text{otherwise.} \end{cases}$$

Observe that in (4.16) the DST-I is a scaled version of the symmetric sine transform SST matrix for $N = 2^m - 1$ defined by,

$$\left[\tilde{S}_{N}^{I}\right]_{mn} = \sqrt{\frac{2}{N+1}}\left\{\sin\left[\frac{(m+1)(n+1)\pi}{N+1}\right]\right\}, \quad m,n = 0,1,\cdots,N-1.$$

Also notice that

$$\left[\tilde{S}_{N}^{I}\right]^{-1} = \left[\tilde{S}_{N}^{I}\right]^{T} = \tilde{S}_{N}^{I}.$$

The basis vectors of the DCTs and DSTs are eigenvectors of a parametrized symmetric tridiagonal Jacobi matrix [222], [523]. The column vectors of the even and odd versions of the DCTs and DSTs are eigenvectors of a symmetric second difference matrix and therefore they are orthogonal [454].

THE DISCRETE FOURIER TRANSFORM OF A SKEWED SEQUENCE

Example 4.5. *Skewed sequence.*

Let the real sequence be $\{x_n, \quad n = 1, 2, \cdots, N-1\}$ *be an* $(N-1)$*-point data sequence. A skewed sequence* s_n *can be constructed as,*

$$s_n = \begin{cases} 0 & n = 0, N, \\ x_n & 0 < n < N, \\ -x_{2N-n} & N < n \leq 2N-1. \end{cases} \tag{4.17}$$

For example, for $x_n = x_1, x_2, x_3$, *then* $N = 4$. *Therefore, we construct the skewed sequence as,* $s_n = 0, x_1, x_2, x_3, 0, -x_3, -x_2, -x_1 \quad$ *for* $n = 0, 1, \cdots, 7$.

Example 4.6. *The Fourier transform of a skewed sequence.*

The Fourier transform of a real skewed-symmetric sequence s_n *is purely imaginary and it is related to the sine transform of the half-sequence. Therefore, the* $2N$*-point discrete Fourier transform (DFT) of the sequence* (4.17) *is given by,*

$$S_f(m) = -2K \sum_{n=1}^{N-1} x_n \sin\left(\frac{mn\pi}{N}\right), \tag{4.18}$$

with K *a constant factor. Thus, the* $2N$*-point DFT of* s_n *is the* $(N-1)$*-point DST of* x_n, *up to a normalization factor. We have defined the* $S_f(m)$ *as,*

$$S_f(m) = -\sum_{n=0}^{2N-1} s_n W_{2N}^{mn}, \tag{4.19}$$

and $W_{2N} = e^{-\left(j\frac{2\pi}{2N}\right)}$ *the* $2N$*th root of unity.*

In other words, the DST-I of the real sequence $\{x_1, x_2, x_3\}$, is equivalent to the DFT of eight real numbers $0, x_1, x_2, x_3, 0, -x_3, -x_2, -x_1$ (odd symmetry), scaled by 0.5.

Example 4.7. *Using MATLAB implement a function to calculate the DST-I of a sequence x via the DFT of a skewed sequence.*

```
% DST-I of a sequence via the DFT of a skewed sequence .

function DST_1 = DSTviaDFTofSkewedSeq(x)
    s = [0, x, 0, -flip(x)];
    FT = -0.5*imag(fft(s));
    N = length(x);
    DST_1= FT(2:N+1);
end
```

THE DISCRETE SINE TRANSFORM VIA THE DFT

The DST-I pair can be expressed as,

$$X_m^{S(1)} = \sqrt{\frac{2}{N+1}} \sum_{n=1}^{N} x_n \sin\left(\frac{nm\pi}{N+1}\right) \quad m = 1, 2, \cdots, N. \tag{4.20}$$

Notice that

$$X_m^{S(1)} = \frac{2}{\sqrt{2(N+1)}} \sum_{n=0}^{2N+1} x_n \sin\left(\frac{2nm\pi}{2(N+1)}\right) \quad m = 1, 2, \cdots, N \tag{4.21}$$

and

$$x_n = \sqrt{\frac{2}{N+1}} \sum_{m=1}^{N} X_m^{S(1)} \sin\left(\frac{mn\pi}{N+1}\right), \quad n = 1, 2, \cdots, N. \tag{4.22}$$

Assuming x_n is real, the DFT of the sequence x_n is given by,

$$X_m^F = \sqrt{\frac{1}{M}} \sum_{n=0}^{M-1} x_n W_M^{nm} \quad m = 0, 1, \cdots, M-1 \tag{4.23}$$

where $W_M^{nm} = e^{-\left(j\frac{2nm\pi}{M}\right)}$. Therefore

$$X_m^F = \sqrt{\frac{1}{M}} \sum_{n=0}^{M-1} x_n \left[\cos\left(\frac{-2nm\pi}{M}\right) - j\sin\left(\frac{2nm\pi}{M}\right)\right]$$

By letting $M = 2(N+1)$ we have

$$X_m^F = \sqrt{\frac{1}{2(N+1)}} \sum_{n=0}^{2N+1} x_n \left[\cos\left(\frac{-2nm\pi}{2(N+1)}\right) - j\sin\left(\frac{2nm\pi}{2(N+1)}\right)\right]. \tag{4.24}$$

Observe that the imaginary parts of (4.24) and (4.21) are equal up to a constant factor, consequently

$$X_m^{S(1)} = -2\mathrm{Im}[\text{ DFT } \{x_n\}] \quad m = 1, 2, \cdots, N, \tag{4.25}$$

with $x_n = 0$ for $m = 0$ and $m \geq N+1$.

Example 4.8. *Using MATLAB implement a function to calculate the N-point DST-I of a sequence x_n via the $(2N+2)$-point DFT.*

```
function [ DST_I ] = DSTviaDFT( x )
% DST-I VIA DFT
    N=length(x);
    X = zeros(1, 2*N+2);
    X(2:N+1)=x;
    imFT = -imag(fft(X));
    DST_I= imFT(2:N+1);
end
```

FAST ALGORITHMS

Wang [520], [521], [522] established the following recursive relationships between the DST and the DCT that can be used to reduce the number of matrices for efficient implementation.

$$[S_N^{II}] = [\bar{I}_N][C_N^{II}][D_N], \tag{4.26}$$

$$[S_N^{III}] = [D_N][C_N^{III}][\bar{I}_N], \tag{4.27}$$

and

$$[S_N^{IV}] = [\bar{I}_N][C_N^{IV}][D_N], \tag{4.28}$$

where $[\bar{I}_N]$ is the opposite identity matrix and $[D_N] = \text{diag}\{(-1)^k\}, k = 0, 1, \cdots, N-1$ is the diagonal matrix with change of sign in odd positions. Observe that fast DST-II and DST-IV algorithms can be obtained from DCT-II and DCT-IV algorithms. Furthermore, in [64] it was shown that DCT-IV can be expressed in terms of a DCT-II by the following matrix multiplication,

$$[C_N^{IV}] = [R_N][C_N^{II}][DC_N], \tag{4.29}$$

where $[DC_N] = \text{diag}\{2\cos\left[\frac{(2k-1)\pi}{4N}\right]\}$ for $k = 1, 2, \cdots, N/2$; and

$$[R_N] = \begin{bmatrix} \frac{1}{2} & 0 & 0 & 0 & \cdots & 0 \\ -\frac{1}{2} & 1 & 0 & 0 & \cdots & 0 \\ \frac{1}{2} & -1 & 1 & 0 & \cdots & 0 \\ -\frac{1}{2} & 1 & -1 & 1 & \cdots & 0 \\ \vdots & \vdots & \vdots & \vdots & \ddots & \vdots \\ -\frac{1}{2} & 1 & -1 & 1 & \cdots & 1 \end{bmatrix}.$$

Also, the following relationships hold [373].

$$
\begin{aligned}
[C^I_{N+1}][\bar{I}_{N+1}] &= [D_{N+1}][C^I_{N+1}],\\
[S^I_{N-1}][\bar{I}_{N-1}] &= [D_{N-1}][S^I_{N-1}],\\
[C^{II}_N][\bar{I}_N] &= [D_N][C^{II}_N],\\
[S^{II}_N][\bar{I}_N] &= [D_N][S^{II}_N],\\
[\bar{I}_N][C^{III}_N] &= [C^{III}_N][D_N],\\
[\bar{I}_N][S^{III}_N] &= [S^{III}_N][D_N],\\
(-1)^{N-1}[C^{IV}_N][\bar{I}_N][D_N] &= [\bar{I}_N][D_N][C^{IV}_N],\\
(-1)^{N-1}[S^{IV}_N][\bar{I}_N][D_N] &= [\bar{I}_N][D_N][S^{IV}_N].
\end{aligned}
\tag{4.30}
$$

Notice that (4.30) limits our discussion to $[S^I_{N-1}]$ and $[\tilde{S}^I_N]$.

Next, we derive some explicit orthonormal matrices for the DST-I for $N = 2,4,8$, in order to develop the signal flowgraphs for the DST. The DST-I matrices of (4.16) are shown following, each row represents a basis vector for the transform.

$$[S^I_1] = 1, \tag{4.31}$$

$$
[S^I_3] = \begin{bmatrix} \sin\frac{\pi}{4} & 1 & \sin\frac{\pi}{4} \\ 1 & 0 & -1 \\ \sin\frac{\pi}{4} & -1 & \sin\frac{\pi}{4} \end{bmatrix}, \tag{4.32}
$$

$$
[S^I_7] = \begin{bmatrix}
\sin\frac{\pi}{8} & \sin\frac{\pi}{4} & \cos\frac{\pi}{8} & 1 & \cos\frac{\pi}{8} & \sin\frac{\pi}{4} & \sin\frac{\pi}{8} \\
\sin\frac{\pi}{4} & 1 & \sin\frac{\pi}{4} & 0 & -\sin\frac{\pi}{4} & -1 & -\sin\frac{\pi}{4} \\
\cos\frac{\pi}{8} & \sin\frac{\pi}{4} & -\sin\frac{\pi}{8} & -1 & -\sin\frac{\pi}{8} & \sin\frac{\pi}{4} & \cos\frac{\pi}{8} \\
1 & 0 & -1 & 0 & 1 & 0 & -1 \\
\cos\frac{\pi}{8} & -\sin\frac{\pi}{4} & -\sin\frac{\pi}{8} & 1 & -\sin\frac{\pi}{8} & -\sin\frac{\pi}{4} & \cos\frac{\pi}{8} \\
\sin\frac{\pi}{4} & -1 & \sin\frac{\pi}{4} & 0 & -\sin\frac{\pi}{4} & 1 & -\sin\frac{\pi}{4} \\
\sin\frac{\pi}{8} & -\sin\frac{\pi}{4} & \cos\frac{\pi}{8} & -1 & \cos\frac{\pi}{8} & -\sin\frac{\pi}{4} & \sin\frac{\pi}{8}
\end{bmatrix}. \tag{4.33}
$$

For the SST we have

$$
[\tilde{S}^I_2] = \sqrt{\frac{2}{3}} \begin{bmatrix} \sin\frac{\pi}{3} & \sin\frac{2\pi}{3} \\ \sin\frac{2\pi}{3} & -\sin\frac{\pi}{3} \end{bmatrix}, \tag{4.34}
$$

$$[\tilde{S}_4^I] = \sqrt{\frac{2}{5}}\begin{bmatrix} \sin\frac{\pi}{5} & \sin\frac{2\pi}{5} & \sin\frac{2\pi}{5} & \sin\frac{\pi}{5} \\ \sin\frac{2\pi}{5} & \sin\frac{\pi}{5} & -\sin\frac{\pi}{5} & -\sin\frac{2\pi}{5} \\ \sin\frac{2\pi}{5} & -\sin\frac{\pi}{5} & -\sin\frac{\pi}{5} & \sin\frac{2\pi}{5} \\ \sin\frac{\pi}{5} & -\sin\frac{2\pi}{5} & \sin\frac{2\pi}{5} & -\sin\frac{\pi}{5} \end{bmatrix}, \tag{4.35}$$

$$[\tilde{S}_8^I] = \frac{\sqrt{2}}{3} \times \begin{bmatrix} \sin\frac{\pi}{9} & \sin\frac{2\pi}{9} & \sin\frac{\pi}{3} & \sin\frac{4\pi}{9} & \sin\frac{4\pi}{9} & \sin\frac{\pi}{3} & \sin\frac{2\pi}{9} & \sin\frac{\pi}{9} \\ \sin\frac{2\pi}{9} & \sin\frac{4\pi}{9} & \sin\frac{\pi}{3} & \sin\frac{\pi}{9} & -\sin\frac{\pi}{9} & -\sin\frac{\pi}{3} & -\sin\frac{4\pi}{9} & -\sin\frac{2\pi}{9} \\ \sin\frac{\pi}{3} & \sin\frac{\pi}{3} & 0 & -\sin\frac{\pi}{3} & -\sin\frac{\pi}{3} & 0 & \sin\frac{\pi}{3} & \sin\frac{\pi}{3} \\ \sin\frac{4\pi}{9} & \sin\frac{\pi}{9} & -\sin\frac{\pi}{3} & -\sin\frac{2\pi}{9} & \sin\frac{2\pi}{9} & \sin\frac{\pi}{3} & -\sin\frac{\pi}{9} & -\sin\frac{4\pi}{9} \\ \sin\frac{4\pi}{9} & -\sin\frac{\pi}{9} & -\sin\frac{\pi}{3} & \sin\frac{2\pi}{9} & \sin\frac{2\pi}{9} & -\sin\frac{\pi}{3} & -\sin\frac{\pi}{9} & \sin\frac{4\pi}{9} \\ \sin\frac{\pi}{3} & -\sin\frac{\pi}{3} & 0 & \sin\frac{\pi}{3} & -\sin\frac{\pi}{3} & 0 & \sin\frac{\pi}{3} & -\sin\frac{\pi}{3} \\ \sin\frac{2\pi}{9} & -\sin\frac{4\pi}{9} & \sin\frac{\pi}{3} & -\sin\frac{\pi}{9} & -\sin\frac{\pi}{9} & \sin\frac{\pi}{3} & -\sin\frac{4\pi}{9} & \sin\frac{2\pi}{9} \\ \sin\frac{\pi}{9} & -\sin\frac{2\pi}{9} & \sin\frac{\pi}{3} & -\sin\frac{4\pi}{9} & \sin\frac{4\pi}{9} & -\sin\frac{\pi}{3} & \sin\frac{2\pi}{9} & -\sin\frac{\pi}{9} \end{bmatrix}. \tag{4.36}$$

Similar to the DCT and the SCT, the DST and SST matrices follow certain symmetries whose center depends on the order (length) of the row vector. The symmetry indicates that the fast forward and inverse algorithms are identical. Also, one can notice that the matrices $[S_{N-1}^I]$, $[S_N^{IV}]$ and $[\tilde{S}_N^I]$ are self-inverse. In other words, they are involutory matrices. The inverse matrices of $[S_N^{II}]$ and $[S_N^{III}]$ are their transpose matrices, which mean that the fast algorithms for the inverse matrices can be obtained by reversing the arrows of the flowgraph of the forward transform.

Let n be the nth element of the kth basis vector of the matrices $[S_{N-1}^I]$ or $[\tilde{S}_N^I]$. Then, we can express the matrices in terms of their basis vectors ψ as

$$[\Psi] = \begin{bmatrix} \psi_{1,n} \\ \psi_{2,n} \\ \vdots \\ \psi_{k,n} \end{bmatrix},$$

Observe that if a basis vector of Ψ is symmetric then $\psi_{k,N-1-n} = \psi_{k,n}, \quad n = 0,1\cdots,N/2$ or $\psi_{k,N-1-n} = -\psi_{k,n}, \quad n = 0,1\cdots,N/2-1$ if it is antisymmetric.

Notice that in $[\tilde{S}_N^I]$ the even-indexed and the odd-indexed basis vectors are symmetric and antisymmetric respectively. Also, notice that the following relations hold.

$$\begin{aligned}
\psi_{k,N-1-n} &= (-1)^k \psi_{k,n}, & n &= 0,1,\cdots,N/2-1,\\
\psi_{2k,N-1-n} &= \psi_{2k,n}, & k,n &= 0,1,\cdots,N/2-1,\\
\psi_{2k+1,N-1-n} &= -\,\psi_{2k+1,n}, & k,n &= 0,1,\cdots,N/2-1.
\end{aligned} \tag{4.37}$$

The symmetry and antisymmetry of the basis vectors of the DST and also the DCT matrices are key features to factorize the matrices into sparse matrices. A factorization scheme for the even/odd transforms (EOT) is reported in [496] and [373] and is shown below,

$$T_N = \begin{bmatrix} E_{\frac{N}{2}} & \bar{E}_{\frac{N}{2}} \\ O_{\frac{N}{2}} & -\bar{O}_{\frac{N}{2}} \end{bmatrix} = \begin{bmatrix} E_{\frac{N}{2}} & 0 \\ 0 & -\bar{O}_{\frac{N}{2}} \end{bmatrix} \begin{bmatrix} I_{\frac{N}{2}} & \tilde{I}_{\frac{N}{2}} \\ \tilde{I}_{\frac{N}{2}} & -I_{\frac{N}{2}} \end{bmatrix}, \tag{4.38}$$

$$= \begin{bmatrix} E_{\frac{N}{2}} & 0 \\ 0 & O_{\frac{N}{2}} \end{bmatrix} \begin{bmatrix} I_{\frac{N}{2}} & \tilde{I}_{\frac{N}{2}} \\ I_{\frac{N}{2}} & -\tilde{I}_{\frac{N}{2}} \end{bmatrix}, \tag{4.39}$$

where $\bar{E}_{\frac{N}{2}} = E_{\frac{N}{2}}\tilde{I}_{\frac{N}{2}}$, $\bar{O}_{\frac{N}{2}} = O_{\frac{N}{2}}\tilde{I}_{\frac{N}{2}}$. $E_{\frac{N}{2}}$ and $O_{\frac{N}{2}}$ are matrices of order $\frac{N}{2}$ with symmetric and antisymmetric properties respectively. Equations (4.38) and (4.39) factorize matrices with order odd. However, the center element is the center of symmetry and cannot be included in this factorization.

THE FAST DST-I VIA THE SST ALGORITHM

Assuming $N = 2^k - 1$ and $M = N + 1$ the coefficients $b_{M,m}$ generated by the SST, ignoring the scaling factor, can be expressed as

$$\tilde{S}_m^I = b_{M,m} = \sum_{n=0}^{M} x_n \sin\left[\frac{(m+1)(n+1)\pi}{M}\right], \quad m = 0,1,\cdots,M-2. \tag{4.40}$$

Coefficients can be split into two subsequences of even $b_{M/2,m}^E$ and odd coefficients $b_{M/2,m}^O$, generated by the even and odd samples x_n respectively.

$$\begin{aligned}
b_{M/2,m}^E &= \sum_{n=0}^{M/2-1} x_{2n} \sin\left[\frac{(m+1)(2n+1)\pi}{2(M/2)}\right],\\
b_{M/2,m}^O &= \sum_{n=0}^{M/2-2} x_{2n+1} \sin\left[\frac{(m+1)(n+1)\pi}{M/2}\right].
\end{aligned} \tag{4.41}$$

It is important to note that $b_{M/2,(M/2-1)} = 0$. Therefore, the $(M-1)$-point sequence can be expressed recursively as,

$$\begin{aligned}
b_{M,m} &= b_{M/2,m}^E + b_{M/2,m}^O, & m &= 0,1,\cdots,M/2-1,\\
b_{M,M-m-2} &= b_{M/2,m}^E - b_{M/2,m}^O, & m &= 0,1,\cdots,M/2-2.
\end{aligned} \tag{4.42}$$

Notice that (4.42) is the decomposition of the $(M-1)$-point DST-I into an $(\frac{M}{2}-1)$-point DST-I and an $(\frac{M}{2})$-point DST-II.

Example 4.9. *Design the DST-I flowgraph for $N=3$ based on the SST algorithm. Since $N=3$ then $M=N+1=4$, from (4.42), for $b_{M,m}$, $m=0,1$, then*

$$b_{4,m} = b^E_{2,m} + b^O_{2,m}.$$

When $m=0$

$$\begin{aligned} b_{4,0} &= \sum_{n=0}^{1} x_{2n} \sin\left[\frac{(2n+1)\pi}{4}\right] + x_1 \sin\left(\frac{\pi}{2}\right) \\ &= (x_0+x_2)\frac{1}{\sqrt{2}} + x_1. \end{aligned} \tag{4.43}$$

When $m=1$

$$\begin{aligned} b_{4,1} &= \sum_{n=0}^{1} x_{2n} \sin\left[\frac{(2n+1)\pi}{2}\right] + x_1 \sin(\pi) \\ &= (x_0-x_2). \end{aligned} \tag{4.44}$$

For $b_{M,M-m-2}$, $m=0$

$$\begin{aligned} b_{4,2} &= b^E_{2,0} - b^O_{2,0}. \\ &= \sum_{n=0}^{1} x_{2n} \sin\left[\frac{(2n+1)\pi}{4}\right] - x_1 \sin\left(\frac{\pi}{2}\right) \\ &= (x_0+x_2)\frac{1}{\sqrt{2}} - x_1. \end{aligned} \tag{4.45}$$

Figure 4.1 shows the signal flowgraph for the fast DST-I computation with $N=3$. Notice that imput samples $\{x_n\}$ are in bit-reversed order and the output coefficients are in reversed order. The inverse transform can be obtained by reversing the direction of the arrows.

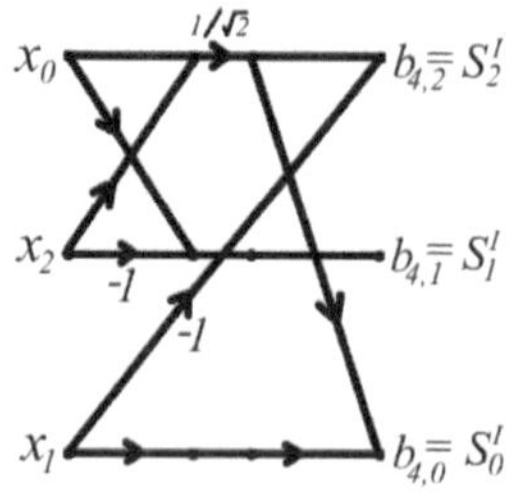

Figure 4.1: Signal flowgraph for the computation of the forward DST-I for $N=3$ via the SST algorithm.

The N-point SST for $N = 2^k$, without scaling factor can be expressed as,

$$\tilde{S}^I_m = b_{N,m} = \sum_{n=0}^{N-1} x_n \sin\left[\frac{(m+1)(n+1)\pi}{N+1}\right], \quad m = 0, 1, \cdots, N-1.$$

Again, we group the terms $2n$ and $2n+1$ to obtain one $N/2$-point DST-II and one $N/2$-point SST as follows,

$$\begin{aligned} b^E_{N/2,m} &= \sum_{n=0}^{N/2-1} x_{2n} \sin\left[\frac{(m+1)(2n+1)2\pi}{N+1}\right], \\ b^O_{N/2,m} &= \sum_{n=0}^{N/2-1} x_{2n+1} \sin\left[\frac{(m+1)(n+1)\pi}{N+1}\right]. \end{aligned} \tag{4.46}$$

Also,

$$\begin{aligned} b_{N,m} &= b^E_{N/2,m} + b^O_{N/2,m}, \quad m = 0, 1, \cdots, N/2-1, \\ b_{N,N-m-1} &= b^E_{N/2,m} - b^O_{N/2,m}, \quad m = 0, 1, \cdots, N/2-1. \end{aligned} \tag{4.47}$$

Notice that the SST decomposition of (4.46) and (4.47) is not recursive. However, it can be factored according to (4.38) because the SST is an EOT.

SPARSE MATRIX FACTORIZATION OF THE DST-I

The modified recursive sparse matrix factorization of the DST-I matrix [41], [92], [524], [560], for $N = 2^k$ is defined in (4.48).

$$S^I_{N-1} = \tilde{P}_{N-1} \begin{bmatrix} \tilde{P}_{\frac{N}{2}} S^{III}_{\frac{N}{2}} & 0 \\ 0 & S^I_{\frac{N}{2}-1} \tilde{I}_{\frac{N}{2}-1} \end{bmatrix} \begin{bmatrix} I_{\frac{N}{2}-1} & & \tilde{I}_{\frac{N}{2}-1} \\ & \sqrt{2} & \\ \tilde{I}_{\frac{N}{2}-1} & & -I_{\frac{N}{2}-1} \end{bmatrix}, \tag{4.48}$$

where $\tilde{P}_{N-1}$ is a permutation matrix to change the output of the transform from BRO to natural order (NO), see Problems 4.12. and 4.13.

A different alternative to (4.48) is the following [48], [388],

$$S^I_{N-1} = \bar{P}_{N-1} \begin{bmatrix} S^{III}_{\frac{N}{2}} & 0 \\ 0 & \tilde{I}_{\frac{N}{2}-1} S^I_{\frac{N}{2}-1} \tilde{I}_{\frac{N}{2}-1} \end{bmatrix} \begin{bmatrix} I_{\frac{N}{2}-1} & & \tilde{I}_{\frac{N}{2}-1} \\ & \sqrt{2} & \\ \tilde{I}_{\frac{N}{2}-1} & & -I_{\frac{N}{2}-1} \end{bmatrix}, \tag{4.49}$$

The matrix $\bar{P}_{N-1}$ reorders an input data vector as

$$\tilde{x}_0 = x_0, \quad \tilde{x}_{n+1} = x_{2n+2}, \quad \tilde{x}_{N-2-n} = x_{2n+1}, \quad n = 0, 1 \cdots, \frac{N}{2} - 1 \tag{4.50}$$

In (4.48) and (4.49) the $(N-1)$-point DST-I is recursively factored using $\frac{N}{2}$-point DST-III and $(\frac{N}{2}-1)$-point DST-I. Figure 4.2 shows the signal flowgraph for the computation of the forward DST-I for $N = 4$ and $N = 8$-point via (4.50).

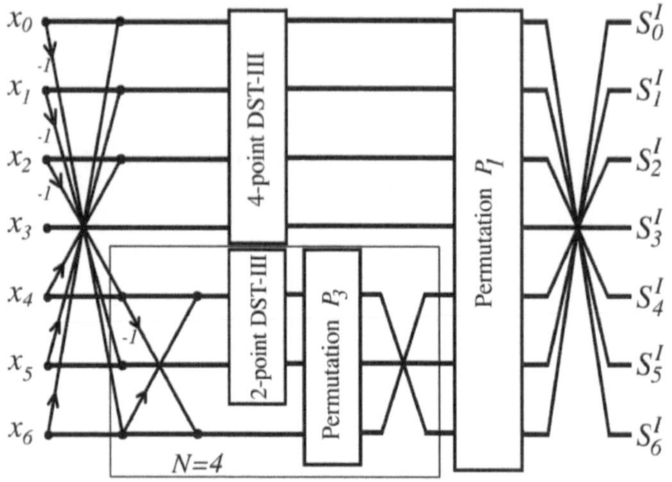

Figure 4.2: Signal flowgraph for the computation of the forward DST-I for $N = 4$ and $N = 8$ via (4.49).

THE SPLIT-RADIX ALGORITHM FOR DST-I

The aim of the split-radix algorithm is to reduce the arithmetic operations. The formulae for split-radix algorithm [464], without normalization factors are given by,

$$\begin{aligned}
S^I_{2m} &= \sum_{n=1}^{N/2-1} (x_n - x_{N-n}) \sin\frac{nm\pi}{N/2}, \quad m = 1, \cdots, N/2-1, \\
S^I_{4m-1} &= a_m - b_m, \quad m = 1, 2, \cdots, N/4, \\
S^I_{4m+1} &= a_m + b_m, \quad m = 0, 1, \cdots, N/4-1, \quad a_0 = 0,
\end{aligned} \tag{4.51}$$

where

$$\begin{aligned}
a_m &= \sum_{n=1}^{N/4-1} \left[(x_n + x_{N-n}) \cos\frac{n\pi}{N} - (x_{N/2-n} + x_{N/2+n}) \sin\frac{n\pi}{N} \right] \sin\frac{nm\pi}{N/4}, \\
b_m &= \sum_{n=0}^{N/4} \left[(x_n + x_{N-n}) \sin\frac{n\pi}{N} + (x_{N/2-n} + x_{N/2+n}) \cos\frac{n\pi}{N} \right] \cos\frac{nm\pi}{N/4} \\
&\quad - x_{N/2} - (x_{N/4} + x_{3N/4}) \cos\frac{\pi}{4}(4m+1).
\end{aligned} \tag{4.52}$$

Example 4.10. *Draw the signal flowgraph for 4-point DST-I based on the split radix algorithm.*

When $N = 4$, $a_m = 0$. Then,

$$S_1^I = b_0$$
$$S_2^I = (x_1 - x_3)\sin\frac{\pi}{2} = (x_3 - x_1)$$
$$S_3^I = -b_1$$
$$b_0 = x_2 + (x_1 + x_3)\sin\frac{\pi}{4}$$
$$b_1 = x_2 - (x_1 + x_3)\sin\frac{\pi}{4}$$

Figure 4.3 shows the signal flowgraph for 4-point DST-I via the split-radix algorithm

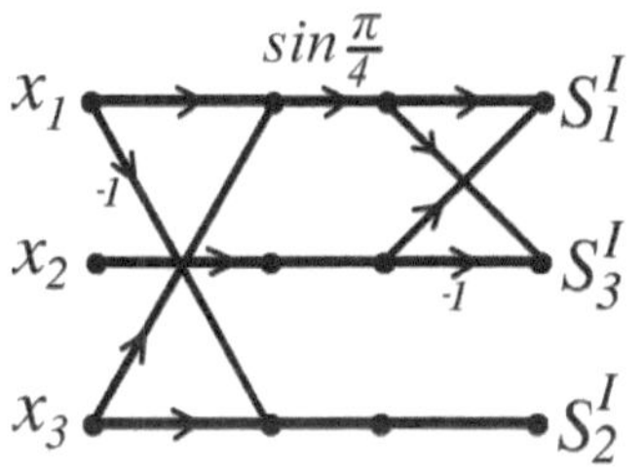

Figure 4.3: Signal flowgraph for the computation of the DST-I for $N = 4$ via the split-radix algorithm.

From (4.50) and (4.52) we can easily see the recursive decomposition of the DST-I into one $(\frac{N}{2} - 1)$-point DST-I, one $(\frac{N}{4} - 1)$-point DST-I and one $(\frac{N}{4} + 1)$-point DCT-I (see Problem 4.14.).

By using the unnormalized DST-I and DCT-I matrices ($\bar{S}^I_{\frac{N}{4}-1}$ and $\bar{C}^I_{\frac{N}{4}+1}$ respectively), the matrix S^I_{N-1} can be recursively factored as,

$$S^I_{N-1} = \tilde{P}_{N-1} \begin{bmatrix} K_{\frac{N}{2}} & 0 \\ 0 & S_{\frac{N}{2}-1}\tilde{I}_{\frac{N}{2}-1} \end{bmatrix} \begin{bmatrix} I_{\frac{N}{2}} & & \tilde{I}_{\frac{N}{2}} \\ & 1 & \\ \tilde{I}_{\frac{N}{2}} & & -I_{\frac{N}{2}} \end{bmatrix}, \tag{4.53}$$

where

$$K_{\frac{N}{2}} = Q_{\frac{N}{2}} \begin{bmatrix} I_{\frac{N}{4}-1} & \tilde{I}_{\frac{N}{4}-1} & 0 & 0 \\ \tilde{I}_{\frac{N}{4}-1} & -I_{\frac{N}{4}-1} & 0 & 0 \\ 0 & 0 & -1 & 0 \\ 0 & 0 & 0 & 1 \end{bmatrix} \begin{bmatrix} \bar{S}^I_{\frac{N}{4}-1} & 0 \\ 0 & \bar{C}^I_{\frac{N}{4}+1} \end{bmatrix} R_{\frac{N}{2}}, \tag{4.54}$$

and

$$K_2 = S_2^{III} = \begin{bmatrix} \sin\frac{\pi}{4} & -1 \\ \sin\frac{\pi}{4} & 1 \end{bmatrix}, \quad Q_4 = \begin{bmatrix} 0 & 0 & 0 & 1 \\ 1 & 0 & 0 & 0 \\ 0 & 1 & 0 & 0 \\ 0 & 0 & 1 & 0 \end{bmatrix}. \tag{4.55}$$

$$\bar{S}^I_{\frac{N}{4}-1} = \tilde{I}_{\frac{N}{4}-1}(B_{\frac{N}{4}-1}S^I_{\frac{N}{4}-1})\tilde{I}_{\frac{N}{4}-1},$$

$$\bar{C}^I_{\frac{N}{4}+1} = \tilde{I}_{\frac{N}{4}+1}(B_{\frac{N}{4}+1}C^I_{\frac{N}{4}+1})\tilde{I}_{\frac{N}{4}-1},$$

with $R_{\frac{N}{2}}$ a rotation vector given by

$$R_{\frac{N}{2}} = \begin{bmatrix} \cos\frac{\pi}{N} & 0 & -\sin\frac{\pi}{N} & 0 \end{bmatrix}, \tag{4.56}$$

APPLICATIONS

DST finds applications in the numerical solutions of differential equations as well as partial differential equations [52], [223], [226], [297]. Also DST has been applied in image processing and signal processing for communication systems, for example, in cepstral analysis in speech processing [156], [336], [558], video coding, [256], [296], pattern recognition [176], [467], [586], medical imaging [346].

Free (GPL) C libraries for fast DSTs (types I–IV) computation in one or more dimensions, of arbitrary size, can be found in [142], [362].

SUMMARY

The DST is discussed in detail in this chapter. The sparse matrix factorization of the transform and explicit form of the orthonormal matrix are discussed. Also the fast algorithms and their flowgraphs and applications are addressed.

PROBLEMS

4.1. Verify the unitarity of the DST.

4.2. Derive the shift in time property for the DST-I, DST-II, DST-III and DST-IV.

4.3. Derive the difference property for DST.

4.4. Derive the convolution property for DST.

4.5. The DST-I basis functions are defined as

$$\theta_{ki} = \theta_{ik} = \sqrt{\frac{2}{N+1}} \sin\left(\frac{ik\pi}{N+1}\right) \quad i,k = 1,2,\cdots,N$$

Show that they are orthonormal.

Hint: Show that

$$\sum_{k=1}^{N} \theta_{lk}\theta_{ik} = \sum_{k=1}^{N} \frac{2}{N+1} \sin\left(\frac{lk\pi}{N+1}\right) \sin\left(\frac{ik\pi}{N+1}\right) = \delta_{il}$$

$$i,l = 1,2,\cdots,N$$

with

$$\delta_{i,l} = \begin{cases} 1 & i = l \\ 0 & i \neq l \end{cases}$$

4.6. Figure 4.1 shows the flowgraph for the DST-I when $N = 3$, via the SST algorithm, develop a similar flowgraph for $N = 7$.
4.7. Prove the relation between DCT-II and DST-II matrices given by (4.26).
4.8. Prove the relation between the DCT-IV. and DST-IV matrices given by (4.28).
4.9. Prove the relation between DCT and DST matrices given by (4.30).
4.10. Prove the even symmetry/antisymmetry of the basis vectors defined in (4.37) for $[\tilde{S}_N^I]$.
4.11. Derive the explicit form of the DST for $N = 16$.
4.12. Equation (4.48) is the modified sparse matrix factorization of DST-I matrix. Draw the corresponding flowgraph for computation of the DST-I for $N = 4$ and list its computational complexity.
4.13. Repeat Problem 4.12. for $N = 8$.
4.14. Repeat Example 4.10 for $N = 8$ and $N = 16$.

PROJECTS

P.4.1. See [108]. The authors have developed a method for noise estimation using DST in audio sequences. Adapt this method to estimate in image denoising.
P.4.2. Develop a Bayesian threshold estimation based on [108].
P.4.3. See P.4.2. Implement a denoising method based on the DST and suitable for H.264/AVC.
P.4.4. See P.4.2. Implement a denoising method based on the DST and suitable for HEVC.
P.4.5. See [176]. The authors have used DST for facial feature extraction, an alternative local linear regression approach (ALLR) and a back propagation neural network (BPNN) to increase the recognition rate. The combination resulted in 100 % recognition rates. Repeat the experiments to obtain the same results.
P.4.6. See [354]. Nirajan has implemented the C-code of the floating-point DCT-I and DST-I into fixed-point code using Q15 format for signed 16-bit fixed-point number representation. The range of numbers was normalized between [-1, 1] and evaluated in various Tensilica processor configurations. The signal-to-noise ratio (SNR) of fixed-point DCT/DST code is between 35 and 76 dBs for different transform sizes and provides approximately 3

to 6 times performance improvement over floating-point code on Tensilica processor cores in terms of clock cycles. Use the same fixed-point code to implement both transforms using FPGAs and compare performances. (See Chapter 6.)

P.4.7. Nikara, Takala and Astola [352] proposed regular fast algorithms for DCT and DST of types II–IV based on the factorization of transform matrices and mapped them onto pipeline architectures. The latency of an ASIC implementation is 94 cycles while operating at 250 MHz frequency. Implement the same algorithms in FPGAs using a lifting scheme (see Chapter 6) and Q15 fixed-point format.

5 The Modified Discrete Cosine Transform

INTRODUCTION

The modified discrete cosine transform (MDCT) is a lapped transform and a type of linear discrete block transformation where the basis functions of the transformation overlap the block boundaries. However, the number of coefficients which result from a series of overlapping block transforms remain the same as any other non-overlapping block transformation.

The MDCT is based on the type-IV discrete cosine transform (DCT-IV) given in (3.36),

Forward:

$$X_m^{C(4)} = \sqrt{\frac{2}{N}} \sum_{n=0}^{N-1} x_n \cos\left[\frac{(2m+1)(2n+1)\pi}{4N}\right], \qquad m = 0, 1, \ldots, N-1;$$

Inverse:

$$x_n = \sqrt{\frac{2}{N}} \sum_{m=0}^{N-1} X_m^{C(4)} \cos\left[\frac{(2m+1)(2n+1)\pi}{4N}\right], \qquad n = 0, 1, \ldots, N-1; \quad (3.36)$$

The normalization factor can be absorbed in the inverse transform and the transform pair can be expressed as

Forward:

$$X_m^{C(4)} = \sum_{n=0}^{N-1} x_n \cos\left[\frac{\pi}{N}(n+0.5)(m+0.5)\right], \qquad m = 0, 1, \ldots, N-1;$$

Inverse:

$$x_n = \frac{2}{N} \sum_{m=0}^{N-1} X_m^{C(4)} \cos\left[\frac{\pi}{N}(n+0.5)(m+0.5)\right], \qquad n = 0, 1, \ldots, N-1;$$

DEFINITIONS

The MDCT uses the concept of time-domain aliasing cancellation (TDAC) and sometimes it is seen as the counterpart of the quadrature mirror filter bank (QMF)

[259], [380], [379], [517]. This transform is designed to be applied on consecutive blocks of larger datasets, in cases where subsequent blocks overlap such that the last half of one block coincides with the first half of the next block. The MDCT is compared to other Fourier-related transforms in that it has half as many outputs as inputs (instead of the same number). Also, it is a linear function, i.e., $f:\mathbb{R}^{2n}\to\mathbb{R}^{N}$, where $\mathbb{R}$ denotes the set of real numbers.

The forward and inverse MDCT pair is shown in (5.1) and (5.2). For simplicity, we drop the superscript $C(4)$

$$X^t_m = \sum_{n=0}^{2N-1} \tilde{x}^t_n \cos\left[\frac{\pi}{N}\left(n+0.5+\frac{N}{2}\right)(m+0.5)\right],$$

$$m = 0,1,\ldots,N-1; \quad (5.1)$$

$$\tilde{x}^t_n = \frac{2}{N} h_n \sum_{m=0}^{N-1} X^t_m \cos\left[\frac{\pi}{N}\left(n+0.5+\frac{N}{2}\right)(m+0.5)\right],$$

$$n = 0,1,\ldots,2N-1, \quad (5.2)$$

where t is the data-block number, $\{\tilde{x}^t_n\}$ is the tth block of $2N$ samples used to calculate the tth transform block $\{X^t_m\}$ of N unique transform coefficients. In (5.2), $\tilde{x}^t_n$ is the tth block containing the inverse modified discrete cosine transform (IMDCT) of $2N$ aliased samples in time domain, obtained from inverse transformation of the N windowed MDCT coefficients in the tth block.

If we assume h_n a symmetric window function of length $2N$ then $\tilde{x}^t_n = h_n x_n$. The window is an identical analysis-synthesis time window, given by,

$$h_n = h_{2N-1-n}, \quad (5.3a)$$

$$h_n^2 + h_{n+N}^2 = 1. \quad (5.3b)$$

This constraint is for perfect reconstruction. Other windows can be applied; however, a sine window is widely used in audio coding as it offers good stop-band attenuation [517]. It also provides attenuation of the block edge effect and allows perfect reconstruction. The sine window mentioned here can be defined as

$$h_n = \sin\left[\frac{\pi}{2N}(n+0.5)\right] \quad n = 0,1,\cdots,2N-1 \quad (5.4)$$

Therefore, we can recover the IMDCT coefficients of (5.2) by

$$\hat{x}_n = \begin{cases} \frac{1}{2}[\tilde{x}_n - \tilde{x}_{N-1-n}] & n = 0,\cdots,N-1 \\ \frac{1}{2}[\tilde{x}_n + \tilde{x}_{3N-1-n}] & n = N,\cdots,2N-1 \end{cases} \quad (5.5)$$

The input signal x_n cannot be perfectly reconstructed from a single block of coefficients. However, it can be perfectly recovered by adding the overlapped IMDCTs of subsequent overlapping blocks as shown in Figure 5.1. This leads to cancellation of errors and hence the original data can be retrieved.

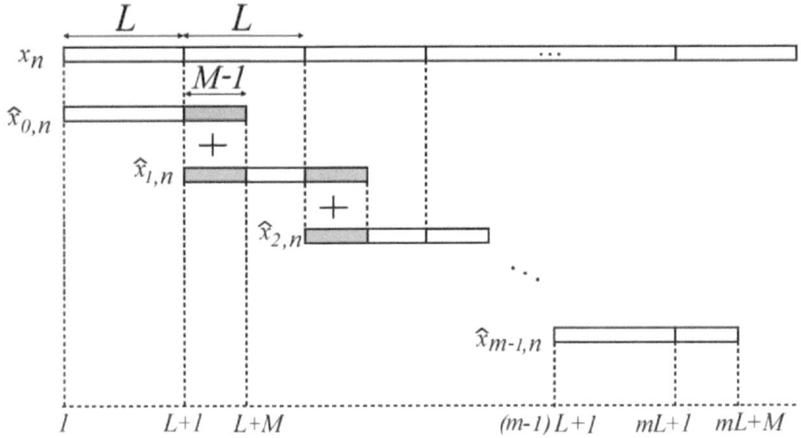

Figure 5.1: Simple overlap and add algorithm [517].

Example 5.1. *Using MATLAB implement the MDCT and the IMDCT shown in* (5.1) *and* (5.2) *to recover the input signal:* $x_n = \{10,12,14,16,14,12,10,8,10,12,14,16,14,12,10,8\}$

```
clear;
% Input sequence
x = [10 12 14 16 14 12 10 8 10 12 14 16 14 12 10 8];

% Length of the block of samples
L = 4   % L = 2N

% Length of the transform block
N= L/2;

% Length of original signal
T = length(x);

% Extended signal
xext=[zeros(1,N),x,zeros(1,2*N-1)];

% Window
n = 0:2*N-1;
h=sin( (pi/2/N).*(n+0.5));

% Number of lapped blocks
NB =floor((T-1)/(2*N)*2)+2;
```

```
%MDCT

for index = 1: N : N*NB
    % Windowing the lapped blocks
    xtilde = xext(index:index+(2*N-1)).*h;
    for m = 0:N-1
        tmp=0;
        for n = 0 : 2*N-1
            tmp= tmp + xtilde(n+1)* ...
                cos((pi/N)*(n+0.5+N/2)*(m+0.5));
        end
        X(m+1, floor(index/N)+1) = tmp;
    end
end

MDCT = X;

% IMDCT

clear n;
clear T;
clear NB;
clear N;
[N, NB] = size(X);

xrec= zeros(1, (NB+1)*N);

% Window
n = 0:2*N-1;
h=sin( (pi/2/N).*(n+0.5));

xblock = zeros(1,2*N);
for index = 1: NB
    Block = X(:,index)';
    for n = 0 : 2*N-1
        tmp =0;
        for m = 0 : N-1
            tmp = tmp + Block(1,m+1)* ...
                cos((pi/N)*(n+0.5+N/2)*(m+0.5));
        end
        xblock(1,n+1)=tmp;
    end
    % Adding
    xrec((index-1)*N+1:(index-1)*N+2*N) = ...
        xrec((index-1)*N+1:(index-1)*N+2*N)+xblock.*h*2/N
end
imdct = xrec(N+1:end-N)
```

MDCT AS THE SHIFTED DISCRETE FOURIER TRANSFORM

The MDCT is equivalent to a shifted discrete Fourier transform (SDFT) [518]. The forward and inverse SDFT are shown below,

$$X_m^{SDFT_{u,v}} = \sum_{n=0}^{2N-1} x_n e^{j2\pi(n+u)(m+v)/2N} \tag{5.6a}$$

$$x_n^{u,v} = \frac{1}{2N}\sum_{m=0}^{2N-1} X_m^{SDFT_{u,v}} e^{-j2\pi(n+u)(m+v)/2N} \tag{5.6b}$$

where u and v represent arbitrary time and frequency domain shifts respectively. Define: $\beta_{n,m,N} = \left[\frac{1}{N}\left(n+0.5+\frac{N}{2}\right)(m+0.5)\right]$. Then, from (5.1)

$$\begin{aligned} X_m &= \sum_{n=0}^{2N-1} \tilde{x}_n \cos\left[\pi\beta_{n,m,N}\right], \\ &= \sum_{n=0}^{N-1} \tilde{x}_n \cos\left[\pi\beta_{n,m,N}\right] + \sum_{n=N}^{2N-1} \tilde{x}_n \cos\left[\pi\beta_{n,m,N}\right] \end{aligned} \tag{5.7}$$

$$\begin{aligned} &= \frac{1}{2}\sum_{n=0}^{N-1} \tilde{x}_n e^{j\pi\beta_{n,m,N}} + \frac{1}{2}\sum_{n=0}^{N-1} \tilde{x}_n e^{-j\pi\beta_{n,m,N}} \\ &+ \frac{1}{2}\sum_{n=N}^{2N-1} \tilde{x}_n e^{j\pi\beta_{n,m,N}} + \frac{1}{2}\sum_{n=N}^{2N-1} \tilde{x}_n e^{-j\pi\beta_{n,m,N}} \end{aligned} \tag{5.8}$$

Replacing the index n in the second and fourth terms of (5.8) with $N-1-n$ and $3N-1-n$ respectively

$$\begin{aligned} X_m &= \frac{1}{2}\sum_{n=0}^{N-1} \tilde{x}_n e^{j\pi\beta_{n,m,N}} + \frac{1}{2}\sum_{n=0}^{N-1} \tilde{x}_{N-1-n} e^{-j\pi(2-\beta_{n,m,N})} \\ &+ \frac{1}{2}\sum_{n=N}^{2N-1} \tilde{x}_n e^{j\pi\beta_{n,m,N}} + \frac{1}{2}\sum_{n=N}^{2N-1} \tilde{x}_{3N-1-n} e^{-j\pi(4-\beta_{n,m,N})} \end{aligned} \tag{5.9a}$$

Notice that $e^{-j\pi(2-\beta_{n,m,N})} = -e^{j\pi\beta_{n,m,N}}$ and $e^{-j\pi(4-\beta_{n,m,N})} = e^{j\pi\beta_{n,m,N}}$. Therefore,

$$\begin{aligned} X_m &= \frac{1}{2}\sum_{n=0}^{N-1} \tilde{x}_n e^{j\pi\beta_{n,m,N}} - \frac{1}{2}\sum_{n=0}^{N-1} \tilde{x}_{N-1-n} e^{j\pi\beta_{n,m,N}} \\ &+ \frac{1}{2}\sum_{n=N}^{2N-1} \tilde{x}_n e^{j\pi\beta_{n,m,N}} + \frac{1}{2}\sum_{n=N}^{2N-1} \tilde{x}_{3N-1-n} e^{j\pi\beta_{n,m,N}} \\ &= \frac{1}{2}\sum_{n=0}^{N-1} [\tilde{x}_n - \tilde{x}_{N-1-n}] e^{j\pi\beta_{n,m,N}} + \frac{1}{2}\sum_{n=N}^{2N-1} [\tilde{x}_n + \tilde{x}_{3N-1-n}] e^{j\pi\beta_{n,m,N}} \end{aligned} \tag{5.9b}$$

Combining (5.5) and (5.9b) we finally obtain

$$X_m = \sum_{n=0}^{2N-1} \hat{x}_n \exp\left[j2\pi \frac{\left(n+0.5+\frac{N}{2}\right)(m+0.5)}{2N} \right] \tag{5.10}$$

which is the SDFT with $u = \frac{N+1}{2}$ and $v = \frac{1}{2}$ for the signal $\hat{x}_n$ formed from the windowed signal $\tilde{x}_n$.

PROPERTIES OF THE MDCT

(1). The MDCT is not an orthogonal transform. Perfect signal reconstruction can be achieved in the overlap-add (OA) process. For the OA window of $2N$ samples, the first N and last N samples of the signal will remain modified. One can easily see this from the fact that performing MDCT and IMDCT of an arbitrary signal x_n reconstructs X_m. Figure 5.1 describes the simple overlap and add algorithm. First, the signal x_n is partitioned into non-overlapping sequences. Then, the DFTs of the sequences $\tilde{x}_{k,n}$ [415] are calculated by multiplying the FFT of x_n with the FFT of h_n. Afterwards, $\hat{x}_n$ is recovered by using the inverse fast Fourier transform (IFFT) [388], the resulting output signal can be reconstructed using overlapping and adding $\hat{x}_n$ as in Figure 5.1. The overlap is based on the idea that a linear convolution is always longer than the original sequences [415].

(2). Non-orthogonality property of the MDCT.

a). If a signal exhibits local symmetry such that

$$\begin{aligned} \tilde{x}_k &= \tilde{x}_{N-k+1} & k &= 0, 1, \cdots, N-1 \\ \tilde{x}_k &= -\tilde{x}_{3N-k-1} & k &= N, N+1, \cdots, 2N-1 \end{aligned}$$

Then its MDCT degenerates to zero; i.e., $X_m = 0$ for $m = 0, 1, \cdots, N-1$. This is an example that MDCT does not fulfill Parseval's theorem, in other words, the time domain energy and the frequency domain energy are not equal.

b). If a signal exhibits local symmetry such that

$$\begin{aligned} \tilde{x}_k &= -\tilde{x}_{N-k+1} & k &= 0, 1, \cdots, N-1 \\ \tilde{x}_k &= \tilde{x}_{3N-k-1} & k &= N, N+1, \cdots, 2N-1, \end{aligned}$$

then the MDCT and the IMDCT will perfectly reconstruct the original time domain samples.

(3). The MDCT becomes an orthogonal transform if the signal length is infinite. This is different from the traditional definition of orthogonality, which requires a square transform matrix.

(4). The MDCT is similar to the orthogonal transforms such as DFT, DCT, DST and it also possesses energy compaction capability.

(5). Invertibility: Since the number of inputs and outputs is not equal it may seem that the MDCT is not invertible. However, the MDCT is perfectly invertible which is achieved by adding the overlapped IMDCTs of subsequent overlapping blocks. This leads the errors to cancel and the original data will be retrieved. This technique is known as time-domain aliasing cancellation (TDAC) [259], [379], [380], [517] (see Example 5.1).

(6). Performing MDCT and then IMDCT with one single frame of time domain samples, the original time samples cannot be perfectly reconstructed, instead the reconstructed samples are normally an alias-embedded version. The MDCT by itself is a lossy process.

(7). The MDCT concentrates 90% of energy within 10% of the normalized frequency scale for most of the test signals for all transforms concerned [519]. The energy compaction property of different transforms becomes more unified with increasing window length.

(8). Window shape has an impact on MDCT energy compaction property. In the case of rectangular window, DCT has always the best energy compaction property since DCT corresponds to an even extension of the signal.

(9). The MDCT is very useful with its TDAC characteristics [379], [380], [517]. However, its mismatch with the DFT domain-based psychoacoustic model must be kept in mind when developing an MDCT-based audio codec with its full potential in terms of coding performance.

(10). The distinct advantage of MDCT lies in its critical sampling property, reduction of block effect and the possibility of adaptive window switching.

(11). In comparison to the orthogonal transforms, the MDCT has a special property, i.e., the input signal cannot be perfectly reconstructed from a single block of MDCT coefficients [116].

(12). In terms of energy compaction property, MDCT does not have any advantage in comparison to DFT and DCT as indicated in Figure 5.2.

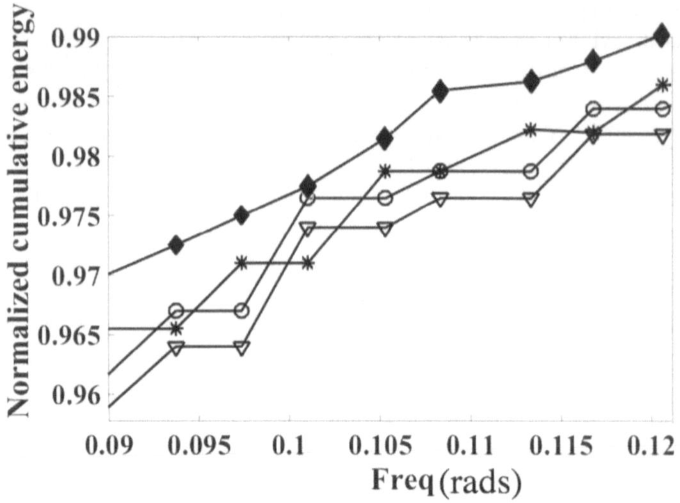

Figure 5.2: Cumulative power spectra of DCT (diamond), $\text{SDFT}_{(N-1)/2,1/1}$ (circle), DFT (star) and MDCT (triangle) with rectangular windows. The window size is 256. The length of the test sequence is 5292000 PCM samples [519].

EVENLY STACKED MDCT

The evenly stacked, block transform MDCT was developed by Princen and Brandly [379] as follows,

$$X_{m,k} = (-1)^{mk}\cos\left(\frac{m\pi}{2}\right)\sum_{r=0}^{P-1} x_{m,mM+r}h_{P-1-r}\cos\left(\frac{2\pi k}{K}(r+n_0)\right)$$
$$+(-1)^{mk}\sin\left(\frac{m\pi}{2}\right)\sum_{r=0}^{P-1} x_{m,mM+r}h_{P-1-r}\sin\left(\frac{2\pi k}{K}(r+n_0)\right),$$
$$k=0,1,\cdots,N-1,\quad M=\frac{N}{2},\quad \frac{N}{2}\le P\le N, \tag{5.11}$$

where $\{x_{m,k}\}$ are the sequences of samples of the mth data block, n_0 is the time shift, m is the mth block time and h_{P-1-r} is a time reversed, finite impulse response filter of length P. Notice that, for a given block time m all the terms in (5.11) vanish except one. The filter or analysis window overlaps between consecutive blocks, $\{X_{m,k}\}$ are the output channel signals. Also, the $\cos\left(\frac{2\pi k}{K}(r+n_0)\right)$ and $\sin\left(\frac{2\pi k}{K}(r+n_0)\right)$ are referred to as the evenly stacked MDCT and modified DST (MDST) transform kernels. For m even the $\{X_{m,k}\}$ are the MDCT of windowed signal segment around time $n=nM$ modified by $(-1)^{mk}$. Similarly, for m odd, $\{X_{k,m}\}$ are the MDST of the windowed signal segment around time $n=nM$ modified by $(-1)^{mk}$. This forms a set of K channels. However, for an evenly stacked system, only $\frac{K}{2}+1$ analysis channels are unique, since the paths are time multiplexed (for m even and odd) with center frequencies $\omega_k = \frac{2\pi k}{K}, \quad k = 0, \cdots K-1$.

The block transform synthesis operation is also possible assuming that no distortion was introduced in $\{X_{m,k}\}$ [379].

$$\hat{x}_{m,r} = \cos\left(\frac{m\pi}{2}\right)\frac{1}{K}\sum_{r=0}^{K-1}(-1)^{mk}X_{m,k}\cos\left(\frac{2\pi k}{K}(r+n_0)\right)$$
$$+\sin\left(\frac{m\pi}{2}\right)\frac{1}{K}\sum_{r=0}^{K-1}(-1)^{mk}X_{m,k}\sin\left(\frac{2\pi k}{K}(r+n_0)\right), \quad (5.12)$$

For m even $\hat{x}_{m,r}$, the time domain aliased samples of the mth data block, are the IMDCT and for m odd it is the inverse modified discrete sine transform (IMDST) of the correspondent modified channel signals.

Assuming perfect reconstruction, then for $n_0 = \frac{K}{4}+\frac{1}{2}$, the time domain aliasing cancels and the original data sequence $\{X_{m,r}\}$ is recovered from two adjacent data blocks by the windowing and OA procedure as follows,

$$x_{m+1,r} = x_{m,\frac{K}{2}+r} = f_{\frac{K}{2}+r}\hat{x}_{m,\frac{K}{2}+r} + f_r\hat{x}_{m+1,r}, \quad r = 0, 1, \cdots, \frac{K}{2}-1, \quad (5.13)$$

where $\{f_r\}$ is the synthesis window function, and the analysis and synthesis window functions have to satisfy the so-called perfect reconstruction condition [379].

MDCT BLOCK TRANSFORM

In general, the fast structures for their implementations and the mathematical properties of the analysis and synthesis filter bank are investigated using block transforms applied to a single windowed input data block $\{x_n\}$, where $n = 1, \cdots, N-1$, with N an even integer representing the samples in a data block; this is also known as a N-point sequence. Using (5.11) and (5.12) the MDCT and the IMDCT evenly (E) stacked block transform pair are, respectively, defined in [379], ignoring the term $(-1)^{mk}$, by

$$X_k^E = \sum_{n=0}^{N-1} x_n \cos\left[\frac{\pi}{N}\left(2n+1+\frac{N}{2}\right)k\right], \quad k = 0, 1, \cdots, N-1, \quad X_{\frac{N}{2}}^E = 0, \quad (5.14)$$

$$\hat{x}_n^E = \frac{1}{N}\sum_{k=0}^{N-1} X_k^E \cos\left[\frac{\pi}{N}\left(2n+1+\frac{N}{2}\right)k\right], \quad n = 0, 1, \cdots, N-1. \quad (5.15)$$

Here, the subscript m of the data block index is omitted, $\hat{x}_n^E$ is used to emphasize that the samples recovered by the IMDCT process do not correspond to the original data samples, in other words, the recovered samples are time domain aliased. It is important to note that the MDST - IMDST pair can be obtained by changing the *cos* function by the *sin* function and knowing that the transform coefficient $X_0^E = 0$ of the MDST, in contrast to (5.14) where the transform coefficient $X_{\frac{N}{2}}^E$ is zero. One of the main differences is the symmetry properties of these transforms.

In evenly stacked systems alternate MDCT/MDST are required. However, the MDCT is recognized as the operation when the time index m of a data block is even and the MDST when m is odd.

SYMMETRY PROPERTIES OF MDCT BLOCK TRANSFORM

The symmetry properties of the evenly stacked MDCT block transform can be investigated by letting $k = N - k$ in (5.14). Therefore, the angle of *cos* can be expressed as $\left[\frac{\pi}{N}(2n+1+\frac{N}{2})(N-k)\right]$. After some simple trigonometric simplifications we obtain,

$$X_{N-k}^E = \sum_{n=0}^{N-1} x_n \cos\left[\pi\left(2n+1+\frac{N}{2}\right)\right]\cos\left[\frac{\pi}{N}\left(2n+1+\frac{N}{2}\right)k\right],$$
$$k = 0, 1, \cdots, \frac{N}{2} - 1. \tag{5.16}$$

Therefore, from the term *cos* $\left[\pi\left(2n+1+\frac{N}{2}\right)\right]$ it is easy to see that

$$X_{N-k}^E = \begin{cases} -X_k^E, & \text{if } \left(2n+1+\frac{N}{2}\right) \text{ is odd}, k = 1, 2, \cdots, \frac{N}{2} - 1, \\ X_k^E, & \text{if } \left(2n+1+\frac{N}{2}\right) \text{ is even}, k = 1, 2, \cdots, \frac{N}{2} - 1, \end{cases} \tag{5.17}$$

In other words, the odd anti-symmetry results when $N/2$ is even and the odd symmetry when $N/2$ is odd. The same analysis is valid for the MDST resulting in a change of signs of the transform block of (5.17) with odd antisymmetry property when $N/2$ is odd and odd symmetry property when $N/2$ is even. From this result we can see that only $\frac{N}{2}$ MDCT or MDST coefficients are unique, whereby $\frac{N}{2}$ coefficients are non-zero.

The time domain aliased data sequence $\{\hat{x}_n^E\}$ recovered by the backward MDCT has the following symmetries

$$\hat{x}_n^E = \hat{x}_{\frac{N}{2}-1-n}^E, \qquad \hat{x}_{\frac{N}{2}+n}^E = \hat{x}_{N-1-n}^E$$
$$n = 0, 1, \cdots, \frac{N}{4} - 1. \tag{5.18}$$

This property exhibits the even symmetries in the first and second half.

For $\{\hat{x}_n^E\}$ recovered by the backward MDST we have,

$$\hat{x}_n^E = -\hat{x}_{\frac{N}{2}-1-n}^E, \qquad \hat{x}_{\frac{N}{2}+n}^E = -\hat{x}_{N-1-n}^E$$
$$n = 0, 1, \cdots, \frac{N}{4} - 1,$$

and exhibits the odd symmetries in the first and second half.

For a proper representation, the MDCT pair of (5.14) and (5.15) is redefined by introducing scaling factors ε_k according to the corresponding symmetry property (5.17) as follows,

$$X_k^E = \sum_{n=0}^{N-1} x_n \cos\left[\frac{\pi}{N}\left(2n+1+\frac{N}{2}\right)k\right], \quad k=0,1,\cdots,\frac{N}{2}-1, \quad X_{\frac{N}{2}}^E = 0, \tag{5.19}$$

$$\hat{x}_n^E = \frac{1}{N}\sum_{k=0}^{\frac{N}{2}-1} \varepsilon_k X_k^E \cos\left[\frac{\pi}{N}\left(2n+1+\frac{N}{2}\right)k\right], \quad n=0,1,\cdots,N-1. \tag{5.20}$$

$$\varepsilon_k = \begin{cases} 1 & \text{for } k=0, \\ 2 & \text{for } k=1,2\cdots,\frac{N}{2}-1. \end{cases}$$

The evenly stacked MDST pair can also be redefined by introducing scaling factors τ_k as,

$$X_k^E = \sum_{n=0}^{N-1} x_n \sin\left[\frac{\pi}{N}\left(2n+1+\frac{N}{2}\right)k\right], \quad k=1,\cdots,\frac{N}{2}, \quad X_0^E = 0,$$

$$\hat{x}_n^E = \frac{1}{N}\sum_{k=1}^{\frac{N}{2}} \tau_k X_k^E \sin\left[\frac{\pi}{N}\left(2n+1+\frac{N}{2}\right)k\right], \quad n=0,1,\cdots,N-1.$$

$$\tau_k = \begin{cases} 1 & \text{for } k=\frac{N}{2}, \\ 2 & \text{for } k=1,2\cdots,\frac{N}{2}-1. \end{cases}$$

MATRIX REPRESENTATION OF THE MDCT

We can represent the MDCT using matrix form. By observing the symmetry properties derived in (5.17), we can easily see that there are only $\frac{N}{2}$ linearly independent rows. Therefore, the dimensions of the MDCT matrix will be $\frac{N}{2} \times N$ or $C^E_{\frac{N}{2}\times N}$. Also, if we let $x = [x_0, x_2, \ldots, x_{N-1}]$ and $X^E = \left[X_0^E, X_1^E, \ldots, X_{\frac{N}{2}-1}^E\right]$ be a single row vector of windowed data and a row vector of transform coefficients respectively, we can write the evenly stacked MDCT block transform pair defined in (5.19) and (5.20) without the normalization factor $\frac{1}{N}$ as,

$$\left[X^E\right]^T = C^E_{\frac{N}{2}\times N}[x]^T, \tag{5.21}$$

$$\left[\hat{x}^E\right]^T = \varepsilon_k C^E_{N\times\frac{N}{2}}\left[X^E\right]^T, \tag{5.22}$$

where T denotes matrix transpositions and $\left[C^E_{\frac{N}{2}\times N}\right]^T = C^E_{N\times\frac{N}{2}}$. Also, the following relations hold [49], [379],

$$C^E_{\frac{N}{2}\times N}\varepsilon_k\left[C^E_{\frac{N}{2}\times N}\right]^T = NI_{\frac{N}{2}},$$

$$\varepsilon_k\left[C^E_{\frac{N}{2}\times N}\right]^T\left[C^E_{\frac{N}{2}\times N}\right] = \frac{N}{2}\begin{bmatrix} I_{\frac{N}{2}}+\tilde{I}_{\frac{N}{2}} & 0 \\ 0 & I_{\frac{N}{2}}+\tilde{I}_{\frac{N}{2}} \end{bmatrix} \tag{5.23}$$

where 0 is a null matrix, $I_{\frac{N}{2}}$ and $\tilde{I}_{\frac{N}{2}}$ are the identity and the opposite identity matrices respectively, both of dimension $\frac{N}{2}$. The time domain aliased sequence can be recovered by inverse MDCT from (5.23) as

$$\left[\hat{x}^E\right]^T = \frac{N}{2}\begin{bmatrix} I_{\frac{N}{2}}+\tilde{I}_{\frac{N}{2}} & 0 \\ 0 & I_{\frac{N}{2}}+\tilde{I}_{\frac{N}{2}} \end{bmatrix} x^T. \tag{5.24}$$

The explicit form of the time domain aliased data samples can be recovered from (5.24). See problem 5.8.

For the evenly stacked block transform MDST (S^E) ignoring the normalization factor $\frac{1}{N}$,

$$\left[X^E\right]^T = S^E_{\frac{N}{2}\times N}[x]^T,$$
$$\left[\hat{x}^E\right]^T = \tau_k S^E_{N\times\frac{N}{2}}\left[X^E\right]^T.$$

The following relationship holds

$$S^E_{\frac{N}{2}\times N}\tau_k\left[S^E_{\frac{N}{2}\times N}\right]^T = NI_{\frac{N}{2}}$$

Hence,

$$\begin{aligned}\tau_k\left[S^E_{\frac{N}{2}\times N}\right]^T S^E_{\frac{N}{2}\times N} &= \frac{N}{2}\begin{bmatrix} I_{\frac{N}{4}} & -\tilde{I}_{\frac{N}{4}} & 0 & 0 \\ -\tilde{I}_{\frac{N}{4}} & I_{\frac{N}{4}} & 0 & 0 \\ 0 & 0 & I_{\frac{N}{4}} & -\tilde{I}_{\frac{N}{4}} \\ 0 & 0 & -\tilde{I}_{\frac{N}{4}} & I_{\frac{N}{4}} \end{bmatrix} \\ &= \frac{N}{2}\begin{bmatrix} I_{\frac{N}{2}}-\tilde{I}_{\frac{N}{2}} & 0 \\ 0 & I_{\frac{N}{2}}-\tilde{I}_{\frac{N}{2}} \end{bmatrix}\end{aligned}$$

where 0 is a null matrix, $I_{\frac{N}{2}}$ and $\tilde{I}_{\frac{N}{2}}$ are the identity and the opposite diagonal identity matrices of order $N/2$ respectively. Notice that,

$$rank\left(C^E_{\frac{N}{2}\times N}\right) = rank\left(S^E_{\frac{N}{2}\times N}\right) = \frac{N}{2}.$$

ODDLY STACKED MDCT

The oddly stacked MDCT, critically sampled, single-sideband SSB filter bank based on the TDAC, was developed as a block transform [232, 380]. The outputs $\{X_{m,k}\}$ of analysis channels are defined by

$$X_{m,k} = \cos(mk\pi)\sum_{r=0}^{N-1} x_{m,r+m\frac{N}{2}} h_{N-r-1} \cos\left[\frac{2\pi}{N}(k+0.5)(r+n_0)\right],$$
$$k = 0,1\cdots,N-1, \quad (5.25)$$

where $\{x_{m,r+m\frac{N}{2}}\}$ are the samples of the $m-$th data block, $\{h_{N-r-1}\}$ is a length N, time reversed low pass window and $cos\left[\frac{2\pi}{N}(k+0.5)(r+n_0)\right]$ the oddly stacked MDCT kernel. Also, after analysis/synthesis MDCT the aliasing distortion is introduced in time domain.

The oddly stacked, block transform synthesis operation is defined by [380]

$$\hat{x}_{m,r} = \frac{2}{N}\sum_{k=0}^{N-1}\cos(mk\pi)X_{m,k}\cos\left[\frac{2\pi}{N}(k+0.5)(r+n_0)\right],$$
$$r = 0,1\cdots,N-1, \quad (5.26)$$

where $\{\hat{x}_{m,r}\}$ are the windowed aliased samples, in time domain, of the mth data block of the backward MDCT. If perfect reconstruction conditions hold, the original data sequence is recovered from two adjacent data sequences $\{\hat{x}_{m,r}\}$ by windowing and OA operations for a phase factor $n_0 = \frac{N}{4} + \frac{1}{2}$.

MDCT BLOCK TRANSFORM

Let $\{x_n\}$ be a single windowed input data block, where $n = 1,\cdots,N-1$ and N an even integer number representing the samples in a data block. The forward MDCT coefficients $\{X_k^o\}$ and the time domain aliased data samples $\hat{x}_n^o$ from IMDCT are computed as [380]

$$X_k^o = \sum_{n=0}^{N-1} x_n \cos\left[\frac{\pi}{2N}\left(2n+1+\frac{N}{2}\right)(2k+1)\right], \quad k = 0,1,\cdots,N-1, \quad (5.27)$$

$$\hat{x}_n^o = \frac{4}{N}\sum_{k=0}^{N-1} X_k^o \cos\left[\frac{\pi}{2N}\left(2n+1+\frac{N}{2}\right)(2k+1)\right], \quad n = 0,1,\cdots,N-1. \quad (5.28)$$

And the forward and inverse MDST pair is defined as,

$$X_k^o = \sum_{n=0}^{N-1} x_n \sin\left[\frac{\pi}{2N}\left(2n+1+\frac{N}{2}\right)(2k+1)\right], \quad k=0,1,\cdots,N-1,$$

$$\hat{x}_n^o = \frac{4}{N}\sum_{k=0}^{\frac{N}{2}-1} X_k^o \sin\left[\frac{\pi}{2N}\left(2n+1+\frac{N}{2}\right)(2k+1)\right], \quad n=0,1,\cdots,N-1.$$

SYMMETRY PROPERTIES OF MDCT BLOCK TRANSFORM

To investigate the symmetry properties of the oddly stacked MDCT block transform, replace k by $k=N-k-1$ into (5.26)

$$X_{N-k-1}^o = \sum_{n=0}^{N-1} x_n \cos\left[\pi\left(2n+1+\frac{N}{2}\right)\right] \times \cos\left[\frac{\pi}{2N}\left(2n+1+\frac{N}{2}\right)(2k+1)\right], \quad k=0,1,\cdots,N-1, \quad (5.29)$$

From (5.29) it is easy to see the symmetry of the forward MDCT coefficients, by observing the argument $\left[\pi\left(2n+1+\frac{N}{2}\right)\right]$,

$$X_{N-k-1}^o = \begin{cases} -X_k^o, & \text{if}\left(2n+1+\frac{N}{2}\right) \text{ is odd}, k=0,1,\cdots,\frac{N}{2}-1, \\ X_k^o, & \text{if}\left(2n+1+\frac{N}{2}\right) \text{ is even}, k=0,1,\cdots,\frac{N}{2}-1, \end{cases} \quad (5.30)$$

In other words, the X_{N-k-1}^o coefficients exhibit either even symmetry property when $\frac{N}{2}$ is odd, or even anti-symmetry when $\frac{N}{2}$ is even. Notice that only $\frac{N}{2}$ coefficients are unique. Also, the aliased time domain sequences have the following local symmetry,

$$\hat{x}_n^o = -\hat{x}_{\frac{N}{2}-1-n}^o \qquad \hat{x}_{\frac{N}{2}+n}^o = \hat{x}_{N-1-n}^o, \quad n=0,1,\cdots,\frac{N}{4}-1. \quad (5.31)$$

Hence the oddly stacked MDCT and IMDCT can be expressed as,

$$X_k^o = \sum_{n=0}^{N-1} x_n \cos\left[\frac{\pi}{2N}\left(2n+1+\frac{N}{2}\right)(2k+1)\right], \quad k=0,1,\cdots,\frac{N}{2}-1, \quad (5.32)$$

$$\hat{x}_n^o = \frac{4}{N}\sum_{k=0}^{\frac{N}{2}-1} X_k^o \cos\left[\frac{\pi}{2N}\left(2n+1+\frac{N}{2}\right)(2k+1)\right], \quad n=0,1,\cdots,N-1. \quad (5.33)$$

The oddly stacked MDCT kernel is defined by,

$$t^o_{k,n} = \cos\left[\frac{\pi}{2N}\left(2n+1+\frac{N}{2}\right)(2k+1)\right], \tag{5.34}$$

substituting $n+N$, and then $n+2N$ for n we obtain the following relationship for the MDCT kernel,

$$t^o_{k,n} = -t^o_{k,n+N}, \quad t^o_{k,n} = t^o_{k,n+2N}, \quad \forall k,$$

this implies that the MDCT transform kernel is an anti-periodic sequences with period $2N$.

Fot the MDST replace $N-k-1$ for k in,

$$X^o_k = \sum_{n=0}^{N-1} x_n \sin\left[\frac{\pi}{2N}\left(2n+1+\frac{N}{2}\right)(2k+1)\right], \quad k=0,1,\cdots,N-1,$$

to get

$$X^o_{N-k-1} = -\sum_{n=0}^{N-1} x_n \cos\left[\pi\left(2n+1+\frac{N}{2}\right)\right]\sin\left[\frac{\pi}{2N}\left(2n+1+\frac{N}{2}\right)(2k+1)\right]$$
$$k=0,1,\cdots,N-1.$$

Also, argument $\left[\pi\left(2n+1+\frac{N}{2}\right)\right]$ can be odd or even depending on the value $\frac{N}{2}$ being even or odd respectively. Hence,

$$X^o_{N-k-1} = \begin{cases} X^o_k, & \text{if}\left(2n+1+\frac{N}{2}\right) \text{ is odd }, k=0,1,\cdots,\frac{N}{2}-1, \\ -X^o_k, & \text{if}\left(2n+1+\frac{N}{2}\right) \text{ is even}, k=0,1,\cdots,\frac{N}{2}-1, \end{cases}$$

The MDST coefficients have either even symmetry when $\frac{N}{2}$ is even, or even anti-symmetry when $\frac{N}{2}$ is odd. The aliased time domain sequences recovered by the backward MDST have the following local symmetry,

$$\hat{x}^o_n = \hat{x}^o_{\frac{N}{2}-1-n} \qquad \hat{x}^o_{\frac{N}{2}+n} = -\hat{x}^o_{N-1-n},$$
$$n=0,1,\cdots,\frac{N}{4}-1.$$

These symmetry and anti-symmetry properties of the MDST allows to express the forward and backward MDST block transforms as [50],

$$X^o_k = \sum_{n=0}^{N-1} x_n \sin\left[\frac{\pi}{2N}\left(2n+1+\frac{N}{2}\right)(2k+1)\right], \quad k=0,1,\cdots,\frac{N}{2}-1,$$

$$\hat{x}^o_n = \frac{4}{N}\sum_{k=0}^{\frac{N}{2}-1} X^o_k \cos\left[\frac{\pi}{2N}\left(2n+1+\frac{N}{2}\right)(2k+1)\right], \quad n=0,1,\cdots,N-1.$$

The oddly stacked MDST kernel is defined by,

$$t^o_{k,n} = \sin\left[\frac{\pi}{2N}\left(2n+1+\frac{N}{2}\right)(2k+1)\right].$$

substituting $n+N$, and then $n+2N$ for n we obtain the following relationship for the MDST kernel,

$$t^o_{k,n} = -t^o_{k,n+N}, \quad t^o_{k,n} = t^o_{k,n+2N}, \quad \forall k,$$

this implies that the MDST transform kernel is an anti-periodic sequence with period $2N$.

MATRIX REPRESENTATION OF THE MDCT

The symmetry properties of (5.30) imply that only $\frac{N}{2}$ rows of the MDCT matrix are linearly independent. Hence, the dimensions of the MDCT matrix will be $\frac{N}{2} \times N$ or $C^o_{\frac{N}{2}\times N}$. Let $x = [x_0, x_2, \ldots, x_{N-1}]$ and $X^o = \left[X^o_0, X^o_1, \ldots, X^o_{\frac{N}{2}-1}\right]$ be a single row vector of windowed data and a row vector of transform coefficients respectively. Therefore, the MDCT and IMDCT block transforms in matrix form can be expressed as

$$[X^o]^T = C^o_{\frac{N}{2}\times N}[x]^T, \tag{5.35}$$

$$[\hat{x}^o]^T = C^o_{N\times\frac{N}{2}}[X^o]^T, \tag{5.36}$$

where the constant factor $\frac{4}{N}$ is ignored, T denotes matrix transpositions and $\left[C^o_{\frac{N}{2}\times N}\right]^T = C^o_{N\times\frac{N}{2}}$. The following relations hold [49, 50],

$$C^o_{\frac{N}{2}\times N}\left[C^o_{\frac{N}{2}\times N}\right]^T = \frac{N}{2}I_{\frac{N}{2}} \tag{5.37}$$

$$\left[C^o_{\frac{N}{2}\times N}\right]^T C^o_{\frac{N}{2}\times N} = \frac{N}{4}\begin{bmatrix} I_{\frac{N}{2}} - \tilde{I}_{\frac{N}{2}} & 0 \\ 0 & I_{\frac{N}{2}} + \tilde{I}_{\frac{N}{2}} \end{bmatrix} \tag{5.38}$$

where 0 is a null matrix, $I_{\frac{N}{2}}$ and $\tilde{I}_{\frac{N}{2}}$ are the identity and the opposite identity matrices respectively, both of dimension $\frac{N}{2}$. The time domain for the oddly stacked IMDCT block transform can be recovered using (5.38),

$$[\hat{x}^o]^T = \frac{N}{4}\begin{bmatrix} I_{\frac{N}{2}} - \tilde{I}_{\frac{N}{2}} & 0 \\ 0 & I_{\frac{N}{2}} + \tilde{I}_{\frac{N}{2}} \end{bmatrix} x^T. \tag{5.39}$$

The explicit form of the time domain aliased data samples can be recovered from (5.39). See problem 5.12.

For the oddly stacked block transform MDST (S^o) ignoring the normalization factor $\frac{N}{4}$,

$$[X^o]^T = S^o_{\frac{N}{2}\times N}[x]^T,$$
$$[\hat{x}^o]^T = S^o_{N\times\frac{N}{2}}[X^o]^T.$$

The following relationship holds

$$S^o_{\frac{N}{2}\times N}\left[S^o_{\frac{N}{2}\times N}\right]^T = \frac{N}{2}I_{\frac{N}{2}}$$

Hence,

$$\left[S^o_{\frac{N}{2}\times N}\right]^T S^o_{\frac{N}{2}\times N} = \frac{N}{4}\begin{bmatrix} I_{\frac{N}{4}} & \tilde{I}_{\frac{N}{4}} & 0 & 0 \\ \tilde{I}_{\frac{N}{4}} & I_{\frac{N}{4}} & 0 & 0 \\ 0 & 0 & I_{\frac{N}{4}} & -\tilde{I}_{\frac{N}{4}} \\ 0 & 0 & -\tilde{I}_{\frac{N}{4}} & I_{\frac{N}{4}} \end{bmatrix}$$
$$= \frac{N}{4}\begin{bmatrix} I_{\frac{N}{2}} + \tilde{I}_{\frac{N}{2}} & 0 \\ 0 & I_{\frac{N}{2}} - \tilde{I}_{\frac{N}{2}} \end{bmatrix}$$

where 0 is a null matrix, $I_{\frac{N}{2}}$ and $\tilde{I}_{\frac{N}{2}}$ are the identity and the opposite diagonal identity matrices of order $N/2$ respectively. Notice that,

$$rank\left(C^o_{\frac{N}{2}\times N}\right) = rank\left(S^o_{\frac{N}{2}\times N}\right) = \frac{N}{2}.$$

APPLICATIONS

The MDCT is employed in most modern lossy audio formats such as MP3, Vorbis, Windows Media Audio, Cook and advanced audio coding (AAC) which are different lossy audio compression codecs [262]. In MP3, the MDCT is not applied to the audio signal directly, but instead it is applied to the output of a 32-band polyphase quadrature filter (PQF) bank. The output of this MDCT is post-processed by an alias reduction formula to reduce the typical aliasing of the PQF bank. Such a combination of a filter bank with an MDCT is called hybrid bank or a subband MDCT. AAC on the other hand, normally uses a pure MDCT; only the MPEG-4, AAC Scalable Sample Rate (AAC-SSR) variant by Sony electronics uses a four-band PQF bank followed by an MDCT. Similar to MP3 the ATRAC uses stacked quadrature mirror filters (QMF) followed by an MDCT. The direct application of MDCT formula require $O(N^2)$ operations, however the number of operations can be reduced to only $O(N\log_2 N)$ complexity. This reduction can be done by recursively factorizing the computation as in the case of FFT. It is also possible to compute MDCT using other transforms such as DFT (FFT) or a DCT combined with $O(N)$ pre-and post-processing steps.

MDCT limits the sources of output distortion at the quantization stage. It is also used as an analysis filter. MDCT performs a series of inner products between the input samples x_n and the analysis filter h_n. This eliminates the blocking artifacts that would cause a problem during the reconstruction of the samples. The inverse MDCT reconstructs the samples without the blocking artifacts. The signal representation in the MDCT is a dominant tool in high-quality audio coding because of its special properties. In addition to the energy compaction capability similar to DCT, the MDCT simultaneously achieves critical sampling, a reduction of the block effect and flexible window switching. MDCT makes use of the concept of time-domain aliasing cancellation (TDAC) whereas the quadrature mirror filter bank (QMF) uses the concept of the frequency-domain aliasing cancellation [259]. This can be viewed as a duality of MDCT and QMF. However, it is to be noted that MDCT also cancels frequency-domain aliasing, whereas QMF does not cancel time-domain aliasing. This means that the MDCT is designed to achieve perfect reconstruction, QMF on the other hand does not produce perfect reconstruction. Overlapped windows allow for better frequency response functions but carry the penalty of additional values in the frequency domain, thus these transforms are not critically sampled. MDCT thus has solved the paradox satisfactorily and is currently the best solution.

The MDCT spectrum of a signal is the Fourier spectrum of the signal mixed with its alias. This compromises the performance of MDCT as a Fourier spectrum analyzer and leads to possible mismatch problems between MDCT- and DFT-based perceptual models. However, MDCT has been applied successfully to perceptual audio compression without major problems if a proper window, such as a sine window, is employed [262]. The TDAC of an MDCT filter bank can only be achieved with the overlap-add process in the time domain [259]. Although MDCT coefficients are quantized in an individual data block, MDCT is usually analyzed in the context of a continuous stream. In the case of discontinuity, such as editing or error concealment, the aliases of the two neighboring blocks in the overlapped area are not able to cancel each other out.

SUMMARY

The MDCT is a linear transformation that takes $2N$inputs and produces N outputs, which is designed to be applied to a sequence of 50%-overlapping blocks of a longer sequence (e.g., audio samples). Because this is non-square (fewer outputs than inputs), the IMDCT is not an "inverse" transformation in the usual sense; it only recovers the original data when IMDCTs of overlapping blocks are added (by "time-domain aliasing cancellation").

The MDCT achieves perfect reconstruction only without quantization, which is never the case in coding applications. On modeling the quantization as a superposition of quantization noise to the MDCT coefficients, the time-domain alias of the input signal will still be cancelled, but the noise components will be extended as additional "noise alias." In order to have 50% window overlap and critical sampling simultaneously, the MDCT time-domain window is twice as long as that of ordinary transforms such as DCT. Because of the increased time-domain window length, the quantization noise is spread to the whole window, thus making pre-echo more likely to be audible. Well-known solutions to this problem are the window switching and temporal noise shaping. In very low bit-rate coding the high-frequency components are often removed. This corresponds to a very steep low-pass filter. Due to the increased window size, the ringing effect caused by high-frequency cutting is longer.

The MDCT is a special case of perfect reconstruction cosine modulated filter bank to avoid audible artifacts in audio codecs. The transform is real, therefore that no phase information is present and it is an example of the class of transforms called time domain aliasing cancellation (TDAC). The MDCT is an orthogonal lapped transform and it has been adopted in several international standards and commercial products such as MPEG-1, MPEG-2, and AC-3 in audio coding. Nevertheless, the computational complexity of this transform can be extensive. In this chapter, we addressed the definitions and properties of the MDCT. Also, we studied the evenly and oddly stacked MDCT as well as the applications, problems and projects.

PROBLEMS

5.1. Obtain the symmetry properties for the MDST similar to those shown in (5.17) for MDCT.

5.2. Obtain the symmetries of the time-aliased data sequence for IMDST similar to those shown in (5.18) for MDCT.

5.3. Verify the symmetry properties (5.18) by proper substitution into (5.15).

5.4. Show that the sequence $\{\hat{x}_n^E\}$ exhibits the even symmetries both in the first and second half for the case of the IMDCT.

5.5. Show that the sequence $\{\hat{x}_n^E\}$ exhibits the odd symmetries both in the first and second half for the case of the IMDST.

5.6. Derive the matrix form of the MDST using a scaling factor τ_k.

5.7. Derive the relation between the MDCT, MDST and DFT.

5.8. Write the explicit form of the time-domain aliased data sampled $\{\tilde{x}_n^E\}$ in terms of the original data samples $\{x_n\}$ using (5.24). Notice that $\tilde{x}^E_{\frac{N}{2}-1-n} = \tilde{x}_n^E$ and $\tilde{x}^E_{N-1-n} = \tilde{x}^E_{\frac{N}{2}+n}$, for $n = 0, 1, \cdots, \frac{N}{4}-1$.

5.9. Similar to (5.27) and (5.28), derive the corresponding forward and backward MDST block transforms.

5.10. Verify the symmetry properties of (5.31) by proper substitution into (5.28).

5.11. Show that the transform kernel of (5.34) is anti-periodic with period N and periodic with period $2N$. (Hint: Replace n by $n+N$ and then n by $n+2N$).

5.12. Write the explicit form of the time-domain aliased data sampled $\{\tilde{x}_n^o\}$ in terms of the original data samples $\{x_n\}$ using (5.39). Notice that $\tilde{x}^o_{\frac{N}{2}-1-n} = -\tilde{x}_n^o$ and $\tilde{x}^o_{N-1-n} = \tilde{x}^o_{\frac{N}{2}+n}$, for all $n = 0, 1, \cdots, \frac{N}{4}-1$.

PROJECTS

P.5.1. In [417] the authors develop an MDCT/IMDCT based on fast rotation-based MDCT computational structure for MP3 audio. Figure 2 of the paper shows the signal graph to compute the 36-point IMDCT. Describe this structure using a hardware description language (HDL) and synthesize it.

P.5.2. In [564] the authors evaluate the IntMDCT implemented by fixed-point arithmetic with quantized lifting coefficients in the MPEG-2 AAC-based lossless audio coding. Describe this structure using an HDL and synthesize it.

P.5.3. In [353] the authors develop an algorithm for fast implementation of forward and inverse MDCT in layer III of MPEG-1 and MPEG-2 audio coding standards. Describe the MDCT/IMDCT structure using an HDL and synthesize it.

P.5.4. Compare the performance (area and speed) of implementations in P.5.1., P.5.2. and P.5.3.

P.5.5. In [51] the authors developed a fast computational structure identical for both the forward and inverseMDCT/MDST. Develop the integer version and compare with the original proposal.

P.5.6. Describe the structure of P.5.5. using an HDL and synthesize it.

P.5.7. Test the identical fast MDCT computational structure of P.5.5. in MPEG layer III (MP3) and the Dolby AC-3 audio codecs.

6 The Integer Discrete Cosine Transform

INTRODUCTION

The DCT-II has been the core of the video coding systems. One problem of the DCT is that some elements of its kernel are irrational numbers; this increases the computational cost and, in some cases, causes a mismatch between the encoder and decoder. The integer discrete cosine transform (ICT or IntDCT) [61], [62] was developed to deal with this problem. The IntDCT is an integer approximation of the discrete cosine transform and can be generated by replacing the real numbered elements of the DCT-II matrix with integers keeping the relative magnitudes and orthogonal relationship among the matrix elements [236]. The resulting matrix can be implemented exclusively with integer arithmetic. It proves to be highly advantageous in cost and speed for hardware implementations. In particular, transforms of sizes larger than 4×4 or 8×8, especially 16×16, 32×32 and now 64×64 [78] are proposed because of their increased applicability to the de-correlation of high resolution video signals. In fact, transform size of 128×128 is proposed for the forthcoming versatile video coding (VVC) [534], [359]. For example, order-16 integer transform is a simple, low computational complexity transform with a high coding efficiency.

Transformation is one popular technique for data compression. The first step is to decorrelate the pixels by applying a transform. The most significant transform coefficients are retained, with the rest set to zero, and inverse transformation is applied to recover the original image. Therefore, it is important that the transform process be simple and fast. The family of orthogonal transforms is well suited for this application because the inverse of an orthogonal matrix is its transpose. The DCT is widely accepted as having a high efficiency [387]. Before H.264/AVC, the transformation in video coding standards was specified in floating point operations. Nevertheless, errors were introduce in the recovered image because of the different floating point implementations of encoders and decoders. In H.264/AVC [204] and HEVC [209] video coding standards, the inverse transforms are specified in integer arithmetic to induce almost the same reconstructed image at the decoder side. The normative is specified in [199]. In 2006 the moving picture experts group (MPEG) approved the replacement specification ISO/IEC 23002-1 [207].

The state of the art HEVC has a hybrid video coding framework as the previous H.26x series. In such a framework, video is first decorrelated by block-wise spatial prediction or temporal prediction with motion estimation (ME) and motion compensation (MC) and then, each prediction error block goes through DCT to exploit spatial redundancy. The superior performance of HEVC originates from the advanced coding tools adopted including quad-tree partition, extended-size IntDCT and additional DST [296]. Even though IntDCT is a vital operation, it is one of the most time-consuming methods together with the rate-distortion optimization (RDO) process. The all zero block (AZB) early determination ahead of transform tends to be useful for saving encoding time caused by redundant transform and quantization [81], [100], [228], [273], [330], [445], [476], [505], [504], [506], [507], [508], [509], [510], [511], [537], [545], [546], [549], [552], [588], [590].

This chapter explains systematic forms to construct the integer discrete cosine transform, the integer discrete Tchebichef transform and the main algorithms used in the late video coding standards as well as hardware implementations, applications, problems and projects of the IntDCT.

INTEGER DCT MATRIX

The elements of the DCT matrix are real numbers. For a 16-point DCT, 8 bits are needed to represent these numbers in order to ensure negligible image reconstruction errors due to finite-length number representation. If the transform matrix elements are integers, then it may be possible to have a smaller number of bit representation and at the same time zero truncation errors. Moreover, the resultant cosine values are difficult to approximate in fixed precision integers, thus producing rounding errors in practical applications. Rounding errors can alter the orthogonality property of the transform. Using the principle of dyadic symmetry [62] order-8 integer cosine transform (ICT) which has zero truncation errors was introduced. This requires a small number, as little as 2 bit representation and efficiency comparable to the DCT [61]. The following are some desirable properties for an ICT matrix:

1. *Coding gain (Energy compaction):* This property refers to the amount of signal energy compacted in a few coefficients. Transforms with high coding gain compact more signal energy and produce a sparser representation of the signal in the transform domain, leading to a more efficient entropy coding.
2. *Orthogonality:* This property allows to preserve the norm (energy) of the signal in the transform domain.
3. *Symmetry (Linear phase):* The linear phase of the basis functions is one of the most important properties in image and video coding.
4. *Integer to integer mapping:* This property refers to exact invertibility and tight dynamic range for lossless coding applications.
5. *DC-Leakage:* This is the amount of DC energy spread in the band-pass and high-pass coefficients. It measures the degree at which the DC energy is concentrated in the DC coefficients. Low DC-Leakage of the transform yields a high decorrelation of the signal and prevents the annoying discontinuities or checkerboard artifacts in the reconstructed signal.
6. *Dyadic-rational coefficients:* The elements of the basis vectors of an ICT are dyadic-rational numbers of the form $a/2^b$, with a and b integer numbers to be easy to implement in hardware and software, to allow the operations to be carried out in finite and small width buses, to avoid high latencies, low memory buffering and to allow parallel implementation.

THE ICT BY DIRECT SCALING

In the direct scaling method, the floating point coefficients of the basis vectors of the DCT are approximated by a integer number. The integer transform T_N is obtained by multiplying the coefficients by a parameter α to control the precision and rounding with the result as follows,

$$T_N(\alpha, C_N) = \text{round}(\alpha \cdot C_N) = T_N(\alpha) \tag{6.1}$$

However, if the parameter α is large the dynamic range of the transform increases, resulting in the need for larger architectures (32 or 64-bit). For example, when $\alpha = 26$ the DCT properties are retained. Hence, a 32-bit architecture is necessary for its implementation.

The integer kernels can be constructed to be a good approximation of the DCT and to keep its properties. This makes the ICT suitable for many video coding standards, such as H.264/AVC [204], AVS [356] and HEVC [208], [209]. ICT can be appropriate for many transform sizes, for example in HEVC the TUs are partitioned adaptively in a quad-tree form. Then, order-8, 16 and 32 ICTs are applied to 4×4, 8×8, 16×16 and 32×32 TUs respectively. Also, there exists a 4×4 integer sine transform (IST) that can be chosen according to the residual TU. In the next subsections, matrices of finite precision approximations to the DCT, recommended in recent video coding standards, are explained.

H.264 4x4 and 8x8 ICT Matrices

In the $H.264$ standard [398] different integer transforms are defined: a 4×4 core transform, 4×4 and 2×2 Hadamard transforms and an 8×8 transform (high profile only). The inverse and re-scaling processes are defined in this standard. Therefore, integer forward and inverse transforms are designed taking into consideration the re-scaling process. In this section, only the 4×4 and the 8×8 ICTs are described.

The following 4×4 forward and inverse ICT matrices were selected as part of the $H.264$ video coding standard and were proposed by H. S. Malvar et al. [308] with an α factor of $\frac{5}{2}$. This approximation is chosen to minimize the complexity while maintaining good compression performance.

$$T_4\left(\frac{5}{2}\right) = \begin{bmatrix} 1 & 1 & 1 & 1 \\ 2 & 1 & -1 & -2 \\ 1 & -1 & -1 & 1 \\ 1 & -2 & 2 & -1 \end{bmatrix} \tag{6.2}$$

The maximum sum of absolute values in any row of (6.2) is 6. Considering the two dimensions, the transform has a dynamic range gain of 6. For this 2-D transform we obtain a total dynamic range gain of 6^2 Since we compute two-dimensional transforms by transforming rows and columns. Therefore, we need $\log_2(6^2) = 5.17 \approx 6$ bits or more than the input signal x_n to store the transform coefficients X_k. The transform dynamic range gain is 5.38 dB for a stationary Gauss–Markov input with correlation coefficient $\rho = 0.9$. The forward ICT approximation minimizes the complexity of the implementation and requires only additions and binary shifts while maintaining a good coding gain. The transform can be computed in integer arithmetic, thus, avoiding inverse transform mismatch problems. The inverse transform is defined as follows [308],

$$T_4^{-1}\left(\frac{5}{2}\right) = \frac{1}{20}\begin{bmatrix} 5 & 4 & 5 & 2 \\ 5 & 2 & -5 & -4 \\ 5 & -2 & -5 & 4 \\ 5 & -4 & 5 & -2 \end{bmatrix} \tag{6.3}$$

Also expressed as

$$T_4^{-1}\left(\frac{5}{2}\right) = \begin{bmatrix} 1 & 1 & 1 & 1/2 \\ 1 & 1/2 & -1 & -1 \\ 1 & -1/2 & -1 & 1 \\ 1 & -1 & 1 & -1/2 \end{bmatrix} \times \text{diag}\left\{\frac{1}{4},\frac{1}{5},\frac{1}{4},\frac{1}{5}\right\}$$

Notice that

$$\left[T_4^{-1}\left(\frac{5}{2}\right)\right] \text{diag}\left\{\frac{1}{4},\frac{1}{5},\frac{1}{4},\frac{1}{5}\right\}\left[T_4\left(\frac{5}{2}\right)\right] = I$$

Figure 6.1 shows the flowgraphs of (a) the fast forward and (b) fast inverse transforms as described above, which are applied to rows and columns of each 4×4 block. Observe the low complexity of the transform. No multiplications are needed, only additions and shifts [308].

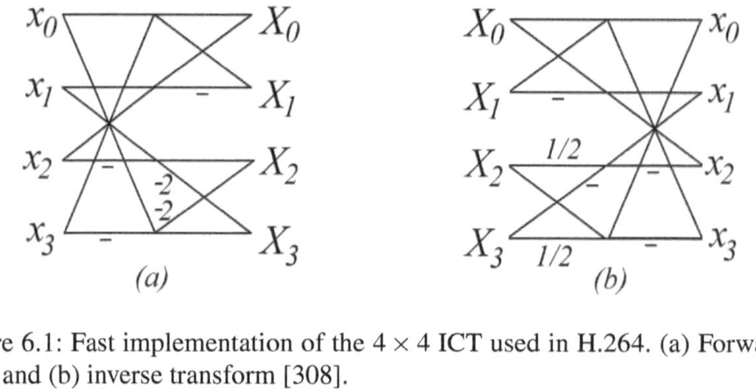

Figure 6.1: Fast implementation of the 4×4 ICT used in H.264. (a) Forward transform and (b) inverse transform [308].

The forward and inverse 8×8 integer transforms were developed similarly to the 4×4 integer transform. However, the 8×8 integer transform cannot be obtained by scaling and rounding an 8–point DCT. The forward and inverse scaling factors are $\frac{1}{2^{22}}$ and $\frac{1}{2^8}$ respectively as the dynamic range is larger than the 4×4 transform [398]. The basis vectors of the core forward transform are shown below,

$$T_8 = \begin{bmatrix} 8 & 8 & 8 & 8 & 8 & 8 & 8 & 8 \\ 12 & 10 & 6 & 3 & -3 & -6 & -10 & -12 \\ 8 & 4 & -4 & -8 & -8 & -4 & 4 & 8 \\ 10 & -3 & -12 & -6 & 6 & 12 & 3 & -10 \\ 8 & -8 & -8 & 8 & 8 & -8 & -8 & 8 \\ 6 & -12 & 3 & 10 & -10 & -3 & 12 & -6 \\ 4 & -8 & 8 & -4 & -4 & 8 & -8 & 4 \\ 3 & -6 & 10 & -12 & 12 & -10 & 6 & -3 \end{bmatrix} \tag{6.4}$$

Each row vector can be normalized by multiplying it by the normalization factor $\frac{1}{\sqrt{\sum_n c_{m,n}^2}}$ where m is the row vector index and n is the nth vector component or column index. Notice that $m, n \in [0, 1, \cdots, 7]$. The inverse transform T_8^{-1} is obtained by dividing $T_8/8$ and is shown below,

$$T_8^{-1} = \begin{bmatrix} 1 & 1 & 1 & 1 & 1 & 1 & 1 & 1 \\ 12/8 & 10/8 & 6/8 & 3/8 & -3/8 & -6/8 & -10/8 & -12/8 \\ 1 & 1/2 & -1/2 & -1 & -1 & -1/2 & 1/2 & 1 \\ 10/8 & -3/8 & -12/8 & -6/8 & 6/8 & 12/8 & 3/8 & -10/8 \\ 1 & -1 & -1 & 1 & 1 & -1 & -1 & 1 \\ 6/8 & -12/8 & 3/8 & 10/8 & -10/8 & -3/8 & 12/8 & -6/8 \\ 1/2 & -1 & 1 & -1/2 & -1/2 & 1 & -1 & 1/2 \\ 3/8 & -6/8 & 10/8 & -12/8 & 12/8 & -10/8 & 6/8 & -3/8 \end{bmatrix} \tag{6.5}$$

The rows can be normalized. The components of basis vectors are in dyadic rational form, that is, in the form $a/2^b$, where a and b are integer numbers.

WMV9 4x4 and 8x8 ICT Matrices

Windows Media 9 series (WMV)9 includes a variety of audio and video codecs, which are key components for authoring and playback of digital media. This standard uses transform matrices for one-dimensional 4×4 and 8×8 integer transformations [130], [449]. The core matrix for the 1D 4×4 integer transformation is shown here.

$$T_4 = \begin{bmatrix} 17 & 17 & 17 & 17 \\ 22 & 10 & -10 & -22 \\ 17 & -17 & -17 & 17 \\ 10 & -22 & 22 & -10 \end{bmatrix} \tag{6.6}$$

The inverse transform is achieved in the same way as the 8×8 matrix and the procedure is shown here.

$$T_8 = \begin{bmatrix} 12 & 12 & 12 & 12 & 12 & 12 & 12 & 12 \\ 16 & 15 & 9 & 4 & -4 & -9 & -15 & -16 \\ 16 & 6 & -6 & -16 & -16 & -6 & 6 & 16 \\ 15 & -4 & -16 & -9 & 9 & 16 & 4 & -15 \\ 12 & -12 & -12 & 12 & 12 & -12 & -12 & 12 \\ 9 & -16 & 4 & 15 & -15 & -4 & 16 & -9 \\ 6 & -16 & 16 & -6 & -6 & 16 & -16 & 6 \\ 4 & -9 & 15 & -16 & 16 & -15 & 9 & -4 \end{bmatrix} \tag{6.7}$$

The inverse transform is achieved by using the row-column wise method as follows. For example, let X be the 8×8 block of transform coefficients. In the first stage, the coefficients are inversely transformed by processing row by row as,

$$D = X \cdot T_8/16. \tag{6.8}$$

Then, in a second stage, the columns of D are inversely transformed to produce the recovered pixels R as,

$$R = T_8^T \cdot D/64, \tag{6.9}$$

where the $[\cdot]^T$ is the transpose operation and R are the recovered pixels.

AVS1 China 8x8 ICT Matrix

Audio and video coding standard (AVS1) is the national standard of China developed to meet a variety of applications such as digital television, laser disc, network streaming media, wireless streaming media, digital audio broadcasting, video surveillance, etc. The compression efficiency of AVS1 is similar to that of the high profile of H.264/AVC. AVS1 uses a separable, integer-precise, 8×8 cosine transform to approximate DCT using a Pre-scaled Integer Transform (PIT) technique [389], [573] to obtain the transform coefficients Y as,

$$Y = (T_8^T \times X \times T_8) \otimes S, \tag{6.10}$$

where the operation $C = A \otimes B$ means $C_{i,k} = A_{i,k} \times B_{i,k}$, the forward scaling operation $\otimes S$ is defined by,

$$S_{m,n} = \frac{1}{\sqrt{\sum_{k=0}^{N-1} T_{8,(m,k)}} \sqrt{\sum_{k=0}^{N-1} T_{8,(k,n)}^T}}, \tag{6.11}$$

and transformation matrix T_8 is,

$$T_8 = \begin{bmatrix} 8 & 10 & 10 & 9 & 8 & 6 & 4 & 2 \\ 8 & 9 & 4 & -2 & -8 & -10 & -10 & -6 \\ 8 & 6 & -4 & -10 & -8 & 2 & 10 & 9 \\ 8 & 2 & -10 & -6 & 8 & 9 & -4 & -10 \\ 8 & -2 & -10 & 6 & 8 & -9 & -4 & 10 \\ 8 & -6 & -4 & 10 & -8 & -2 & 10 & -9 \\ 8 & -9 & 4 & 2 & -8 & 10 & -10 & 6 \\ 8 & -10 & 10 & -9 & 8 & -6 & 4 & -2 \end{bmatrix} \tag{6.12}$$

The inverse transform is defined according to the forward transform process as

$$Y = (T_8 \times X \otimes S' \times T_8^T), \tag{6.13}$$

where $\otimes S'$ is the inverse scaling process. When the scaling process is done with the quantization process, multiplications can be saved. Hence, the forward transform can be carried out using only shifts and additions,

$$Y = (T_8 \times X \otimes \times T_8 + (1 << 4)) >> 5, \tag{6.14}$$

afterward, the scaling and quantization step size Q can be easily combined to yield

$$\hat{Y} = (Y \otimes CS)//Q, \tag{6.15}$$

where $A//B$ means $A_{m,n}/B_{m,n}$ and CS is a constant matrix that combines S and S' [566] only for the forward transform operation. The inverse transform and dequantization process is defined by,

$$\hat{X} = (T_8 \times (\hat{Y} \otimes Q) \times T_8^T), \tag{6.16}$$

Observe that the recovery of the signal does not include the scaling operation and the division operations are avoided. For low-complexity inverse transform and dequantization, shifting operations are applied separately in horizontal and vertical directions at the inverse transform side to keep immediate data with a 16-bit representation.

H.265 4x4, 8x8, 16x16 and 32x32 ICT Matrices

In the $H.265$ video coding standard, also known as high efficiency video coding (HEVC) or $MPEG-H$ Part 2 [209], a scaling value of $\alpha = 2^{(6+\frac{M}{2})}$ was used [54] to calculate the 4×4 transform matrix, where the matrix size is $N = 4$ and $M = log_2(N) = 2$. Afterwards, a careful manual tuning on the resulting coefficients was carried out to give a good compromise between orthogonality, closeness to DCT and norm measure. This process was used to calculate a 32×32 integer transform from which coefficients can be obtained for sub-matrices of size 16×16, 8×8 and 4×4 [209], [296], [387], [533].

$$T_4\left(2^{128}\right) \approx \left[\begin{array}{cc|cc} 64 & 64 & 64 & 64 \\ 83 & 36 & -36 & -83 \\ 64 & -64 & -64 & 64 \\ 36 & -83 & 83 & -36 \end{array}\right] \tag{6.17}$$

However, cascading of the 2D forward and inverse transform will results in a scaling of α. Consequently, to preserve the norm through the two-dimensional forward and inverse transforms, the product of all scale factors are equal to $\left[\frac{1}{2^{(6+\frac{M}{2})}}\right]^4$. When the residual video has a bit depth of B bits, the range will be $[-2^B+1, 2^B-1]$, requiring $(B+1)$ bits to represent it. The worst case will be when all the residual samples have values of -2^B at the input of the first stage of transformation. The maximum value of an output sample will be $-2^B \times N \times 64$ which corresponds to the inner product of the first basis vector with the input vector consisting of all values equal to -2^B. Therefore, the output samples can fit within 16 bits (i.e., maximum values of -2^{15}), a scaling of $1/\left(2^B \times 2^M \times 2^6 \times 2^{-15}\right) = 2^{-(B+M-9)}$ is required to be applied to the first output. For example, for $B = 8$ and $M = 2$ the scaling factor after the first stage of transformation is 2^{-1} or one shift to the right.

The output of first transform is a matrix of coefficients with all the elements in the first row equal to -2^{15} and zeros the elements of the remaining rows. Then, this matrix is multiplied by $[T_4]^T$ to analyze the signal along the second dimension. The result is a transform matrix with only a DC coefficient of value $-2^{15} \times [2^M \times 2^6]$ and the AC components equal to zero. Hence, the required scaling factor at the output of the second stage of transformation is $2^{-(M+6)}$ to fit in the required 16 bits.

To recover an approximation of the input signal, the output of the forward transform (in our example is a matrix with only the DC component equal to -2^{15}) is multiplied by $[T_4]^T$. The resulting matrix has the first column with elements equal to $-2^{15} \times 2^6$. Therefore, the scaling factor required here is 2^{-6} to fit in 16 bits. After scaling, the resulting matrix with elements of the first column equal to -2^{15} is multiplied by $[T_4]$. This yields a matrix with all elements equal to $-2^{15} \times 2^6$. Therefore, the scaling factor at the output of the second transformation stage is $2^{-(21-B)}$ in order to fit the output values in the range $[-2^B, 2^B - 1]$. Notice that here it is -2^B instead of $-2^B + 1$ or $2^B - 1$ since it is a power of 2. The basis matrices for the 8×8, 16×16 and 32×32 ICTs are shown in (6.18), (6.19) and (6.20) respectively [209].

$$T_8 = \left[\begin{array}{cccc|cccc}
64 & 64 & 64 & 64 & 64 & 64 & 64 & 64 \\
89 & 75 & 50 & 18 & -18 & -50 & -75 & -89 \\
83 & 36 & -36 & -83 & -83 & -36 & 36 & 83 \\
75 & -18 & -89 & -50 & 50 & 89 & 18 & -75 \\
64 & -64 & -64 & 64 & 64 & -64 & -64 & 64 \\
50 & -89 & 18 & 75 & -75 & -18 & 89 & -50 \\
36 & -83 & 83 & -36 & -36 & 83 & -83 & 36 \\
18 & -50 & 75 & -89 & 89 & -75 & 50 & -18
\end{array}\right] \tag{6.18}$$

$$T_{16} = \left[\begin{array}{cccccccc|cccccccc}
64 & 64 & 64 & 64 & 64 & 64 & 64 & 64 & 64 & 64 & 64 & 64 & 64 & 64 & 64 & 64 \\
90 & 87 & 80 & 70 & 57 & 43 & 25 & 9 & -9 & -25 & -43 & -57 & -70 & -80 & -87 & -90 \\
89 & 75 & 50 & 18 & -18 & -50 & -75 & -89 & -89 & -75 & -50 & -18 & 18 & 50 & 75 & 89 \\
87 & 57 & 9 & -43 & -80 & -90 & -70 & -25 & 25 & 70 & 90 & 80 & 43 & -9 & -57 & -87 \\
83 & 36 & -36 & -83 & -83 & -36 & 36 & 83 & 83 & 36 & -36 & -83 & -83 & -36 & 36 & 83 \\
80 & 9 & -70 & -87 & -25 & 57 & 90 & 43 & -43 & -90 & -57 & 25 & 87 & 70 & -9 & -80 \\
75 & -18 & -89 & -50 & 50 & 89 & 18 & -75 & -75 & 18 & 89 & 50 & -50 & -89 & -18 & 75 \\
70 & -43 & -87 & 9 & 90 & 25 & -80 & -57 & 57 & 80 & -25 & -90 & -9 & 87 & 43 & -70 \\
64 & -64 & -64 & 64 & 64 & -64 & -64 & 64 & 64 & -64 & -64 & 64 & 64 & -64 & -64 & 64 \\
57 & -80 & -25 & 90 & -9 & -87 & 43 & 70 & -70 & -43 & 87 & 9 & -90 & 25 & 80 & -57 \\
50 & -89 & 18 & 75 & -75 & -18 & 89 & -50 & -50 & 89 & -18 & -75 & 75 & 18 & -89 & 50 \\
43 & -90 & 57 & 25 & -87 & 70 & 9 & -80 & 80 & -9 & -70 & 87 & -25 & -57 & 90 & -43 \\
36 & -83 & 83 & -36 & -36 & 83 & -83 & 36 & 36 & -83 & 83 & -36 & -36 & 83 & -83 & 36 \\
25 & -70 & 90 & -80 & 43 & 9 & -57 & 87 & -87 & 57 & -9 & -43 & 80 & -90 & 70 & -25 \\
18 & -50 & 75 & -89 & 89 & -75 & 50 & -18 & -18 & 50 & -75 & 89 & -89 & 75 & -50 & 18 \\
9 & -25 & 43 & -57 & 70 & -80 & 87 & -90 & 90 & -87 & 80 & -70 & 57 & -43 & 25 & -9
\end{array}\right] \tag{6.19}$$

Observe that the basis vectors located in even rows are symmetric and in odd rows are anti-symmetric with respect to the vertical line. Therefore, the second half of the matrix is easy to find from the first half. Also, notice how the T_8 is embedded in the even rows of this matrix.

$$T_{32} = \left[\begin{array}{cccccccccccccccc|c}
64 & 64 & 64 & 64 & 64 & 64 & 64 & 64 & 64 & 64 & 64 & 64 & 64 & 64 & 64 & 64 & \cdots \\
90 & 90 & 88 & 85 & 82 & 78 & 73 & 67 & 61 & 54 & 46 & 38 & 31 & 22 & 13 & 4 & \cdots \\
90 & 87 & 80 & 70 & 57 & 43 & 25 & 9 & -9 & -25 & -43 & -57 & -70 & -80 & -87 & -90 & \cdots \\
90 & 82 & 67 & 46 & 22 & -4 & -31 & -54 & -73 & -85 & -90 & -88 & -78 & -61 & -38 & -13 & \cdots \\
89 & 75 & 50 & 18 & -18 & -50 & -75 & -89 & -89 & -75 & -50 & -18 & 18 & 50 & 75 & 89 & \cdots \\
88 & 67 & 31 & -13 & -54 & -82 & -90 & -78 & -46 & -4 & 38 & 73 & 90 & 85 & 61 & 22 & \cdots \\
87 & 57 & 9 & -43 & -80 & -90 & -70 & -25 & 25 & 70 & 90 & 80 & 43 & -9 & -57 & -87 & \cdots \\
85 & 46 & -13 & -67 & -90 & -73 & -22 & 38 & 82 & 88 & 54 & -4 & -61 & -90 & -78 & -31 & \cdots \\
83 & 36 & -36 & -83 & -83 & -36 & 36 & 83 & 83 & 36 & -36 & -83 & -83 & -36 & 36 & 83 & \cdots \\
82 & 22 & -54 & -90 & -61 & 13 & 78 & 85 & 31 & -46 & -90 & -67 & 4 & 73 & 88 & 38 & \cdots \\
80 & 9 & -70 & -87 & -25 & 57 & 90 & 43 & -43 & -90 & -57 & 25 & 87 & 70 & -9 & -80 & \cdots \\
78 & -4 & -82 & -73 & 13 & 85 & 67 & -22 & -88 & -61 & 31 & 90 & 54 & -38 & -90 & -46 & \cdots \\
75 & -18 & -89 & -50 & 50 & 89 & 18 & -75 & -75 & 18 & 89 & 50 & -50 & -89 & -18 & 75 & \cdots \\
73 & -31 & -90 & -22 & 78 & 67 & -38 & -90 & -13 & 82 & 61 & -46 & -88 & -4 & 85 & 54 & \cdots \\
70 & -43 & -87 & 9 & 90 & 25 & -80 & -57 & 57 & 80 & -25 & -90 & -9 & 87 & 43 & -70 & \cdots \\
67 & -54 & -78 & 38 & 85 & -22 & -90 & 4 & 90 & 13 & -88 & -31 & 82 & 46 & -73 & -61 & \cdots \\
64 & -64 & -64 & 64 & 64 & -64 & -64 & 64 & 64 & -64 & -64 & 64 & 64 & -64 & -64 & 64 & \cdots \\
61 & -73 & -46 & 82 & 31 & -88 & -13 & 90 & -4 & -90 & 22 & 85 & -38 & -78 & 54 & 67 & \cdots \\
57 & -80 & -25 & 90 & -9 & -87 & 43 & 70 & -70 & -43 & 87 & 9 & -90 & 25 & 80 & -57 & \cdots \\
54 & -85 & -4 & 88 & -46 & -61 & 82 & 13 & -90 & 38 & 67 & -78 & -22 & 90 & -31 & -73 & \cdots \\
50 & -89 & 18 & 75 & -75 & -18 & 89 & -50 & -50 & 89 & -18 & -75 & 75 & 18 & -89 & 50 & \cdots \\
46 & -90 & 38 & 54 & -90 & 31 & 61 & -88 & 22 & 67 & -85 & 13 & 73 & -82 & 4 & 78 & \cdots \\
43 & -90 & 57 & 25 & -87 & 70 & 9 & -80 & 80 & -9 & -70 & 87 & -25 & -57 & 90 & -43 & \cdots \\
38 & -88 & 73 & -4 & -67 & 90 & -46 & -31 & 85 & -78 & 13 & 61 & -90 & 54 & 22 & -82 & \cdots \\
36 & -83 & 83 & -36 & -36 & 83 & -83 & 36 & 36 & -83 & 83 & -36 & -36 & 83 & -83 & 36 & \cdots \\
31 & -78 & 90 & -61 & 4 & 54 & -88 & 82 & -38 & -22 & 73 & -90 & 67 & -13 & -46 & 85 & \cdots \\
25 & -70 & 90 & -80 & 43 & 9 & -57 & 87 & -87 & 57 & -9 & -43 & 80 & -90 & 70 & -25 & \cdots \\
22 & -61 & 85 & -90 & 73 & -38 & -4 & 46 & -78 & 90 & -82 & 54 & -13 & -31 & 67 & -88 & \cdots \\
18 & -50 & 75 & -89 & 89 & -75 & 50 & -18 & -18 & 50 & -75 & 89 & -89 & 75 & -50 & 18 & \cdots \\
13 & -38 & 61 & -78 & 88 & -90 & 85 & -73 & 54 & -31 & 4 & 22 & -46 & 67 & -82 & 90 & \cdots \\
9 & -25 & 43 & -57 & 70 & -80 & 87 & -90 & 90 & -87 & 80 & -70 & 57 & -43 & 25 & -9 & \cdots \\
4 & -13 & 22 & -31 & 38 & -46 & 54 & -61 & 67 & -73 & 78 & -82 & 85 & -88 & 90 & -90 & \cdots
\end{array}\right] \qquad (6.20)$$

If the transform size increases the complexity also increases. HEVC makes use of the fast DCT algorithm proposed by Chen, Smith and Fralick in [73] to minimize the complexity.

AVS2 China 4x4, 8x8, 16x16 and 32x32 ICT Matrices

In the second generation AVS standard, referred to as AVS2, the primary application target is ultra high definition (UHD) video, supporting the efficient compression of ultra high resolution (4K and higher), high dynamic range (HDR) videos. AVS2 supports efficient coding of 3D video, multi-view and virtual reality (VR) video; efficient loss and lossless coding of stereo and multi-channel audio; efficient coding of monitor video and new media services for tri-networks integration [24].

Tests by radio and television metering center of state administration of radio, film, and television (SARFT) show that AVS2 is twice as efficient as the previous generation standards AVS1 [356] and H.264/AVC [204], and exceeds the same generation standard HEVC [208], [209]. The transform matrices for $N = 4, 8, 16$ and 32 are shown from (6.21) to (6.24).

$$T_4 = \left[\begin{array}{rr|rr} 32 & 32 & 32 & 32 \\ 42 & 17 & -17 & -42 \\ 32 & -32 & -32 & 32 \\ 17 & -42 & 42 & -17 \end{array}\right] \tag{6.21}$$

$$T_8 = \left[\begin{array}{rrrr|rrrr} 32 & 32 & 32 & 32 & 32 & 32 & 32 & 32 \\ 44 & 38 & 25 & 9 & -9 & -25 & -38 & -44 \\ 42 & 17 & -17 & -42 & -42 & -17 & 17 & 42 \\ 38 & -9 & -44 & -25 & 25 & 44 & 9 & -38 \\ 32 & -32 & -32 & 32 & 32 & -32 & -32 & 32 \\ 25 & -44 & 9 & 38 & -38 & -9 & 44 & -25 \\ 17 & -42 & 42 & -17 & -17 & 42 & -42 & 17 \\ 9 & -25 & 38 & -44 & 44 & -38 & 25 & -9 \end{array}\right] \tag{6.22}$$

$$T_{16} = \left[\begin{array}{rrrrrrrr|rrrrrrrr} 32 & 32 & 32 & 32 & 32 & 32 & 32 & 32 & 32 & 32 & 32 & 32 & 32 & 32 & 32 & 32 \\ 45 & 43 & 40 & 35 & 29 & 21 & 13 & 4 & -4 & -13 & -21 & -29 & -35 & -40 & -43 & -45 \\ 44 & 38 & 25 & 9 & -9 & -25 & -38 & -44 & -44 & -38 & -25 & -9 & 9 & 25 & 38 & 44 \\ 43 & 29 & 4 & -21 & -40 & -45 & -35 & -13 & 13 & 35 & 45 & 40 & 21 & -4 & -29 & -43 \\ 42 & 17 & -17 & -42 & -42 & -17 & 17 & 42 & 42 & 17 & -17 & -42 & -42 & -17 & 17 & 42 \\ 40 & 4 & -35 & -43 & -13 & 29 & 45 & 21 & -21 & -45 & -29 & 13 & 43 & 35 & -4 & -40 \\ 38 & -9 & -44 & -25 & 25 & 44 & 9 & -38 & -38 & 9 & 44 & 25 & -25 & -44 & -9 & 38 \\ 35 & -21 & -43 & 4 & 45 & 13 & -40 & -29 & 29 & 40 & -13 & -45 & -4 & 43 & 21 & -35 \\ 32 & -32 & -32 & 32 & 32 & -32 & -32 & 32 & 32 & -32 & -32 & 32 & 32 & -32 & -32 & 32 \\ 29 & -40 & -13 & 45 & -4 & -43 & 21 & 35 & -35 & -21 & 43 & 4 & -45 & 13 & 40 & -29 \\ 25 & -44 & 9 & 38 & -38 & -9 & 44 & -25 & -25 & 44 & -9 & -38 & 38 & 9 & -44 & 25 \\ 21 & -45 & 29 & 13 & -43 & 35 & 4 & -40 & 40 & -4 & -35 & 43 & -13 & -29 & 45 & -21 \\ 17 & -42 & 42 & -17 & -17 & 42 & -42 & 17 & 17 & -42 & 42 & -17 & -17 & 42 & -42 & 17 \\ 13 & -35 & 45 & -40 & 21 & 4 & -29 & 43 & -43 & 29 & -4 & -21 & 40 & -45 & 35 & -13 \\ 9 & -25 & 38 & -44 & 44 & -38 & 25 & -9 & -9 & 25 & -38 & 44 & -44 & 38 & -25 & 9 \\ 4 & -13 & 21 & -29 & 35 & -40 & 43 & -45 & 45 & -43 & 40 & -35 & 29 & -21 & 13 & -4 \end{array}\right] \tag{6.23}$$

$$
T_{32} = \left[\begin{array}{cccccccccccccccc|c}
32 & 32 & 32 & 32 & 32 & 32 & 32 & 32 & 32 & 32 & 32 & 32 & 32 & 32 & 32 & 32 & \cdots \\
45 & 45 & 44 & 43 & 41 & 39 & 36 & 34 & 30 & 27 & 23 & 19 & 15 & 11 & 7 & 2 & \cdots \\
45 & 43 & 40 & 35 & 29 & 21 & 13 & 4 & -4 & -13 & -21 & -29 & -35 & -40 & -43 & -45 & \cdots \\
45 & 41 & 34 & 23 & 11 & -2 & -15 & -27 & -36 & -43 & -45 & -44 & -39 & -30 & -19 & -7 & \cdots \\
44 & 38 & 25 & 9 & -9 & -25 & -38 & -44 & -44 & -38 & -25 & -9 & 9 & 25 & 38 & 44 & \cdots \\
44 & 34 & 15 & -7 & -27 & -41 & -45 & -39 & -23 & -2 & 19 & 36 & 45 & 43 & 30 & 11 & \cdots \\
43 & 29 & 4 & -21 & -40 & -45 & -35 & -13 & 13 & 35 & 45 & 40 & 21 & -4 & -29 & -43 & \cdots \\
43 & 23 & -7 & -34 & -45 & -36 & -11 & 19 & 41 & 44 & 27 & -2 & -30 & -45 & -39 & -15 & \cdots \\
42 & 17 & -17 & -42 & -42 & -17 & 17 & 42 & 42 & 17 & -17 & -42 & -42 & -17 & 17 & 42 & \cdots \\
41 & 11 & -27 & -45 & -30 & 7 & 39 & 43 & 15 & -23 & -45 & -34 & 2 & 36 & 44 & 19 & \cdots \\
40 & 4 & -35 & -43 & -13 & 29 & 45 & 21 & -21 & -45 & -29 & 13 & 43 & 35 & -4 & -40 & \cdots \\
39 & -2 & -41 & -36 & 7 & 43 & 34 & -11 & -44 & -30 & 15 & 45 & 27 & -19 & -45 & -23 & \cdots \\
38 & -9 & -44 & -25 & 25 & 44 & 9 & -38 & -38 & 9 & 44 & 25 & -25 & -44 & -9 & 38 & \cdots \\
36 & -15 & -45 & -11 & 39 & 34 & -19 & -45 & -7 & 41 & 30 & -23 & -44 & -2 & 43 & 27 & \cdots \\
35 & -21 & -43 & 4 & 45 & 13 & -40 & -29 & 29 & 40 & -13 & -45 & -4 & 43 & 21 & -35 & \cdots \\
34 & -27 & -39 & 19 & 43 & -11 & -45 & 2 & 45 & 7 & -44 & -15 & 41 & 23 & -36 & -30 & \cdots \\
32 & -32 & -32 & 32 & 32 & -32 & -32 & 32 & 32 & -32 & -32 & 32 & 32 & -32 & -32 & 32 & \cdots \\
30 & -36 & -23 & 41 & 15 & -44 & -7 & 45 & -2 & -45 & 11 & 43 & -19 & -39 & 27 & 34 & \cdots \\
29 & -40 & -13 & 45 & -4 & -43 & 21 & 35 & -35 & -21 & 43 & 4 & -45 & 13 & 40 & -29 & \cdots \\
27 & -43 & -2 & 44 & -23 & -30 & 41 & 7 & -45 & 19 & 34 & -39 & -11 & 45 & -15 & -36 & \cdots \\
25 & -44 & 9 & 38 & -38 & -9 & 44 & -25 & -25 & 44 & -9 & -38 & 38 & 9 & -44 & 25 & \cdots \\
23 & -45 & 19 & 27 & -45 & 15 & 30 & -44 & 11 & 34 & -43 & 7 & 36 & -41 & 2 & 39 & \cdots \\
21 & -45 & 29 & 13 & -43 & 35 & 4 & -40 & 40 & -4 & -35 & 43 & -13 & -29 & 45 & -21 & \cdots \\
19 & -44 & 36 & -2 & -34 & 45 & -23 & -15 & 43 & -39 & 7 & 30 & -45 & 27 & 11 & -41 & \cdots \\
17 & -42 & 42 & -17 & -17 & 42 & -42 & 17 & 17 & -42 & 42 & -17 & -17 & 42 & -42 & 17 & \cdots \\
15 & -39 & 45 & -30 & 2 & 27 & -44 & 41 & -19 & -11 & 36 & -45 & 34 & -7 & -23 & 43 & \cdots \\
13 & -35 & 45 & -40 & 21 & 4 & -29 & 43 & -43 & 29 & -4 & -21 & 40 & -45 & 35 & -13 & \cdots \\
11 & -30 & 43 & -45 & 36 & -19 & -2 & 23 & -39 & 45 & -41 & 27 & -7 & -15 & 34 & -44 & \cdots \\
9 & -25 & 38 & -44 & 44 & -38 & 25 & -9 & -9 & 25 & -38 & 44 & -44 & 38 & -25 & 9 & \cdots \\
7 & -19 & 30 & -39 & 44 & -45 & 43 & -36 & 27 & -15 & 2 & 11 & -23 & 34 & -41 & 45 & \cdots \\
4 & -13 & 21 & -29 & 35 & -40 & 43 & -45 & 45 & -43 & 40 & -35 & 29 & -21 & 13 & -4 & \cdots \\
2 & -7 & 11 & -15 & 19 & -23 & 27 & -30 & 34 & -36 & 39 & -41 & 43 & -44 & 45 & -45 & \cdots
\end{array}\right] \tag{6.24}
$$

It is important to note the symmetry of the basis vectors in all integer matrices. For example, let t_i be the ith basis vector of an order-$2N$ ICT, it is required that vectors $t_{2n} \quad n \in 0, 1, \cdots$ are symmetric while the remaining $t_{2n+1} \quad n \in 0, 1, \cdots$ are anti-symmetric.

Versatile Video Coding (VVC)

The new video coding standard, called versatile video coding (VVC) [215], [534], will be released in 2020 by the joint video experts team (JVET) (on future video coding). This standard is intended to significantly improve the current HEVC [208], [209] and will aid for the deployment of higher-quality video services and emerging applications such as 360° omnidirectional immersive multimedia and HDR video. JVET developed the experimental software called joint exploration model (JEM). The exploration software was built on top of HEVC-HM and can be downloaded from [140]. One of the modifications is the adaptive multiple core transform (AMT) that includes transform matrices quantized more accurately, applicable for block sizes less than or equal to 64×64 and the mode dependent non-separable secondary transform (MDNSST) in intra coding. In intra prediction, switchable primary transforms are used, mostly DCT/DST variants. Also, secondary transforms targeting specific cases of prediction residual characteristics. The DCT/DST transform kernels proposed for the VVC [359] are,

$$\text{DCT-II:} \quad \left[C_N^{II}\right]_{mn} = k_m \cdot \sqrt{\frac{2}{N}} \cdot \cos\left[\frac{m \cdot (2n+1) \cdot \pi}{2N}\right]$$

$$k_m = \begin{cases} \sqrt{\frac{2}{N}} & \text{if } m = 0 \\ 1 & \text{if } m \neq 0 \end{cases}$$

$$\text{DCT-V:} \quad \left[C_N^{V}\right]_{mn} = k_m \cdot k_n \cdot \sqrt{\frac{2}{2N-1}} \cdot \cos\left[\frac{mn2\pi}{2N-1}\right]$$

$$k_m = \begin{cases} \sqrt{\frac{2}{N}} & \text{if } m = 0 \\ 1 & \text{if } m \neq 0 \end{cases}, \quad k_n = \begin{cases} \sqrt{\frac{2}{N}} & \text{if } n = 0 \\ 1 & \text{if } n \neq 0 \end{cases}$$

$$\text{DCT-VIII:} \quad \left[C_N^{VIII}\right]_{mn} = \sqrt{\frac{4}{2N+1}} \cdot \cos\left[\frac{(2m+1) \cdot (2n+1) \cdot \pi}{4N+2}\right]$$

$$\text{DST-I:} \quad \left[S_N^{I}\right]_{mn} = \sqrt{\frac{2}{N+1}} \cdot \sin\left[\frac{(m+1) \cdot (n+1) \cdot \pi}{N+1}\right]$$

$$\text{DST-VII:} \quad \left[S_N^{I}\right]_{mn} = \sqrt{\frac{4}{2N+1}} \cdot \sin\left[\frac{(2m+1) \cdot (n+1) \cdot \pi}{2N+1}\right]$$

For $m, n = 0, 1, \cdots, N-1$.

The mode-dependent transform selection, for intra prediction are grouped into 3 sets. Set 0 includes DST-VII and DCT-VIII, set 1 includes DST-VII and DST-I and set 2 includes DST-VII and DCT-VIII. Notice that, in HEVC the DST-VII has been adopted for coding 4×4 luma intra prediction residuals because it compacts the residual energy more efficiently than DCT-II [170]. Now the encoder is able to select a transform from several candidates to code a block because the statistics of residual blocks varies largely even for the same prediction mode. However, the computational complexity is high. Moreover, the inclusion of non-separable secondary 2-D transform (NSST) schemes to break the limitations of conventional separable 2-D transform and further compact the energy of the residuals blocks also increases the complexity [584]. The standard will process large block-size transform with high-frequency zeroing. The maximum transform size proposed is of up to 128×128.

THE ICT BY KERNEL COMPONENT REPLACEMENT

An ICT matrix can be of the form $[T] = [K][J]$, where $[T]$ is the orthogonal ICT matrix, and $[K]$ is a diagonal matrix whose elements scale the rows of the matrix $[J]$ so that the relative magnitudes of elements of the ICT matrix $[T]$ are similar to those in the DCT matrix. The matrix $[J]$ is orthogonal with elements that are all integers.

The ICT can be generated from DCT-II by kernel component replacement matrix and keeping the relative magnitudes and orthogonal relationship among the matrix elements [61], [63]. The kernel of the order-N DCT-II, depicted in (3.20) is $\left[C_N^{II}\right]_{mn}$ with the nth component of the mth DCT basis vector and $m \in [0, N-1]$. In [61], the following three steps are proposed to convert an order-8 DCT kernel into an ICT kernel,

1. Let $N = 8$ therefore, the 8×8 DCT-II matrix can be decomposed as

$$\left[C_8^{II}\right]_{mn} = [k_0 J_0, k_1 J_1, \cdots, k_7 J_7]^T, \tag{6.25}$$

   where J_{mn} is the nth element of the J_m basis vector and k_m is a scaling constant such that $|k_m \cdot J_m| = 1$. For simplicity, we shall denote $\left[C_8^{II}\right]_{mn}$ by $[C]_{mn}$. The 8×8 matrix is

$$[C]_{mn} = \begin{bmatrix} 0.3536 & 0.3536 & 0.3536 & 0.3536 & 0.3536 & 0.3536 & 0.3536 & 0.3536 \\ 0.4904 & 0.4157 & 0.2778 & 0.0975 & -0.0975 & -0.2778 & -0.4157 & -0.4904 \\ 0.4619 & 0.1913 & -0.1913 & -0.4619 & -0.4619 & -0.1913 & 0.1913 & 0.4619 \\ 0.4157 & -0.0975 & -0.4904 & -0.2778 & 0.2778 & 0.4904 & 0.0975 & -0.4157 \\ 0.3536 & -0.3536 & -0.3536 & 0.3536 & 0.3536 & -0.3536 & -0.3536 & 0.3536 \\ 0.2778 & -0.4904 & 0.0975 & 0.4157 & -0.4157 & -0.0975 & 0.4904 & -0.2778 \\ 0.1913 & -0.4619 & 0.4619 & -0.1913 & -0.1913 & 0.4619 & -0.4619 & 0.1913 \\ 0.0975 & -0.2778 & 0.4157 & -0.4904 & 0.4904 & -0.4157 & 0.2778 & -0.0975 \end{bmatrix}. \tag{6.26}$$

   Notice that $[C]_{1,0} = -[C]_{1,7} = -[C]_{3,2} = [C]_{3,5} = -[C]_{5,1} = [C]_{5,6} = -[C]_{7,3} = [C]_{7,4}$. The magnitudes of the corresponding components $J_{1,0}, J_{1,7}, J_{3,2}, J_{3,5}, J_{5,1}, J_{5,6}, J_{7,3}, J_{7,4}$ can be replaced by a single variable 'a'. This can be done for all eight basis vectors to express the transform as ICT {'a', 'b', 'c', 'd', 'e', 'f'} as shown in Table 6.1 and choosing k_m such that variables 'd' and 'f' are unity. Then, the remaining variables represent irrational numbers whose values can be approximated to 5.027, 4.2620, 2.8478 and 2.4142.
2. To find the orthogonality of the basis vectors of Table 6.1, it is necessary that each basis vector has at least one dyadic symmetry. A vector of 2^k elements $[a_0, \cdots, a_{2^k-1}]$ is said to have ith dyadic symmetry if and only if $a_l = s \cdot a_{l \oplus i}$ where $\oplus$ is the exclusive OR operation, l is in the range $[0, 2^k - 1]$ and i in the range $[1, 2^k - 1]$, $s = 1$ when symmetry is even (E) and $s = -1$ when the symmetry is odd (O). Notice that seven possible dyadic symmetries can be obtained with a vector of eight elements. Table 6.1 shows that the 8 basis vectors have at least one dyadic symmetry and Table 6.2 shows the dyadic symmetry in each basis vector. According to [61], two vectors are orthogonal if they have the same type of dyadic symmetry and one is even and the other is odd.

Table 6.1
The eight scaled basis vectors [61]

m	J_m							
0	1	1	1	1	1	1	1	1
1	a	b	c	d	$-d$	$-c$	$-b$	$-a$
2	e	f	$-f$	$-e$	$-e$	$-f$	f	e
3	b	$-d$	$-a$	$-c$	c	a	d	$-b$
4	1	-1	-1	1	1	-1	-1	1
5	c	$-a$	d	b	$-b$	$-d$	a	$-c$
6	f	$-e$	e	$-f$	$-f$	e	$-e$	f
7	d	$-c$	b	$-a$	a	$-b$	c	$-d$

Table 6.2
Sth dyadic symmetry in basis vectors. *E* is even, *O* is odd and – no dyadic symmetry [61]

	Dyadic S symmetry in J_m						
m	1	2	3	4	5	6	7
0	E	E	E	E	E	E	E
1	–	–	–	–	–	–	O
2	–	–	O	–	–	–	E
3	–	–	–	–	–	–	O
4	O	O	E	E	O	O	E
5	–	–	–	–	–	–	O
6	–	–	O	–	–	–	E
7	–	–	–	–	–	–	O

For example, J_0 and J_1 are always orthogonal to each other for all a, b, c and d because they have even and odd 7th dyadic symmetry respectively. J_0 and J_2 are always orthogonal to each other for all e and f because they have even and odd 3rd dyadic symmetry respectively. Also, J_1 and J_2 are orthogonal to each other for all a, b, c, d, e and f because they have odd and even 7th dyadic symmetry

Table 6.3
Conditions to ensure orthogonality of the basis vectors [61].

m						
1	2	3	4	5	6	7
*3	*2	*3	*2	*3	*2	*3
	*3	*1	*3	*1	*3	*4
		*3	*2	*3	*4	*3
			*3	*4	*3	*1
				*3	*2	*3
					*3	*1
						*3

*1 if $a \cdot b = a \cdot c + b \cdot d + c \cdot d$.
*2 must be orthogonal due to the 3rd dyadic symmetry.
*3 must be orthogonal due to 7th dyadic symmetry.
*4 must be orthogonal as their dot product is zero.

Table 6.4
Fifteen order-8 ICT coefficients and $a \leq 255$ suitable to be implemented using 8 bits [61].

ICT $\{a,b,c,d,e,f\}$ coefficients	
$\{230,201,134,46,3,1\}$	$\{55,48,32,11,3,1\}$
$\{175,153,102,35,3,1\}$	$\{205,180,120,41,3,1\}$
$\{120,105,70,24,3,1\}$	$\{140,123,82,28,3,1\}$
$\{185,162,108,37,3,1\}$	$\{215,189,126,43,3,1\}$
$\{250,219,146,50,3,1\}$	$\{75,66,44,15,3,1\}$
$\{65,57,38,13,3,1\}$	$\{235,207,138,47,3,1\}$
$\{10,9,6,2,3,1\}$	$\{5,3,2,1,3,1\}$
	$\{4,2,2,0,4,2\}2$

respectively. Table 6.3 summarizes the conditions that constants a, b, c, d, e and f must satisfy to ensure orthogonality of the basis vectors.

3. Set up boundary conditions. In (3.20) we see that $a \geq b \geq c \geq d$ and $e \geq f$, the condition $a \cdot b = a \cdot c + b \cdot d + c \cdot d$ ensures that $[C]_{mn}$ is orthogonal and we can restrict a, b, c, d, e and f to be integers.

Other order-8 ICTs transform can be derived, because the relation $a \cdot b - a \cdot c - b \cdot d - c \cdot d = 0$ has an infinite number of solutions with e, f arbitrary. For example,

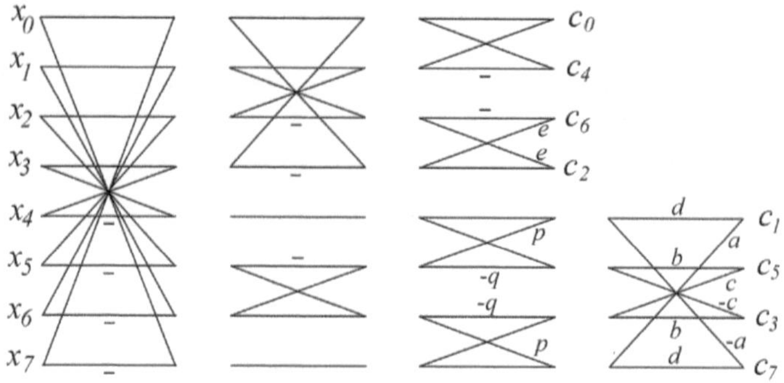

Figure 6.2: Signal flowgraph for the computation of the fast order-8 ICT, where $p = (b+c)/2a$ and $q = (a-d)/2c$ [61].

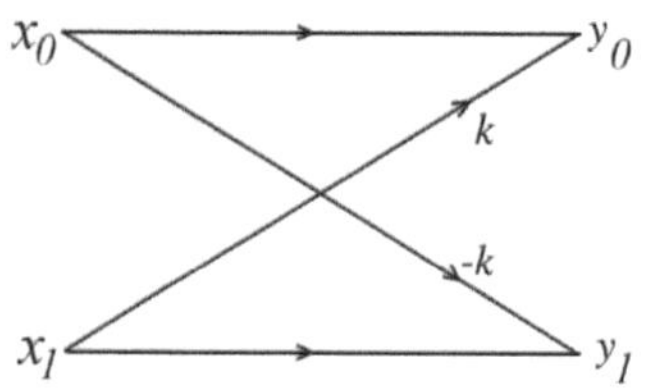

Figure 6.3: Lattice structure.

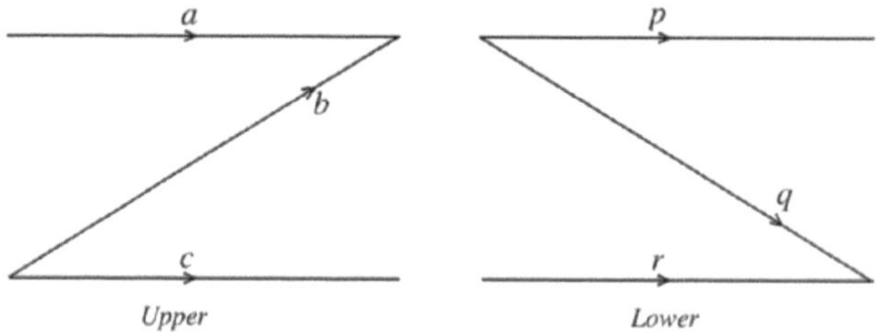

Figure 6.4: Upper and lower structures.

Table 6.4 [61] presents fifteen order-8 ICTs. Note that $a \leq 255$ which is suitable to be represented in at most 8-bits.

Fast Algorithm

The [ICT] matrix is of the form [ICT]=$[K][J]$ as shown in (6.4). Matrix $[K]$ is a diagonal matrix and $[J]$ is a matrix that contains the n basis vectors given in Table 6.1

with integer elements only. In this case, one good candidate to be implemented as an order-8 DCT is ICT$\{10,9,6,2,3,1\}$. Figure 6.2 shows the signal flowgraph for the computation of the fast algorithm.

THE LIFTING MATRIX AND THE LIFTING STEP

Before going further, we will briefly explain the lifting scheme with an example.

Example 6.1. *Consider the lattice structure of Figure 6.3 described by the input-output relationship of* (6.27),

$$\begin{bmatrix} y_0 \\ y_1 \end{bmatrix} = \begin{bmatrix} 1 & k \\ -k & 1 \end{bmatrix} \begin{bmatrix} x_0 \\ x_1 \end{bmatrix} \tag{6.27}$$

The goal is to decompose the matrix into a product of simple matrices, one upper and one lower triangular matrix to separate the branches of the lattice and not let both branches interact with each other at once. Then, we can express the full matrix as follows

$$\begin{bmatrix} 1 & k \\ -k & 1 \end{bmatrix} = \begin{bmatrix} a & b \\ 0 & c \end{bmatrix} \begin{bmatrix} p & 0 \\ q & r \end{bmatrix} \tag{6.28}$$

Figure 6.4 shows the separation into upper and lower branches.

In (6.28) *we have 4 equations and 6 variables. Therefore, there are degrees of freedom that we can exploit, in other words, we can solve 4 variables only and we are free to choose 2 variables appropriately. Then, comparing both sides of* (6.28) *we have*

$$\begin{aligned} a\cdot p + b\cdot q &= 1 \\ c\cdot q &= -k \\ b\cdot r &= k \\ c\cdot r &= 1 \end{aligned}$$

We exploit the degrees of freedom by choosing simple values for some entries, for example, to avoid the multiplications we can set $p = 1$ or $r = 1$ or both $p = r = 1$ in this case we have

$$\begin{aligned} a + b\cdot q &= 1 \\ c\cdot q &= -k \\ b &= k \\ c &= 1 \end{aligned}$$

We have found that $b = k$, $c = 1$. With these results, we can solve for the rest of variables: $q = -k$ and $a = 1 + k^2$. We replace the variables in (6.28) *to obtain,*

$$\begin{bmatrix} 1 & k \\ -k & 1 \end{bmatrix} = \begin{bmatrix} 1+k^2 & k \\ 0 & 1 \end{bmatrix} \begin{bmatrix} 1 & 0 \\ -k & r \end{bmatrix} \tag{6.29}$$

Notice the simplicity of these results that yields a cascade of two lifting stages named lifting steps. We can redraw the diagram of Figure 6.5 as shown below

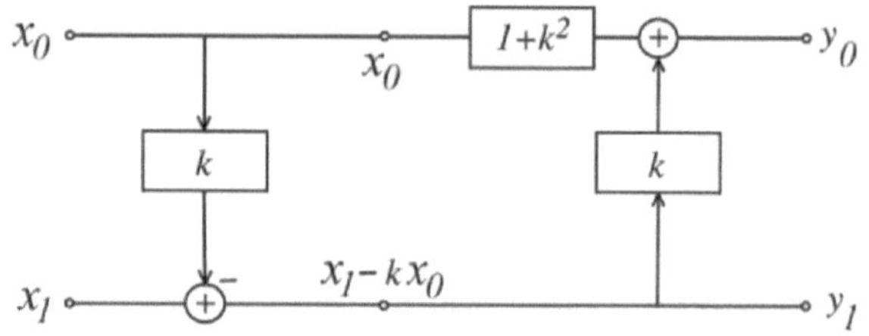

Figure 6.5: Two lifting steps.

The first step combines the upper path with a lower path to modify the lower path and the second stage combines the lower path with the upper path to modify the upper path.

Example 6.2. *This example shows the (unnormalized) polyphase Haar matrix and its decomposition in lower and upper matrices.*

$$\begin{bmatrix} 1 & -\frac{1}{2} \\ 1 & \frac{1}{2} \end{bmatrix} = \begin{bmatrix} 1 & 0 \\ 1 & 1 \end{bmatrix} \begin{bmatrix} 1 & -\frac{1}{2} \\ 0 & 1 \end{bmatrix} \tag{6.30}$$

Example 6.3. *The Givens rotations can also be decomposed into two lifting steps (for $\theta \neq k\pi + \frac{\pi}{2}$ with k an integer).*

$$\begin{bmatrix} \cos\theta & -\sin\theta \\ \sin\theta & \cos\theta \end{bmatrix} = \begin{bmatrix} 1 & 0 \\ \frac{\sin\theta}{\cos\theta} & 1 \end{bmatrix} \begin{bmatrix} \cos\theta & -\sin\theta \\ 0 & \frac{1}{\cos\theta} \end{bmatrix} \tag{6.31}$$

Also, we can factorize the upper triangular matrix to get a scaling diagonal matrix named scaled lifting structure,

$$\begin{bmatrix} \cos\theta & -\sin\theta \\ \sin\theta & \cos\theta \end{bmatrix} = \begin{bmatrix} 1 & 0 \\ \frac{\sin\theta}{\cos\theta} & 1 \end{bmatrix} \begin{bmatrix} 1 & -\sin\theta\cos\theta \\ 0 & 1 \end{bmatrix} \begin{bmatrix} \cos\theta & 0 \\ 0 & \frac{1}{\cos\theta} \end{bmatrix} \tag{6.32}$$

Furthermore, we can do it in three lifting steps without scaling (for $\theta \neq k\pi$ with k an integer),

$$\begin{bmatrix} \cos\theta & -\sin\theta \\ \sin\theta & \cos\theta \end{bmatrix} = \begin{bmatrix} 1 & \frac{(\cos\theta-1)}{\sin\theta} \\ 0 & 1 \end{bmatrix} \begin{bmatrix} 1 & 0 \\ \sin\theta & 1 \end{bmatrix} \begin{bmatrix} 1 & \frac{(\cos\theta-1)}{\sin\theta} \\ 0 & 1 \end{bmatrix} \tag{6.33}$$

By using the method of Example 6.1 we can build lifting matrices to construct integer discrete transforms [58], [469], [571]. A lifting matrix has 1s on its main diagonal and only one element off diagonal is non-zero [58], [469]. Therefore, the matrix can be expressed only by the off-diagonal element s at the ith row and kth column as $L_{i,k}(s) \quad (i \neq k$ and $i,k = 0,1,\cdots,N-1)$ where N is the order of the lifting matrix [571].

Example 6.4. *Express $L_{2,1}(3.1)$ in terms of a lifting matrix and its inverse,*

$$A = \begin{bmatrix} 1 & 0 & 0 & 0 \\ 0 & 1 & 0 & 0 \\ 0 & 3.1 & 1 & 0 \\ 0 & 0 & 0 & 1 \end{bmatrix} \tag{6.34}$$

Notice that $L_{i,k}^{-1}(s) = L_{i,k}(-s)$. Hence, $L_{2,1}^{-1}(3.1) = L_{2,1}(-3.1)$. In other words,

$$A^{-1} = \begin{bmatrix} 1 & 0 & 0 & 0 \\ 0 & 1 & 0 & 0 \\ 0 & -3.1 & 1 & 0 \\ 0 & 0 & 0 & 1 \end{bmatrix} \tag{6.35}$$

In the examples 6.1 to 6.4, one can easily see that if a matrix can be factored into a product of lifting matrices its inverse is also a product of lifting matrices.

A lifting step is the multiplication of a lifting matrix with a vector. The operation $y = L_{i,k}(s)x$, where x and y are column vectors of length N, is called a lifting step.

Example 6.5. *Express the lifting matrix $L_{2,1}(s)$ of size $N = 4$ into a lifting step and express the lifting step in a general compact form.*

The lifting step is,

$$\begin{bmatrix} y_0 \\ y_1 \\ y_2 \\ y_3 \end{bmatrix} = \begin{bmatrix} 1 & 0 & 0 & 0 \\ 0 & 1 & 0 & 0 \\ 0 & s & 1 & 0 \\ 0 & 0 & 0 & 1 \end{bmatrix} \begin{bmatrix} x_0 \\ x_1 \\ x_2 \\ x_3 \end{bmatrix} \tag{6.36}$$

$$\begin{bmatrix} y_0 \\ y_1 \\ y_2 \\ y_3 \end{bmatrix} = \begin{bmatrix} x_0 \\ x_1 \\ sx_1 + x_2 \\ x_3 \end{bmatrix}. \tag{6.37}$$

Or in compact form

$$y_i = sx_k + x_i, \quad y_p = x_p, \quad p \neq i \tag{6.38}$$

Example 6.6. *Consider another the problem where $N = 2$ and $L_{1,0}(3.1)x$, express the lifting step in terms of the lifting matrix and draw the corresponding forward and inverse lifting steps.*

$$\begin{bmatrix} y_0 \\ y_1 \end{bmatrix} = \begin{bmatrix} 1 & 0 \\ 3.1 & 1 \end{bmatrix} \begin{bmatrix} x_0 \\ x_1 \end{bmatrix} \tag{6.39}$$

To reverse the operation we invert the matrix to obtain

$$\begin{bmatrix} x_0 \\ x_1 \end{bmatrix} = \begin{bmatrix} 1 & 0 \\ -3.1 & 1 \end{bmatrix} \begin{bmatrix} y_0 \\ y_1 \end{bmatrix} \tag{6.40}$$

Figure 6.6 (a) shows the lifting step and its corresponding inverse (b).

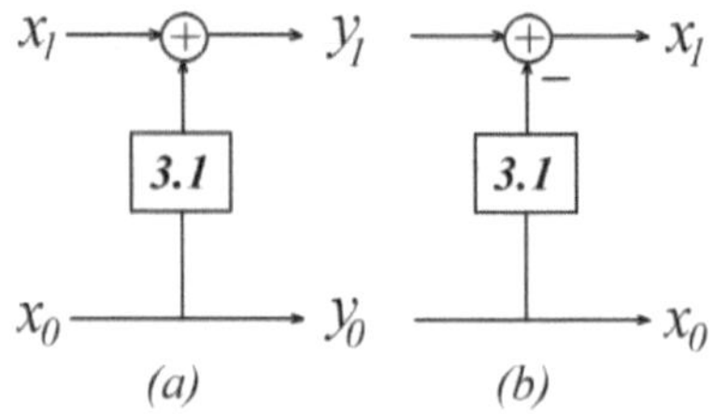

Figure 6.6: The lifting step (a) and its corresponding inverse (b) [469].

One advantage of the lifting step is the simplicity to construct fast transforms that map integers to integers by placing a quantization operation (floor, ceil, round, or other quantization operation) $[\cdot]$, after the element $L_{1,0}(3.1)$, as shown in Figure 6.7.

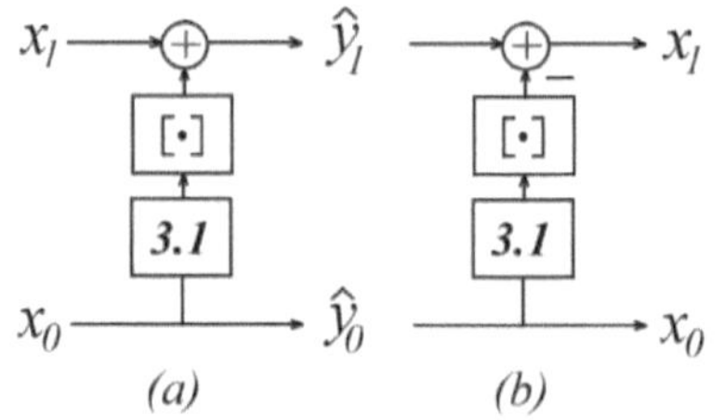

Figure 6.7: Integer to integer mapping with exact invertibility [469].

The compact form of the lifting step (6.38) shows that if s is not an integer number we need to carry out a floating point multiplication. Hence, s is desired to be of dyadic rational form $a/2^b$, where a and b are integer numbers. This number is represented by $RB(s)$ [571]. Therefore, s can be approximated by a number of bits and the resulting number may not be unique. For example, the number 3.1 can be approximated by 3, $13/4$, $26/8$, $52/16$, etc.

The lifting matrix can be now represented as $L_{i,k}(RB(s))$ with inverse lifting matrix $L_{i,k}(-RB(s))$ as shown in Figure 6.6. Then, in (6.38) the coefficients y_i can be approximated by $\hat{y}_i$ as,

$$\hat{y}_i = \lfloor s x_k \rfloor + x_i, \quad \hat{y}_p = x_p, \quad p \neq i \tag{6.41}$$

Notice that this is a nonlinear transform. However, it is invertible with inverse given by,

$$x_i = \hat{y}_i - \lfloor s \hat{y}_k \rfloor, \quad x_p = \hat{y}_p, \quad p \neq i \tag{6.42}$$

Observe that the nonlinear equation (6.41) needs no floating point multiplications and is invertible. The inverse matrix does not need floating point multiplication (6.42)).

It is also known that any matrix with determinant 1 can be factored into products of lifting matrices [39], [58], [469], [571]. If D is a diagonal matrix with determinant 1, it can be expressed as block diagonal matrices with each block of the form,

$$\begin{bmatrix} c & 0 \\ 0 & \frac{1}{c} \end{bmatrix} \tag{6.43}$$

This block is known as scaling and it can be represented into lifting steps as,

$$\begin{bmatrix} c & 0 \\ 0 & \frac{1}{c} \end{bmatrix} = \begin{bmatrix} 1 & c-1 \\ 0 & 1 \end{bmatrix} \begin{bmatrix} 1 & 0 \\ 1 & 1 \end{bmatrix} \begin{bmatrix} 1 & \frac{1}{c}-1 \\ 0 & 1 \end{bmatrix} \begin{bmatrix} 1 & 0 \\ -c & 1 \end{bmatrix}. \tag{6.44}$$

Also, two more lifting forms can be derived

$$\begin{bmatrix} c & 0 \\ 0 & \frac{1}{c} \end{bmatrix} = \begin{bmatrix} 1 & c-c^2 \\ 0 & 1 \end{bmatrix} \begin{bmatrix} 1 & 0 \\ -\frac{1}{c} & 1 \end{bmatrix} \begin{bmatrix} 1 & c-1 \\ 0 & 1 \end{bmatrix} \begin{bmatrix} 1 & 0 \\ 1 & 1 \end{bmatrix},$$

or

$$\begin{bmatrix} c & 0 \\ 0 & \frac{1}{c} \end{bmatrix} = \begin{bmatrix} 1 & 0 \\ -1 & 1 \end{bmatrix} \begin{bmatrix} 1 & 1-\frac{1}{c} \\ 0 & 1 \end{bmatrix} \begin{bmatrix} 1 & 0 \\ c & 1 \end{bmatrix} \begin{bmatrix} 1 & \frac{1}{c^2}-\frac{1}{c} \\ 0 & 1 \end{bmatrix}$$

We can always merge one of the four lifting steps with the last lifting step for the factorization and only three lifting steps will be needed to avoid scaling. In general, it was shown in [39] that any matrix with determinant integer and different from zero can be factored into lifting matrices and integer matrices. In other words, any invertible matrix can be factored into a product of lifting matrices and diagonal matrices with elements that may not be integers. This result is important to construct integer transforms. Moreover, a diagonal matrix D of order $N/2$, with determinant 1 can be factored as,

$$D_{N/2} = \left\{ \prod_{k=0}^{N/4-1} \left[L_{2k,2k+1}(\alpha_{2k} - 1) L_{2k+1,2k} L(1) \times L_{2k,2k+1}\left(\frac{1}{\alpha_{2k}} - 1\right) L_{2k+1,2k}(-\alpha_{2k}) \right] \right\} \cdot \left\{ \prod_{k=1}^{N/4-1} \left[L_{2k-1,2k}(\alpha_{2k-1} - 1) L_{2k,2k-1}(1) \times L_{2k-1,2k}\left(\frac{1}{\alpha_{2k-1}} - 1\right) L_{2k,2k-1}(-\alpha_{2k-1}) \right] \right\} \quad (6.45)$$

where $L_{i,k}(s)$ are the lifting matrices of order $N/2$. To avoid floating point multiplications, the non-zero coefficients in (6.45) can be replaced by numbers of the form $a/2^b$ as,

$$\bar{D}_{N/2} = \left\{ \prod_{k=0}^{N/4-1} \left[L_{2k,2k+1}(RB(\alpha_{2k}) - 1) L_{2k+1,2k} L(1) \times L_{2k,2k+1}\left(RB\left(\frac{1}{\alpha_{2k}}\right) - 1\right) L_{2k+1,2k}(-RB(\alpha_{2k})) \right] \right\} \cdot \left\{ \prod_{k=1}^{N/4-1} \left[L_{2k-1,2k}(RB(\alpha_{2k-1}) - 1) L_{2k,2k-1}(1) \times L_{2k-1,2k}\left(RB\left(\frac{1}{\alpha_{2k-1}}\right) - 1\right) L_{2k,2k-1}(-RB(\alpha_{2k-1})) \right] \right\}. \quad (6.46)$$

The inverse matrix is

$$\bar{D}_{N/2}^{-1} = \left\{ \prod_{k=1}^{N/4-1} \left[L_{2k,2k-1}(RB(\alpha_{2k-1})) L_{2k-1,2k}\left(-RB\left(\frac{1}{\alpha_{2k-1}}\right) + 1\right) \times L_{2k,2k-1}(-1) L_{2k-1,2k}(-RB(\alpha_{2k-1}) + 1) \right] \right\} \cdot \left\{ \prod_{k=0}^{N/4-1} L_{2k+1,2k}(RB(\alpha_{2k})) L_{2k,2k+1}\left(-RB\left(\frac{1}{\alpha_{2k}}\right) + 1\right) \times L_{2k+1,2k}(-1) L_{2k,2k+1}(-RB(\alpha_{2k}) + 1) \right\} \quad (6.47)$$

The Integer DCT-II by Lifting Matrix

The kernel of the DCT-II described in (3.20), can be defined as

$$\left[C_N^{II}\right]_{mn} = \left[\cos\left(\frac{m(2n+1)\pi}{2N}\right)\right]_{m,n=0,1,\ldots,N-1}. \tag{6.48}$$

We can drop the subscript m,n for the sake of visual clarity and express the recursive factorization of (6.48) as,

$$\left[C_N^{II}\right] = P_N \begin{bmatrix} I_{N/2} & 0 \\ 0 & U_{N/2} \end{bmatrix} \begin{bmatrix} C_{N/2}^{II} & 0 \\ 0 & C_{N/2}^{II} \end{bmatrix} \times \begin{bmatrix} I_{N/2} & 0 \\ 0 & D_{N/2} \end{bmatrix} \begin{bmatrix} I_{N/2} & \tilde{I}_{N/2} \\ I_{N/2} & -\tilde{I}_{N/2} \end{bmatrix} \tag{6.49}$$

where I and $\tilde{I}$ are the identity and the identity matrix with reversed rows respectively, the subscripts N and $N/2$ are the dimension of the matrices. The matrices U, P and D are defined as follows,

$$U_{N/2} = \begin{bmatrix} \frac{1}{2} & 0 & 0 & \cdots & 0 & 0 \\ -\frac{1}{2} & 1 & 0 & \cdots & 0 & 0 \\ \frac{1}{2} & -1 & 1 & \cdots & 0 & 0 \\ & \cdots & & \cdots & & \\ \frac{1}{2} & -1 & 1 & \cdots & 1 & 0 \\ -\frac{1}{2} & 1 & -1 & \cdots & -1 & 1 \end{bmatrix} \tag{6.50}$$

$$P_N = \begin{bmatrix} 1 & 0 & \cdots & 0 & 0 & 0 & \cdots & 0 \\ 0 & 0 & \cdots & 0 & 1 & 0 & \cdots & 0 \\ 0 & 1 & \cdots & 0 & 0 & 0 & \cdots & 0 \\ 0 & 0 & \cdots & 0 & 0 & 1 & \cdots & 0 \\ & & \cdots & & & \cdots & & 0 \\ 0 & 0 & \cdots & 1 & 0 & 0 & \cdots & 0 \\ 0 & 0 & \cdots & 0 & 0 & 0 & \cdots & 1 \end{bmatrix} \tag{6.51}$$

$$D_{N/2} = \operatorname{diag}\left(2\cos\frac{\pi}{2N}, 2\cos\frac{3\pi}{2N}, \cdots, 2\cos\frac{(N-1)\pi}{2N}\right) \tag{6.52}$$

Observe that any vector multiplication with matrix $D_{N/2}$ yields a floating point result. Hence, we want to transform $D_{N/2}$ into a product of lifting matrices to approximate the elements in dyadic rational form $a/2^b$ where a and b are integers. However, the determinant of $D_{N/2}$ is not 1. Hence, we can separate this matrix into the product of two diagonal matrices as $D_{N/2} = E_{N/2}F_{N/2}$ where

$$E_{N/2} = \operatorname{diag}\left(\sqrt{2}, 1, \cdots, 1\right) \tag{6.53}$$

$$F_{N/2} = \operatorname{diag}\left(\sqrt{2}\cos\frac{\pi}{2N}, 2\cos\frac{3\pi}{2N}, \cdots, 2\cos\frac{(N-1)\pi}{2N}\right) \tag{6.54}$$

Also, (6.53) can be approximated by,

$$\bar{E}_{N/2} = \text{diag}\left(RB(\sqrt{2}), 1, \cdots, 1\right) \tag{6.55}$$

The determinant of (6.54) is 1. Hence, we can represent this matrix into a product of lifting matrices as in (6.45) with,

$$\alpha_0 = \sqrt{2}\cos\frac{\pi}{2N}, \quad \alpha_k = 2\alpha_{k-1}\cos\left[\frac{(2k+1)\pi}{2N}\right]$$
$$k = 1, 2, \cdots, N/2-1, \tag{6.56}$$

then use (6.46) to avoid floating point multiplications of (6.54) to get $\bar{F}_{N/2}$. Therefore, (6.52) is now $\bar{D}_{N/2} = \bar{E}_{N/2}\bar{F}_{N/2}$.

Consequently, the matrix $\left[C_N^{II}\right]$ can be approximated by,

$$\left[\bar{C}_N^{II}\right] = P_N \begin{bmatrix} I_{N/2} & 0 \\ 0 & U_{N/2} \end{bmatrix} \begin{bmatrix} \bar{C}_{N/2}^{II} & 0 \\ 0 & \bar{C}_{N/2}^{II} \end{bmatrix} \times \begin{bmatrix} I_{N/2} & 0 \\ 0 & \bar{D}_{N/2} \end{bmatrix} \begin{bmatrix} I_{N/2} & \tilde{I}_{N/2} \\ I_{N/2} & -\tilde{I}_{N/2} \end{bmatrix} \tag{6.57}$$

Finally, we can decompose recursively $\bar{C}_{N/2}^{II}$ until order 1. The transform is not unique and depends on the function RB. The inverse matrix is,

$$\left[\bar{C}_N^{II}\right]^{-1} = \begin{bmatrix} \frac{1}{2}I_{N/2} & \frac{1}{2}I_{N/2} \\ \frac{1}{2}\tilde{I}_{N/2} & -\frac{1}{2}\tilde{I}_{N/2} \end{bmatrix} \begin{bmatrix} I_{N/2} & 0 \\ 0 & \bar{F}_{N/2}^{-1}\bar{E}_{N/2}^{-1} \end{bmatrix} \times \begin{bmatrix} \left[\bar{C}_{N/2}^{II}\right]^{-1} & 0 \\ 0 & \left[\bar{C}_{N/2}^{II}\right]^{-1} \end{bmatrix} \begin{bmatrix} I_{N/2} & 0 \\ 0 & U_{N/2}^{-1} \end{bmatrix} P_N^{-1}, \tag{6.58}$$

where $\bar{E}_{N/2}^{-1} = \text{diag}\left(\frac{1}{RB(\sqrt{2})}, 1, \cdots, 1\right)$, $\bar{F}_{N/2}^{-1}$ is obtained by (6.47) and

$$U_{N/2}^{-1} = \begin{bmatrix} 2 & 0 & 0 & \cdots & 0 & 0 \\ 1 & 1 & 0 & \cdots & 0 & 0 \\ 0 & 1 & 1 & \cdots & 0 & 0 \\ & \cdots & & & \cdots & \\ 0 & 0 & 0 & \cdots & 1 & 0 \\ 0 & 0 & 0 & \cdots & 1 & 1 \end{bmatrix} \tag{6.59}$$

LOW-COMPLEXITY ICT (LCICT)

The low-complexity order-64 integer cosine transform (LCICT) for HEVC [78] is based on the recursive integer cosine transform (RICT) [135]. The design allows recursive implementation from order-8 to order-64 implementing fully factorizable structures with the potential to be generalized to arbitrary orders. The structure is also based on the assumption that kernel (6.48) can be approximated by the product $D_N \cdot \bar{C}_N$ with D_N a diagonal matrix containing normalization factors, $\bar{C}_N$ a kernel of integer components and N the order of the transform. Besides, the approximation takes into consideration the symmetry of the basis vectors as explained at the end of §6. Then,

$$\left[\bar{C}_{2N}\right] = P_{2N}\begin{bmatrix} \bar{C}_N & 0 \\ 0 & G_N \end{bmatrix} A_{2N} = P_{2N}\begin{bmatrix} \bar{C}_N & 0 \\ 0 & G_N \end{bmatrix}\begin{bmatrix} I_N & \tilde{I}_N \\ \tilde{I}_N & -I_N \end{bmatrix}$$

Notice that $\bar{C}_N$ can be further factorized in the same way. It is required that G_N be factorized recursively [135] to have a full factorization and full recursivity of the algorithm. In the RICT [135], G_N is the DCT-IV, it is recursive and factored by the following recursive equation,

$$G_{2N} = S_{2N}\begin{bmatrix} G'_N & 0 \\ 0 & G'_N \end{bmatrix} B_{2N},$$

where

$$S_{2N} = \begin{bmatrix} D_{N,0} & D_{N,1}\tilde{I}_N \\ -\tilde{I}_N D_{N,1} & \tilde{I}_N D_{N,0}\tilde{I}_N \end{bmatrix} \quad , \quad B_{2N} = \begin{bmatrix} B_{N,0} \\ B_{N,1} \end{bmatrix}$$

and

$$\begin{aligned} D_{N,0} &= \operatorname{diag}\left(d_{2N,0}, d_{2N,2}, \cdots, d_{2N,2N-2}\right) \\ D_{N,1} &= \operatorname{diag}\left(d_{2N,1}, d_{2N,3}, \cdots, d_{2N,2N-1}\right) \end{aligned}$$

where $d_{2N,i}$ are the integer parameters. Let the rows and columns indices be $m \in \{0,1,\cdots,N-1\}$ and $n \in \{0,1,\cdots,2N-1\}$ respectively, matrix $B_{N,k}$ is defined as,

$$B_{N,k}(m,n) = \begin{cases} (-1)^{km}, & \text{for } n = 2m; \\ (-1)^{k(m+1)}, & \text{for } n = 2m+1; \\ 0, & \text{otherwise.} \end{cases}$$

The LCICT reduces the complexity of RICT by introducing other ICT structures and uses the recursive equation to factorize G_{2N} for $n \geq 4$ into S_{2N}, G'_N and B_{2N}. The inverse transform is found by transposition of the forward transform. Also, the LCICT does not require that G'_N be recursively factorized by the recursive equation. This allows to introduce other ICT structures [78] by optimizing their integer parameters and constraining the row vectors to be orthogonal.

The structure of the order-8 $\bar{C}_8$ [78] is constructed by dyadic symmetry principle [61]. Figure 6.8 shows the flowgraph of the order-32 LCICT (or $\bar{C}_{32}$). The inputs are represented by $\{x_n\}$, the outputs by $\{X_n\}$ the intermediate results by $\{z_n\}$. Dashed branches denote multiplication by -1. The values of parameters (multipliers) are listed on the right, where 2^n and $1/2^n$ denote that the corresponding multiplications are implemented as bitwise shifts.

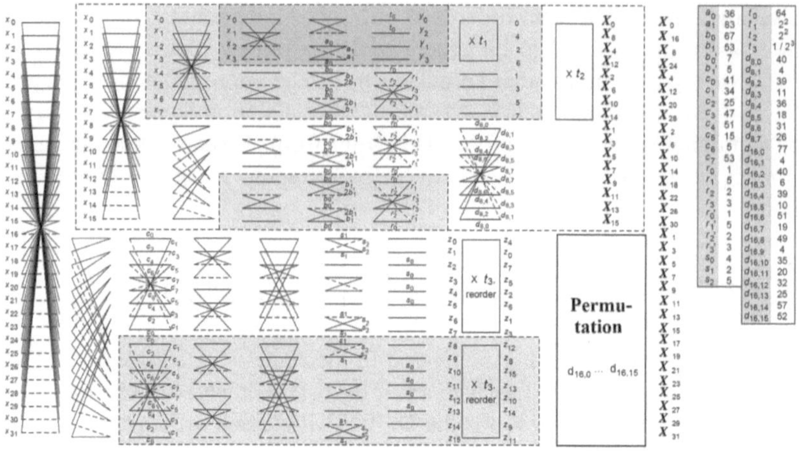

Figure 6.8: Signal flowgraph of the order-32 LCICT. The inputs are represented by $\{x_n\}$, the outputs by $\{X_n\}$ the intermediate results by $\{z_n\}$. Dashed branches denote multiplication by -1. The values of parameters (multipliers) are listed on the right, where 2^n and $1/2^n$ denote that the corresponding multiplications are implemented as bitwise shifts [78].

Figure 6.8 shows the flowgraph of the order-64 LCICT. The inputs are represented by $\{x_n\}$ and the outputs by $\{X_n\}$. Dashed branches denote multiplication by -1. The values of parameters (multipliers) are listed on the right, where $1/2^n$ denote that the multiplications are implemented as bitwise shifts.

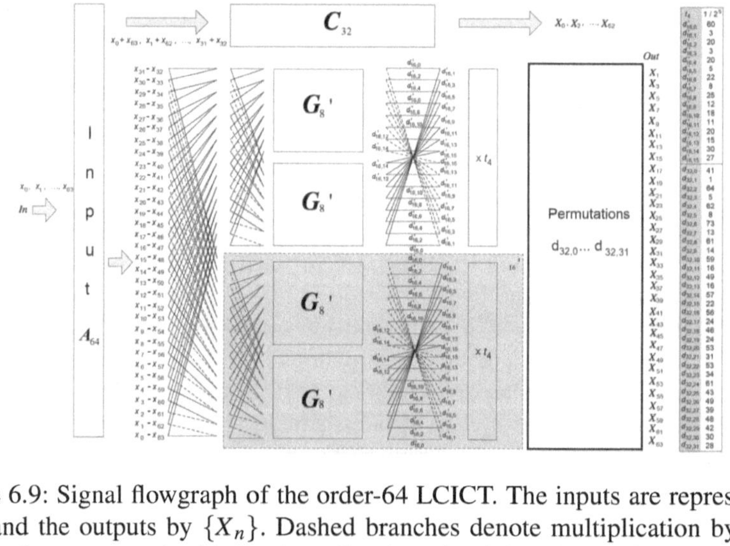

Figure 6.9: Signal flowgraph of the order-64 LCICT. The inputs are represented by $\{x_n\}$ and the outputs by $\{X_n\}$. Dashed branches denote multiplication by -1. The values of parameters (multipliers) are listed on the right, where $1/2^n$ denote that the multiplications are implemented as bitwise shifts [78].

Table 6.5 shows that the parameters $\{d'_{16,n}\}$ and $\{d_{32,n}\}$ in the order-64 RICT are determined in the same way as the ones in LCICT.

Table 6.5
Parameters in G_{32} in the order-64 RICT [78].

$d'_{16,0}$	$d'_{16,1}$	$d'_{16,2}$	$d'_{16,3}$	$d'_{16,4}$	$d'_{16,5}$	$d'_{16,6}$	$d'_{16,7}$
42	2	20	3	20	5	23	8
$d'_{16,8}$	$d'_{16,9}$	$d'_{16,10}$	$d'_{16,11}$	$d'_{16,12}$	$d'_{16,13}$	$d'_{16,14}$	$d'_{16,15}$
17	8	15	9	24	18	11	10
$d_{32,0}$	$d_{32,1}$	$d_{32,2}$	$d_{32,3}$	$d_{32,4}$	$d_{32,5}$	$d_{32,6}$	$d_{32,7}$
32	1	90	7	89	12	70	12
$d_{32,8}$	$d_{32,9}$	$d_{32,10}$	$d_{32,11}$	$d_{32,12}$	$d_{32,13}$	$d_{32,14}$	$d_{32,15}$
90	20	102	28	58	19	84	33
$d_{32,16}$	$d_{32,17}$	$d_{32,18}$	$d_{32,19}$	$d_{32,20}$	$d_{32,21}$	$d_{32,22}$	$d_{32,23}$
83	36	54	28	92	53	77	50
$d_{32,24}$	$d_{32,25}$	$d_{32,26}$	$d_{32,27}$	$d_{32,28}$	$d_{32,29}$	$d_{32,30}$	$d_{32,31}$
58	41	71	55	70	57	24	21

SQUARE WAVE TRANSFORM

The square wave transform based on the DCT was proposed by Haweel [177]. The $N \times N$ transform matrix is generated by applying only the signum function operator to the basis vectors of the DCT matrix. The resulting transform requires only additions/subtractions operations, without transcendental operations ($\sin(\theta)$ and $\cos(\theta)$) and is obtained as follows,

$$T_{SDCT}(m,n) = \frac{1}{\sqrt{N}} \operatorname{sign}\left(T_{DCT}(m,n)\right), \tag{6.60}$$

where $T_{SDCT}(m,n)$ and $T_{DCT}(m,n)$ are the elements of the generated signed discrete cosine transform and the DCT respectively. sign$(\cdot)$ is the signum function defined as,

$$\text{sign}(x) = \begin{cases} +1 & \text{if } x > 0, \\ 0 & \text{if } x = 0, , \\ -1 & \text{if } x < 0. \end{cases} \tag{6.61}$$

For example, the forward 8×8 $T_{SDCT}(m,n)$, generated from the DCT-II matrix of (6.48), is

$$T_{SDCT} = \frac{1}{\sqrt{8}} \begin{bmatrix} 1 & 1 & 1 & 1 & 1 & 1 & 1 & 1 \\ 1 & 1 & 1 & 1 & -1 & -1 & -1 & -1 \\ 1 & 1 & -1 & -1 & -1 & -1 & 1 & 1 \\ 1 & -1 & -1 & -1 & 1 & 1 & 1 & -1 \\ 1 & -1 & -1 & 1 & 1 & -1 & -1 & 1 \\ 1 & -1 & 1 & 1 & -1 & -1 & 1 & -1 \\ 1 & -1 & 1 & -1 & -1 & 1 & -1 & 1 \\ 1 & -1 & 1 & -1 & 1 & -1 & 1 & -1 \end{bmatrix} \tag{6.62}$$

The inverse matrix is

$$T_{SDCT}^{-1} = \frac{1}{\sqrt{8}} \begin{bmatrix} 1 & 2 & 1 & 2 & 1 & 0 & 1 & 0 \\ 1 & 2 & 1 & 0 & -1 & -2 & -1 & 0 \\ 1 & 0 & -1 & -2 & -1 & 0 & 1 & 2 \\ 1 & 0 & -1 & 0 & 1 & 2 & -1 & -2 \\ 1 & 0 & -1 & 0 & 1 & -2 & -1 & 2 \\ 1 & 0 & -1 & 2 & -1 & 0 & 1 & -2 \\ 1 & -2 & 1 & 0 & -1 & 2 & -1 & 0 \\ 1 & -2 & 1 & -2 & 1 & 0 & 1 & 0 \end{bmatrix}, \tag{6.63}$$

and the coefficients are integers numbers $0, \pm 1, \pm 2$. Only shifts and sign bit changes are needed to implement the inverse transform. The DCT periodicity, spectral structure and energy compaction properties [177] are retained. The transform is suitable for real time applications.

Note that T_{SDCT} is not orthogonal. Therefore, it is not energy preserving, this is one drawback of this transform. To overcome the orthogonality problem, Bouguezel, Ahmad and Swamy [43] proposed a fast 8×8 transform aimed at image compression and based on the T_{SDCT} [43], [177]. The forward transform (C) is obtained by setting to zero the entries that make (6.62) non-orthogonal and keeping the symmetry of the basis vectors. The forward and inverse matrices are given by,

$$C = DT, \tag{6.64}$$

$$C^{-1} = C^T = T^T D, \tag{6.65}$$

where ^{T} denotes matrix transpose operation and

$$D = \text{diag}\left(\frac{1}{2\sqrt{2}}, \frac{1}{2}, \frac{1}{2\sqrt{2}}, \frac{1}{\sqrt{2}}, \frac{1}{2\sqrt{2}}, \frac{1}{2}, \frac{1}{2\sqrt{2}}, \frac{1}{\sqrt{2}}\right) \tag{6.66}$$

and

$$T = \begin{bmatrix} 1 & 1 & 1 & 1 & 1 & 1 & 1 & 1 \\ 1 & 1 & 0 & 0 & 0 & 0 & -1 & -1 \\ 1 & 1 & -1 & -1 & -1 & -1 & 1 & 1 \\ 0 & 0 & -1 & 0 & 0 & 1 & 0 & 0 \\ 1 & -1 & -1 & 1 & 1 & -1 & -1 & 1 \\ 1 & -1 & 0 & 0 & 0 & 0 & 1 & -1 \\ 1 & -1 & 1 & -1 & -1 & 1 & -1 & 1 \\ 0 & 0 & 0 & -1 & 1 & 0 & 0 & 0 \end{bmatrix}, \tag{6.67}$$

Then (6.67) is decomposed into the product of three matrices, $T = T_3 \times T_2 \times T_1$, to reduce up to 18 the number of additions, where,

$$T_1 = \begin{bmatrix} 1 & 0 & 0 & 0 & 0 & 0 & 0 & 1 \\ 0 & 1 & 0 & 0 & 0 & 0 & 1 & 0 \\ 0 & 0 & 1 & 0 & 0 & 1 & 0 & 0 \\ 0 & 0 & 0 & 1 & 1 & 0 & 0 & 0 \\ 0 & 0 & 0 & 1 & -1 & 0 & 0 & 0 \\ 0 & 0 & 1 & 0 & 0 & -1 & 0 & 0 \\ 0 & 1 & 0 & 0 & 0 & 0 & -1 & 0 \\ 1 & 0 & 0 & 0 & 0 & 0 & 0 & -1 \end{bmatrix}, \tag{6.68}$$

$$T_2 = \begin{bmatrix} 1 & 0 & 0 & 1 & 0 & 0 & 0 & 0 \\ 0 & 0 & 0 & 0 & 0 & 0 & 1 & 1 \\ 0 & 1 & 1 & 0 & 0 & 0 & 0 & 0 \\ 0 & 0 & 0 & 0 & 0 & -1 & 0 & 0 \\ 0 & 1 & -1 & 0 & 0 & 0 & 0 & 0 \\ 0 & 0 & 0 & 0 & 0 & 0 & -1 & 1 \\ 1 & 0 & 0 & -1 & 0 & 0 & 0 & 0 \\ 0 & 0 & 0 & 0 & -1 & 0 & 0 & 0 \end{bmatrix}, \tag{6.69}$$

$$T_3 = \begin{bmatrix} 1 & 0 & 1 & 0 & 0 & 0 & 0 & 0 \\ 0 & 1 & 0 & 0 & 0 & 0 & 0 & 0 \\ 0 & 0 & 0 & 0 & 1 & 0 & 1 & 0 \\ 0 & 0 & 0 & 1 & 0 & 0 & 0 & 0 \\ 1 & 0 & -1 & 0 & 0 & 0 & 0 & 0 \\ 0 & 0 & 0 & 0 & 0 & 1 & 0 & 0 \\ 0 & 0 & 0 & 0 & -1 & 0 & 1 & 0 \\ 0 & 0 & 0 & 0 & 0 & 0 & 0 & 1 \end{bmatrix}. \tag{6.70}$$

For compression applications D can be included in the quantization matrix or it can be approximated as in (6.55).

ROUND-OFF APPROXIMATION OF IntDCT

Cintra and Bayer [90] proposed a DCT approximation for image compression using the following round-off operation in each coefficient $c_{m,n}$ of kernel (6.48),

$$\mathrm{round}(c_{m,n}) = \mathrm{sign}(c_{m,n}) \left\lfloor \left| c_{m,n} \right| + 0.5 \right\rfloor \tag{6.71}$$

where sign(·) is the signum function (6.61), $\lfloor \cdot \rfloor$ is the floor function and $|\cdot|$ is the absolute value. The resulting matrix is,

$$C_0 = \begin{bmatrix} 1 & 1 & 1 & 1 & 1 & 1 & 1 & 1 \\ 1 & 1 & 1 & 0 & 0 & -1 & -1 & -1 \\ 1 & 0 & 0 & -1 & -1 & 0 & 0 & 1 \\ 1 & 0 & -1 & -1 & 1 & 1 & 0 & -1 \\ 1 & -1 & -1 & 1 & 1 & -1 & -1 & 1 \\ 1 & -1 & 0 & 1 & -1 & 0 & 1 & -1 \\ 0 & -1 & 1 & 0 & 0 & 1 & -1 & 0 \\ 0 & -1 & 1 & -1 & 1 & -1 & 1 & 0 \end{bmatrix}. \tag{6.72}$$

which is a coarse approximation of the DCT matrix $\left[C_N^{II}\right]_{mn}$ and related by $\left[\hat{C}_N^{II}\right]_{mn} = \frac{1}{2}\left[C_0\right]_{mn}$. Notice that,

$$C_0 \cdot C_0^T = \begin{bmatrix} 8 & 0 & 0 & 0 & 0 & 0 & 0 & 0 \\ 0 & 6 & 0 & 0 & 0 & 0 & 0 & 0 \\ 0 & 0 & 4 & 0 & 0 & 0 & 0 & 0 \\ 0 & 0 & 0 & 6 & 0 & 0 & 0 & 0 \\ 0 & 0 & 0 & 0 & 8 & 0 & 0 & 0 \\ 0 & 0 & 0 & 0 & 0 & 6 & 0 & 0 \\ 0 & 0 & 0 & 0 & 0 & 0 & 4 & 0 \\ 0 & 0 & 0 & 0 & 0 & 0 & 0 & 6 \end{bmatrix} \neq I \tag{6.73}$$

Therefore $C_0^T \neq C_0^{-1}$ to adjust this result the orthogonalization matrix [90] $S = \sqrt{\left(C_0 \cdot C_0^T\right)^{-1}}$ is necessary, therefore,

$$S = \mathrm{diag}\left(\sqrt{\frac{1}{8}}, \sqrt{\frac{1}{6}}, \frac{1}{2}, \sqrt{\frac{1}{6}}, \sqrt{\frac{1}{8}}, \sqrt{\frac{1}{6}}, \frac{1}{2}, \sqrt{\frac{1}{6}}\right) \tag{6.74}$$

Hence, $S \cdot C_0 \cdot C_0^T \cdot S = I$. Then, the orthogonal approximation to the DCT becomes,

$$\hat{C} = S \cdot C_0 \tag{6.75}$$

Other similar methods to derive low-complexity DCT approximations can be found in [36], [44], [45], [91], [135], [371], [376] among others.

THE TCHEBICHEF POLYNOMIALS

The Tchebichef polynomials are a family of orthogonal polynomials with major applications to solve ordinary and partial differential equations [320]. The polynomials of the first and second kind are obtained from the Tchebichef differential equation as follows,

$$(1-x^2)\frac{d^2y}{dx^2} - x\frac{dy}{dx} + n^2y = 0, \quad n = 0, 1, \cdots \tag{6.76}$$

Let $x = \cos\theta$ then $dx = -\sin\theta\, d\theta$ and $d\theta/dx = -1/sin\theta$. Then, using the chain and the product rules,

$$\frac{dy}{dx} = \frac{dy}{d\theta}\left(\frac{d\theta}{dx}\right) = -\frac{1}{\sin\theta}\left(\frac{dy}{d\theta}\right) \tag{6.77}$$

$$\begin{aligned}\frac{d^2y}{dx^2} &= \frac{dy}{d\theta}\frac{d}{dx}\left(\frac{d\theta}{dx}\right) + \frac{d\theta}{dx}\frac{d}{dx}\left(\frac{dy}{d\theta}\right) \\ &= \frac{dy}{d\theta}\frac{d\theta}{dx}\frac{d}{d\theta}\left(\frac{d\theta}{dx}\right) + \frac{d\theta}{dx}\frac{d\theta}{dx}\frac{d}{d\theta}\left(\frac{dy}{d\theta}\right) \\ &= \frac{dy}{d\theta}\left(-\frac{1}{\sin\theta}\right)\frac{d}{d\theta}\left(-\frac{1}{\sin\theta}\right) - \left(\frac{1}{\sin\theta}\right)\left(\frac{1}{\sin\theta}\right)\frac{d}{d\theta}\left(\frac{d^2y}{d\theta}\right) \\ &= \frac{1}{\sin^2\theta}\left[-\frac{\cos\theta}{\sin\theta}\left(\frac{dy}{d\theta}\right) + \frac{d^2y}{d\theta^2}\right]\end{aligned} \tag{6.78}$$

Replace (6.77) and (6.78) in (6.76); also, we assumed that $x = \cos\theta$ and using $(1-\cos^2\theta) = \sin^2\theta$ we obtain,

$$\sin^2\theta\left[\frac{1}{\sin^2\theta}\left(\frac{\cos\theta}{\sin\theta}\right)\left(\frac{dy}{d\theta}\right) + \frac{d^2y}{d\theta^2}\right] - \cos\theta\left[\frac{-1}{\sin\theta}\left(\frac{dy}{d\theta}\right)\right] + n^2y = 0 \tag{6.79}$$

Notice that (6.79) reduces to the following linear ordinary differential equation with constant coefficients

$$\frac{d^2y}{d\theta^2} + n^2y = 0$$

with general solution

$$y(\theta) = A\cos(n\cos^{-1}x) + B\sin(n\cos^{-1}x), \tag{6.80}$$

or

$$y(\theta) = AT_n(x) + BU_n(x) \quad |x| < 1, \tag{6.81}$$

where $T_n(x)$ and $U_n(x)$ are the Tchebichef polynomials of the first and second kind of order n respectively. We can generate the nth Tchebishef polynomial of the first kind $T_n(x)$ using the fact that $x = \cos\theta$ and $T_n(\cos\theta) = \cos n\theta$,

1. $T_0(x) = 1$
2. $T_1(x) = \cos\theta = x$
3. $T_2(x) = \cos 2\theta = 2\cos^2\theta - 1 = 2x^2 - 1$
4. $T_3(x) = \cos 3\theta = 4\cos^3\theta - 3\cos\theta = 4x^3 - 3x$
5. $T_4(x) = \cos 4\theta = 8\cos^4\theta - 8\cos^2\theta + 1 = 8x^4 - 8x^2 + 1$

In general, we can obtain the following recursive relations:

$$T_{n+1}(x) = 2xT_n(x) - T_{n-1}(x) \tag{6.82}$$

$$U_{n+1}(x) = 2xU_n(x) - U_{n-1}(x) \tag{6.83}$$

ORTHOGONALITY

The Tchebichef polynomials $T_n(x)$ are orthogonal with respect to $(1-x^2)^{-1/2}$ and $U_n(x)$ with respect to $(1-x^2)^{1/2}$ in the interval $[-1,1]$

$$\int_{-1}^{1} T_n(x)T_m(x)(1-x^2)^{-1/2}dx = N\delta_{n,m}, \tag{6.84}$$

with $N = \pi/2$ for $n = m \neq 0$, $N = 0$ for $n \neq m$ and $N = \pi$ for $n = m = 0$

$$\int_{-1}^{1} U_n(x)U_m(x)(1-x^2)^{1/2}dx = N\delta_{n,m}, \tag{6.85}$$

with $N = \pi/2$ for $n = m$ and $N = 0$ for $n \neq m$.

DISCRETE TCHEBICHEF POLYNOMIALS

A kth order discrete Tchebichef polynomial is defined as [201],

$$t_{k,n} = (1-N)_k \cdot {}_3F_2(-k, -n, 1+k; 1, 1-N; 1) \qquad k, n = 0, 1, 2, \cdots, N-1, \tag{6.86}$$

where

$${}_3F_2(a_1, a_2, a_3; b_1, b_2; z) = \sum_{n=0}^{\infty}\left[\frac{(a_1)_k(a_2)_k(a_3)_k}{(b_1)_k(b_2)_k}\right]\frac{z^k}{k!}, \tag{6.87}$$

is the hypergeometric function and $(a)_k = a(a+1)\cdots(a+k-1)$. The discrete Tchebichef polynomials can be obtained as follows [360],

$$t_{k,n} = \left[\frac{2k-1}{k}t_{k,1}\right]t_{k-1,n} - \left[\frac{k-1}{k}(N^2-(k-1)^2)\right]t_{k-2,n} \qquad k = 0, 1, \cdots, N-1, \tag{6.88}$$

for $t_{0,n} = 1$ and $t_{1,n} = 2n - N + 1$.

Orthogonality

The discrete Tchebichef polynomials are orthogonal and satisfy the following condition

$$\sum_{i=0}^{N-1} t_{i,n} t_{m,n} = \rho_{m,N} \cdot \delta_{i,m}, \tag{6.89}$$

with $\delta_{i,m}$ the Kronecker function ($\delta_{i,m} = 1$ if $i = m$ and 0 otherwise) and $\rho_{k,N}$ the squared norm given by

$$\rho_{k,N} = \frac{(N+k)!}{(2k+1)(N-k-1)!} \tag{6.90}$$

Then set $\{t_{k,n}\}$ is an orthogonal basis with respect to the unit weight. The basis can be normalized by,

$$\tilde{t}_{k,n} = \frac{t_{k,n}}{\sqrt{\rho_{k,N}}}, \quad k = 0, 1, \cdots N-1, \tag{6.91}$$

Even Symmetry

The Tchebichef polynomials satisfy the even property [335],

$$\tilde{t}_{k,N-1-n} = (-1)^k \tilde{t}_{k,n}, \quad k = 0, 1, \cdots, N-1 \tag{6.92}$$

This property is useful to reduce the number of multiplications when implementing a fast transform.

Nakagaki and Mukundan [340] proposed a more elegant form to generate the discrete Tchebichef polynomial, given by the following recursive formula,

$$\tilde{t}_{k,n} = (\alpha_1 n + \alpha_2)\tilde{t}_{k-1,n} + \alpha_3 \tilde{t}_{k-2,n} \tag{6.93}$$

$$\tilde{t}_{0,n} = \frac{1}{\sqrt{N}} \tag{6.94}$$

$$\tilde{t}_{1,n} = (2n+1-N)\sqrt{\frac{3}{N(N^2-1)}} \tag{6.95}$$

where

$$\alpha_1 = \frac{2}{k}\sqrt{\frac{4k^2-1}{N^2-k^2}}$$

$$\alpha_2 = \frac{1-N}{k}\sqrt{\frac{4k^2-1}{N^2-k^2}}$$

$$\alpha_3 = \frac{1-k}{k}\sqrt{\frac{2k+1}{2k-3}}\sqrt{\frac{N^2-(k-1)^2}{N^2-k^2}}$$

THE DISCRETE TCHEBICHEF TRANSFORM

The discrete Tchebishef transform (DTT) is based on the discrete Tchebishef polynomials. The properties of the DTT (i.e., decorrelation and energy compaction) are equivalent to those for the DCT [189]. The DTT has been used in some areas such as image and video compression [201], [340], [360], [378], image analysis and recognition [279], [335] compressed sensing [278] and blind integrity verification of medical images [193]. One advantage of the DTT is that the transform conforms naturally to an integer representation without the need of approximation or intermediate scaling by 2 as in the ICT case.

Let $\{x\}_{m,n} \quad m,n=0,1,\cdots,N-1$, be the intensity values from a discrete image of size $N\times N$ pixels. The 2D DTT of the image denoted by $X_{p,q}, \quad p,q=0,1, \quad N-1$ is given by [340], [360]

$$X_{p,q}=\sum_{m=0}^{N-1}\sum_{n=0}^{N-1} t_{p,m}\cdot t_{q,n}\cdot x_{m,n} \quad p,q=0,1,\cdots,N-1. \tag{6.96}$$

The transform kernel in (6.96) is separable. Therefore, the transform coefficients are obtained as,

$$X_{p,q}=\sum_{m=0}^{N-1} t_{p,m}\sum_{n=0}^{N-1}\cdot t_{q,n}\cdot x_{m,n}. \tag{6.97}$$

In matrix form,

$$X=T_N\cdot x\cdot T_N^T. \tag{6.98}$$

T_N is the N-point unidimensional DTT matrix given by,

$$T_N=\begin{bmatrix} \tilde{t}_{0,0} & \tilde{t}_{0,1} & \cdots & \tilde{t}_{0,N-1} \\ \tilde{t}_{1,0} & \tilde{t}_{1,1} & \cdots & \tilde{t}_{1,N-1} \\ \vdots & \vdots & \ddots & \vdots \\ \tilde{t}_{N-1,0} & \tilde{t}_{N-1,1} & \cdots & \tilde{t}_{N-1,N-1}. \end{bmatrix} \tag{6.99}$$

The pixels can be recovered as

$$x=T_N^T\cdot X\cdot T_N. \tag{6.100}$$

The following MATLAB code shows the implementation of the 1D N-point DTT of (6.93)

```
clear;
N = 4;
F = zeros(N,N);
n = 0:N-1;
k = 0:N-1;
F(1:1, 1:N) =  1/sqrt(N);
F(2:2, 1:N) =  (2*n+1-N)*sqrt(3/(N*(N^2 - 1)));
```

```
for k = 2:N-1
    k2 = k*k;
    N2 = N*N;
    alpha1 = (2/k)*sqrt((4*k2 - 1)/(N2-k2));
    alpha2 = ((1-N)/k)*sqrt((4*k2 - 1)/(N2-k2));
    alpha3 = ((1-k)/k)*sqrt((2*k+1)/(2*k - 3))*...
        sqrt((N2-(k - 1)^2)/(N2-k2));

    F(k+1:k+1,1:N) = (alpha1*n+alpha2).*...
        F(k:k,1:N)+alpha3.*F(k-1:k-1,1:N);
end
```

4X4 AND 8X8 DISCRETE TCHEBICHEF TRANSFORMS

Using (6.93), (6.94) and (6.95) for $N = 4$, $k \in \{0,1,2,3\}$ and $n \in \{0,1,2,3\}$ we have,

$$F_4 = \left[\begin{array}{cc|cc} \frac{1}{2} & \frac{1}{2} & \frac{1}{2} & \frac{1}{2} \\ -\frac{3\sqrt{5}}{10} & -\frac{\sqrt{5}}{10} & \frac{\sqrt{5}}{10} & \frac{3\sqrt{5}}{10} \\ \frac{1}{2} & -\frac{1}{2} & -\frac{1}{2} & \frac{1}{2} \\ -\frac{\sqrt{5}}{10} & \frac{3\sqrt{5}}{10} & -\frac{3\sqrt{5}}{10} & \frac{\sqrt{5}}{10} \end{array}\right] \tag{6.101}$$

Notice that $\sqrt{5}/10 = 1/\sqrt{20}$. Therefore, (6.101) can be expressed as the product of a diagonal matrix D_4 and an integer matrix T_4 as follows,

$$F_4 = \operatorname{diag}\left(\frac{1}{2}, \frac{1}{\sqrt{20}}, \frac{1}{2}, \frac{1}{\sqrt{20}}\right) \left[\begin{array}{cc|cc} 1 & 1 & 1 & 1 \\ -3 & -1 & 1 & 3 \\ 1 & -1 & -1 & 1 \\ -1 & 3 & -3 & 1 \end{array}\right] = D_4 \cdot T_4 \tag{6.102}$$

For $N = 8$, $k \in \{0,1,\cdots,7\}$ and $n \in \{0,1,\cdots,7\}$ we have,

$$F_8 = \frac{1}{2}\operatorname{diag}\left(\frac{1}{\sqrt{2}}, \frac{1}{\sqrt{42}}, \frac{1}{\sqrt{42}}, \frac{1}{\sqrt{66}}, \frac{1}{\sqrt{154}}, \frac{1}{\sqrt{546}}, \frac{1}{\sqrt{66}}, \frac{1}{\sqrt{858}}\right)$$

$$\times \left[\begin{array}{cccc|cccc} 1 & 1 & 1 & 1 & 1 & 1 & 1 & 1 \\ -7 & -5 & -3 & -1 & 1 & 3 & 5 & 7 \\ 7 & 1 & -3 & -5 & -5 & -3 & 1 & 7 \\ -7 & 5 & 7 & 3 & -3 & -7 & -5 & 7 \\ 7 & -13 & -3 & 9 & 9 & -3 & -13 & 7 \\ -7 & 23 & -17 & -15 & 15 & 17 & -23 & 7 \\ 1 & -5 & 9 & -5 & -5 & 9 & -5 & 1 \\ -1 & 7 & -21 & 35 & -35 & 21 & -7 & 1 \end{array}\right] = D_8 \cdot T_8 \tag{6.103}$$

Observe that T_N is an integer orthogonal but not orthonormal matrix and the orthonormalization matrix D_N is a diagonal matrix with floating point representation.

INTEGER DISCRETE TCHEBICHEF TRANSFORM

Ishwar, Meher and Swamy [201] proposed a multiplier-free algorithm to represent T_4. For processing the input intensity values along one dimension, (6.98) is expressed as $X = \left[T_4 \cdot (T_4 \cdot x)^T\right]^T$ where output of the 1D transform is given by $Y = T_4 \cdot x$. Also T_4 in (6.102) can be factored into two sparse matrices,

$$T_{4,1} = \left[\begin{array}{rr|rr} 1 & 1 & 1 & 1 \\ 0 & -1 & 1 & 0 \\ 1 & -1 & -1 & 1 \\ -1 & 0 & 0 & 1 \end{array}\right], \quad T_{4,2} = \left[\begin{array}{rr|rr} 0 & 0 & 0 & 0 \\ -1 & 0 & 0 & 1 \\ 0 & 0 & 0 & 0 \\ 0 & 1 & -1 & 0 \end{array}\right] \tag{6.104}$$

where $T_4 = T_{4,1} + 3T_{4,2}$. The symmetry property of (6.92) allows to group the input vector $x = [x_0, x_1, x_2, x_3]^T$ as $u_0 = x_3 + x_0$, $v_0 = x_3 - x_0$, $u_1 = x_2 + x_1$ and $v_1 = x_2 - x_1$. Then, output coefficients can be separated into even Y_e and odd Y_o parts as,

$$Y = \begin{bmatrix} Y_e \\ Y_o \end{bmatrix} = \begin{bmatrix} Y_0 \\ Y_2 \\ Y_1 \\ Y_3 \end{bmatrix} = \begin{bmatrix} u_0 + u_1 \\ u_0 - u_1 \\ 3v_0 + v_1 \\ v_0 - 3v_1 \end{bmatrix}.$$

The multiplication operations can be expressed in terms of one-bit left shift as $3v_0 = v_0 + (<< v0)$ and $3v_1 = v_1 + (<< v1)$ where $<<$ is the one-bit left shift operator. Figure 6.10 shows the signal flowgraph for this algorithm [201].

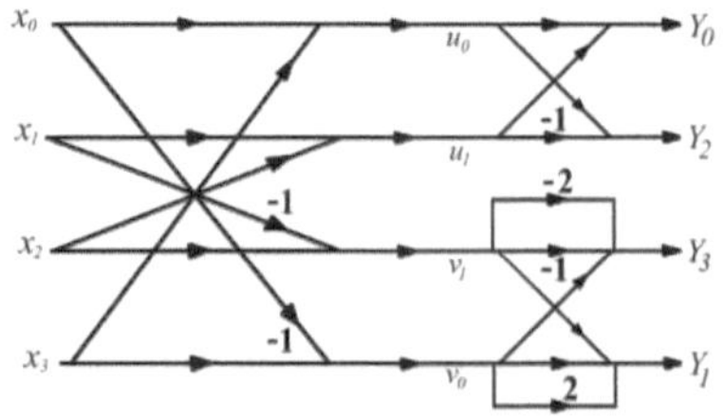

Figure 6.10: Signal flowgraph for the 4×4 DTT [201].

Oliveira, Cintra and Bayer [360] proposed an integer representation of the DTT as in (6.1) and to reduce the dynamic ranges of F_4 and F_8 they normalize the rows of the matrices according to multiplication by

$$D_4 = \text{diag}\left(2, \frac{\sqrt{20}}{3}, 2, , \frac{\sqrt{20}}{3}\right) \tag{6.105}$$

and

$$D_8 = \operatorname{diag}\left(\sqrt{8}, \frac{\sqrt{168}}{7}, \frac{\sqrt{168}}{7}, \frac{\sqrt{264}}{7}, \frac{\sqrt{568}}{13}, \frac{\sqrt{2184}}{23}, \frac{\sqrt{264}}{9}, \frac{\sqrt{3432}}{35}\right), \quad (6.106)$$

respectively. Therefore, the parametric matrices are defined by,

$$F_N(\alpha) = \operatorname{round}(\alpha \cdot D_N \cdot F_N) \quad N \in \{4, 8\}, \quad (6.107)$$

where round$(\cdot)$ is the round-off operation defined in (6.71) and $F_N(\alpha)$ is the low-complexity matrix that can be used to approximate the DTT by orthogonalization or quasi-orthogonalization [36], [90], [91]. Hence, the approximation $\hat{F}_N(\alpha)$ can be obtained by,

$$\hat{F}_N(\alpha) = S_N(\alpha) \cdot F_N(\alpha) \quad (6.108)$$

where $S_N(\alpha) = \sqrt{\operatorname{diag}\{F_N(\alpha) \cdot F_N^T(\alpha)\}^{-1}}$ and diag$(\cdot)$ is the diagonal matrix with diagonal entries of the resulting matrix. If $F_N(\alpha) \cdot F_N^T(\alpha)$ is a diagonal matrix $\hat{F}_N(\alpha)$ is orthogonal, otherwise it is a near-orthogonal matrix [91] and the approximation is considered quasi-orthogonal as long as the deviation from diagonality $\delta(A)$ is small, where

$$\delta(A) = 1 - \frac{\|\operatorname{diag}(A)\|_F}{\|A\|_F} \quad (6.109)$$

and $\|\cdot\|_F$ is the Frobenius norm for matrices. For compression purposes a deviation from diagonality below $1 - (\frac{2}{\sqrt{5}}) \approx 0.1056$ indicates quasi-orthogonality [177]. Afterward, a multi-criteria optimization problem is solved to obtain the optimal scaling parameter α^* that yields the optimal low-complexity DTT matrix $F_N^* \triangleq \hat{F}_N^*(\alpha^*)$ [360]:

$$\alpha^* = \underset{0<\alpha<5/2}{\operatorname{argmax}} \{C_g(\hat{F}_N(\alpha)), \eta(\hat{F}_N(\alpha))\} \quad N \in \{4, 8\}$$

where C_g and η are the coding gain and transform efficiency explained in Section §6. The resulting optimal interval of values for α^* when $N = 4$ is $(3/1, 5/2)$ and $(23/14, 69/34)$ for $N = 8$. The low-complexity matrices for $\alpha = 2$ are as follows,

$$F_4^* = \left[\begin{array}{cc|cc} 1 & 1 & 1 & 1 \\ -2 & -1 & 1 & 2 \\ 1 & -1 & -1 & 1 \\ -1 & 2 & -2 & 1 \end{array}\right], \quad (6.110)$$

and

$$F_8^* = \left[\begin{array}{cccc|cccc} 2 & 2 & 2 & 2 & 2 & 2 & 2 & 2 \\ -2 & -1 & -1 & 0 & 0 & 1 & 1 & 2 \\ 2 & 0 & -1 & -1 & -1 & -1 & 0 & 2 \\ -2 & 1 & 2 & 1 & -1 & -2 & -1 & 2 \\ 1 & -2 & 0 & 1 & 1 & 0 & -2 & 1 \\ -1 & 2 & -1 & -1 & 1 & 1 & -2 & 1 \\ 0 & -1 & 2 & -1 & -1 & 2 & -1 & 0 \\ 0 & 0 & -1 & 2 & -2 & 1 & 0 & 0 \end{array}\right]. \tag{6.111}$$

The optimal approximations ($\hat{F}_N^* \triangleq \hat{F}_N^*(\alpha^*)$) are computed by (6.108) as,

$$\hat{F}_4^* = \mathrm{diag}\left(\frac{1}{2}, \frac{1}{\sqrt{10}}, \frac{1}{2}, \frac{1}{\sqrt{10}}\right) \cdot F_4^*$$

and

$$\hat{F}_8^* = \mathrm{diag}\left(\frac{1}{\sqrt{32}}, \frac{1}{\sqrt{12}}, \frac{1}{\sqrt{12}}, \frac{1}{\sqrt{20}}, \frac{1}{\sqrt{12}}, \frac{1}{\sqrt{14}}, \frac{1}{\sqrt{12}}, \frac{1}{\sqrt{10}}\right) \cdot F_8^*$$

Here, $F_4^* \cdot (F_4^*)^T$ is a diagonal matrix, therefore $\hat{F}_4^*$ is orthonormal unlike $\hat{F}_8^*$ that is not orthogonal. Therefore, We are not sure that the inverse matrix has a low complexity as the forward matrix. However, the Frobenius norm of matrix $F_8^* \cdot (F_8^*)^T$ is 0.0137 which indicates a small deviation from diagonality. Then, we can assume that $(\hat{F}_8^*)^{-1} \approx (F_8^*)^T \cdot [\mathrm{diag}(8,3,3,5,3,7/2,3,5/2)]^{-1}$, Therefore $(F_8^*)^{-1}$ can be replaced with $(F_8^*)^T$ and use the same low-complexity algorithm to implement the forward and inverse transforms . The sparse matrix factorizations are [360],

$$F_8^* = P \cdot A_2 \cdot A_1 \cdot B_8,$$

where P is the following cost-free permutation matrix,

$$P = \begin{bmatrix} 1 & 0 & 0 & 0 & 0 & 0 & 0 & 0 \\ 0 & 0 & 0 & 0 & 1 & 0 & 0 & 0 \\ 0 & 0 & 0 & 0 & 0 & 0 & 1 & 0 \\ 0 & 0 & 1 & 0 & 0 & 0 & 0 & 0 \\ 0 & 0 & 0 & 0 & 0 & 0 & 0 & 1 \\ 0 & 0 & 0 & 1 & 0 & 0 & 0 & 0 \\ 0 & 1 & 0 & 0 & 0 & 0 & 0 & 0 \\ 0 & 0 & 0 & 0 & 0 & 1 & 0 & 0 \end{bmatrix},$$

A_2 and A_1 are the following additive matrices with bit-shifting operations,

$$A_2 = \begin{bmatrix} 1 & 1 & 1 & 0 & 0 & 0 & 0 & 0 & 0 & 0 \\ 0 & 1 & -2 & 0 & 0 & 0 & 0 & 0 & 0 & 0 \\ 0 & 0 & 0 & 1 & 0 & 0 & 0 & 0 & 0 & 0 \\ 0 & 0 & 0 & 0 & 1 & 0 & 0 & 0 & 0 & 0 \\ 0 & 0 & 0 & 0 & 0 & 1 & 0 & 0 & 0 & 0 \\ 0 & 0 & 0 & 0 & 0 & 0 & 1 & 1 & 0 & 1 \\ 0 & 0 & 0 & 0 & 0 & 0 & 0 & -1 & 0 & 1 \\ 0 & 0 & 0 & 0 & 0 & 0 & -1 & 0 & 1 & 0 \end{bmatrix},$$

$$A_1 = \begin{bmatrix} 0 & 0 & 1 & 0 & 0 & 0 & 0 & 0 \\ 1 & 0 & 0 & 1 & 0 & 0 & 0 & 0 \\ 0 & 1 & 0 & 0 & 0 & 0 & 0 & 0 \\ 0 & -1 & 2 & -1 & 0 & 0 & 0 & 0 \\ 2 & 0 & -1 & -1 & 0 & 0 & 0 & 0 \\ 0 & 0 & 0 & 0 & 2 & -1 & 0 & 0 \\ 0 & 0 & 0 & 0 & 1 & 1 & 0 & 0 \\ 0 & 0 & 0 & 0 & 0 & 1 & 1 & 0 \\ 0 & 0 & 0 & 0 & 0 & 0 & 2 & -1 \\ 0 & 0 & 0 & 0 & 0 & 0 & 0 & -2 \end{bmatrix},$$

and B_8 is the butterfly structure,

$$B_8 = \begin{bmatrix} 1 & 0 & 0 & 0 & 0 & 0 & 0 & 1 \\ 0 & 1 & 0 & 0 & 0 & 0 & 1 & 0 \\ 0 & 0 & 1 & 0 & 0 & 1 & 0 & 0 \\ 0 & 0 & 0 & 1 & 1 & 0 & 0 & 0 \\ 0 & 0 & 0 & 1 & -1 & 0 & 0 & 0 \\ 0 & 0 & 1 & 0 & 0 & -1 & 0 & 0 \\ 0 & 1 & 0 & 0 & 0 & 0 & -1 & 0 \\ 1 & 0 & 0 & 0 & 0 & 0 & 0 & -1 \end{bmatrix}.$$

The signal flowgraph for F_8^* according to the previous matrix factorization is shown in Figure 6.11. The algorithm uses only 24 additions and 6 bit-shifting operations.

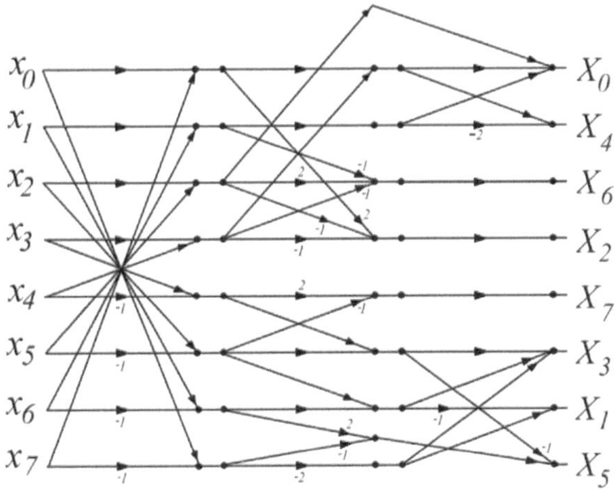

Figure 6.11: Signal flowgraph for the 8 × 8 DTT [360].

METRICS TO EVALUATE A TRANSFORM

To evaluate a transform, the input signal x is modeled as a first-order stationary Markov process with zero-mean, unit variance and adjacent inter-element correlation coefficient $\rho \in [0,1]$. The covariance matrix at the input is defined as,

$$[R_x]_{m,n} = \rho^{|m-n|}, \tag{6.112}$$

where the covariance matrix R_x is symmetric and Toeplitz. The covariance matrix of the transformed vector $y = Cx$ is defined by,

$$R_y = C \cdot R_x \cdot C^T. \tag{6.113}$$

MEAN-SQUARE ERROR

For a 1D input vector x of length M, the error between the true DCT transform C_N and its approximation is $\bar{C}_N$ is $e = \left(C_N - \bar{C}_N\right) x \triangleq Dx$. The MSE is defined as [52],

$$\begin{aligned} MSE \triangleq \frac{1}{M} E\left[x^T \cdot D^T \cdot D \cdot x\right] &= \frac{1}{M} \text{Trace}\left\{D \cdot E[x \cdot x^T] \cdot D^T\right\} \\ &= \frac{1}{M} \text{Trace}\left\{D \cdot R_x \cdot D^T\right\}, \end{aligned} \tag{6.114}$$

where Trace$\{\cdot\}$ is the trace of a matrix which is defined as the sum of its diagonal elements and R_x is the covariance matrix of the input signal x.

CODING GAIN

The coding gain is one of the most important metrics. This parameter represents the capability of a transform to pack energy into few coefficients. In other words, transforms with higher coding gain pack more energy into fewer coefficients [52]. The coding gain is defined by the covariance matrix R_y as,

$$C_g = 10 \log_{10} \frac{\frac{1}{M} \sum_{i=0}^{M-1} \sigma_{x_i}^2}{\left(\prod_{i=0}^{M-1} \sigma_{x_i}^2 \|f_i\|^2\right)^{\frac{1}{M}}}, \tag{6.115}$$

where M is the number of subbands or transform coefficients, $\sigma_{x_i}^2$ is the variance of the ith subband or transform coefficient and $\|f_i\|^2$ is the 2-norm of the ith basis function of the transform matrix.

TRANSFORM EFFICIENCY

This metric measures the capability of a transform to decorrelate the input signal. For example, the optimal KLT has an efficiency $\eta = 100\%$ for any value of ρ while the DCT-II has transform efficiency $\eta = 93.9911\%$ for $\rho = 0.95$. The transform efficiency, for the covariance matrix R_y is defined as,

$$\eta = \frac{\sum_{i=0}^{M-1} |r_{i,i}|}{\sum_{i=0}^{M-1} \sum_{k=0}^{M-1} |r_{i,k}|} \times 100, \quad i \neq k \tag{6.116}$$

where $r_{i,k}$ are the elements of R_y [52].

VARIANCE DISTRIBUTION

The auto-covariance matrix of a random vector in the transform domain is

$$\begin{aligned}[A_{tr}] &= E[XX^T] \\ &= [\phi]_N \left(E[xx^T]\right) [\phi]_N^T \\ &= [\phi]_N [A] [\phi]_N^T \end{aligned} \tag{6.117}$$

where $[\phi]_N$ is any $N \times N$ orthogonal matrix. The diagonal elements of $[A_{tr}]$ are the variances of the transform coefficients $X_0, X_1, \cdots, X_{N-1}$.

ENERGY PACKING EFFICIENCY (EPE)

Energy packing efficiency EPE is a performance measure defined as the energy portion contained in the first M of N transform coefficients [254], i.e.,

$$EPE(M) = \frac{\sum_{k=0}^{M-1} E[X_k^2]}{\sum_{k=0}^{N-1} E[X_k^2]} \tag{6.118}$$

The EPE for other transforms was evaluated by Yip and Rao [390], who concluded that the EPE is invariant with respect to those discrete transforms that display block spectral structure. The EPE is directly equivalent to the ratio of the sum of the first M variances to the sum of all the N variances [390]. Thus EPE is an alternative to the variance distribution in terms of the performance criteria.

RESIDUAL CORRELATION

The residual correlation metric measures how well a discrete transform can decorrelate a given sequence. The KLT is an optimal transform, as it completely decorrelates a Markov-1 sequence. Other discrete transforms, although suboptimal, can be judged by how well or how closely they approximate the KLT in decorrelating the sequence. Whereas the auto-covariance matrix in the KLT domain is diagonal, the covariance matrices for the other discrete transforms have non-zero off-diagonal elements. Hamidi and Pearl [168] proposed a measure called fractional correlation (FC) left undone by a transform. Comparison of DCT-II with DFT [168] and with DHT [168] shows that the DCT is superior to both DFT and DHT based on the residual correlation for all N (sequence length) and ρ (adjacent correlation coefficient) for Markov-1 signals. The FC as defined by Hamidi and Pearl [168] can be developed as follows. Let $[A_{tr}]_{i,i}$ be the diagonal matrix representing the diagonal elements of $[A_{tr}]$ defined in (6.117). The auto-covariance matrix $[A^{'}]$ can be obtained by inverse transforming $[A_{tr}]_{i,i}$ such that

$$[A^{'}] = [\phi]_N^T [A_{tr}]_{i,i} [\phi]_N. \tag{6.119}$$

The FC left undone by the transform is then defined as

$$FC = \frac{\left|[A] - [A^{'}]\right|^2}{\left|[A] - [I_N]\right|^2} \tag{6.120}$$

$|\cdot|^2$ denotes the Hilbert–Schmidt weak norm defined as

$$\left|[A_N]\right|^2 \equiv \frac{1}{N} \sum_{i=0}^{N-1} \sum_{k=0}^{N-1} (a_{i,j})^2 = \frac{1}{N} \sum_{i=0}^{N-1} \lambda_i, \tag{6.121}$$

where $\{\lambda_i\}$ $\quad i = 0, 1, \cdots N-1$ are the eigenvalues of $[A_N]$.

IMPLEMENTATIONS

Meher, Park and Bohanty [322] proposed a flexible hardware architecture for computing the IntDCT of lengths $N \in \{4, 8, 16, 32\}$ suitable for HEVC video compression standard in the context of scalability and reusability architecture for 1D and 2D IntDCT with the same throughput of processing irrespective of transform size. The proposed reusable architecture for the implementation of DCT of any of the prescribed lengths is shown in Figure 6.12(a).

The input to one $(N/2)$-point DCT unit is fed through $(N/2)$ 2 : 1 multiplexers (MUXes) that select either $[a(0), ..., a(N/21)]$ or $[x(0), ..., x(N/21)]$, depending on whether it is used for N-point DCT computation or for the DCT of a lower size. The other $(N/2)$-point DCT unit takes the input $[x(N/2), \cdots, x(N1)]$ when it is used for the computation of DCT of $N/2$- point or a lower size, otherwise, the input is reset by an array of $(N/2)$ AND gates to disable this $(N/2)$-point DCT unit. The output of this $(N/2)$-point DCT unit is multiplexed with that of the output adder unit (OAU), which is preceded by the SAUs and IAU of the structure. The N AND gates before IAU are used to disable the IAU, SAU, and OAU when the architecture is used to compute the $(N/2)$-point DCT or a lower size. The input of the control

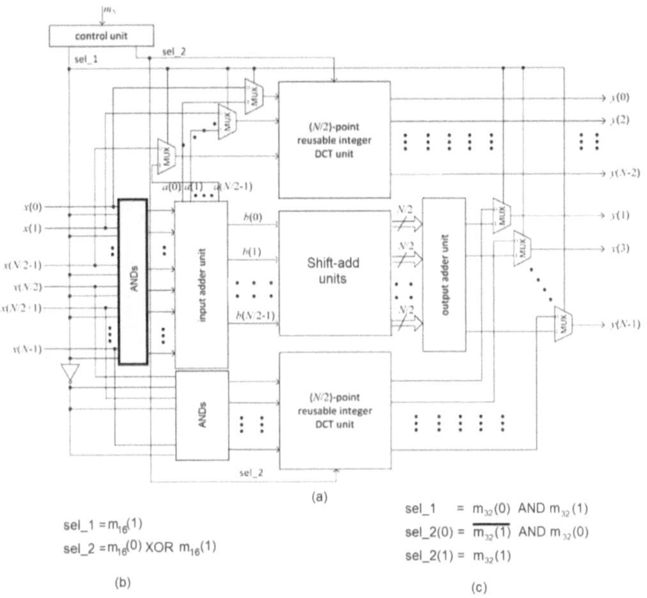

Figure 6.12: Reusable architecture of integer DCT (a) for $N = 8, 16$, and 32. (b) Control unit for $N = 16$. (c) Control unit for $N = 32$ [322].

unit, m_N is used to decide the size of DCT computation. The proposed structure can compute one 32-point DCT, two 16-point DCTs, four eight-point DCTs, and eight four-point DCTs, while the throughput remains the same as 32 DCT coefficients per cycle irrespective of the desired transform size.

Meher, Park and Bohanty [322] also proposed a folded and a full parallel architecture structure to compute the 2D DCT. The folded structure is shown in Figure 6.13(a). The architecture consists of one N-point 1D DCT module and a transposition buffer. The 4×4 transposition buffer is shown in Figure 6.13(b) and consists of 16 registers arranged in four rows and four columns. $N \times N$ transposition buffer can store N values in any column of registers by setting the appropriate signals $EN_i, \quad i = 0, 1, \ldots, N-1$.

The full parallel architecture consists of two N-point 1D DCT modules and a transposition buffer. The 4×4 transposition buffer consists of 16 register cells (RC) arranged in four rows and four columns. $N \times N$ transposition buffer can store N values in a cycle either row-wise or column-wise by selecting the inputs by the MUXes at the input of RCs. To read the output from the buffer, N number of $(2N-1):1$ MUXs are used, where outputs of the ith row and the ith column of RCs are fed as input to the ith MUX. Figure 6.14(a) shows full parallel 2D DCT architecture, (b)

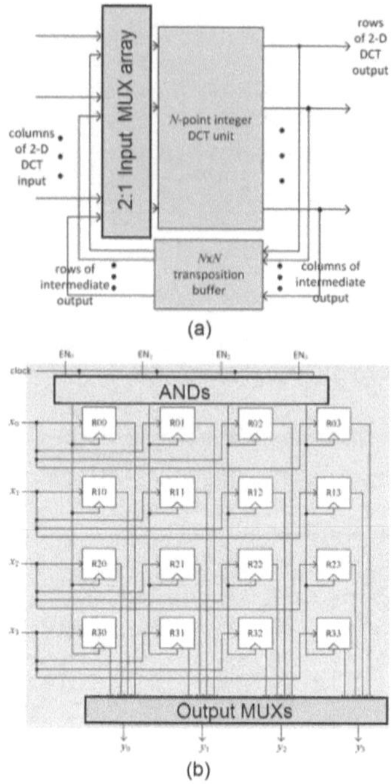

Figure 6.13: Folded structure of $(N \times N)$-point 2D integer DCT. (a) Folded 2D architecture and (b) 4×4 transposition buffers [322].

shows the structure of the transposition buffer for input size 4×4. Figure 6.14(a) (c) and (d) are the register cells RCs_{ik} and the 7:1 MUX for 4×4 transposition buffer.

Kalali and Hamzaoglu [241], and Kalali et al. [240] developed a 2D IDCT / IDST using the field programmable gate array (FPGA) of Xilinx XC6VLX550T FF1760 and the high-level synthesis tool Vivaldo from Xilinx for the HEVC video compression standard. The hardware uses a butterfly structure for column and row transforms. IDCT inputs are selected depending on the TU size (4×4, 8×8, 16×16 or 32×32). Then, IDCT and IDST multiplications are performed in the datapaths using only adders and shifters.

The resulting coefficients are stored in a transpose memory after 1D column IDCT (see Figure 6.15). Then, they are used as input for 1D row IDCT. The transpose memory is implemented using 32 Block RAMs (BRAM). 4, 8, 16 and 32 BRAMs are used for 4×4, 8×8, 16×16 and 32×32 transform unit (TU) sizes, respectively. In Figure 6.15, the numbers in each box show the BRAM that coefficient is stored.

Mert, Kalali and Hamzaoglu [324] developed a future video coding (FVC) baseline 2D transforms hardware, using FPGAs and Verilog hardware design language, for TUs of sizes 4×4 and 8×8, which uses the heuristic cumulative benefit (Hcub) multiple constant multiplication (MCM) algorithm [225]. The hardware of the baseline is shown in Figure 6.16.

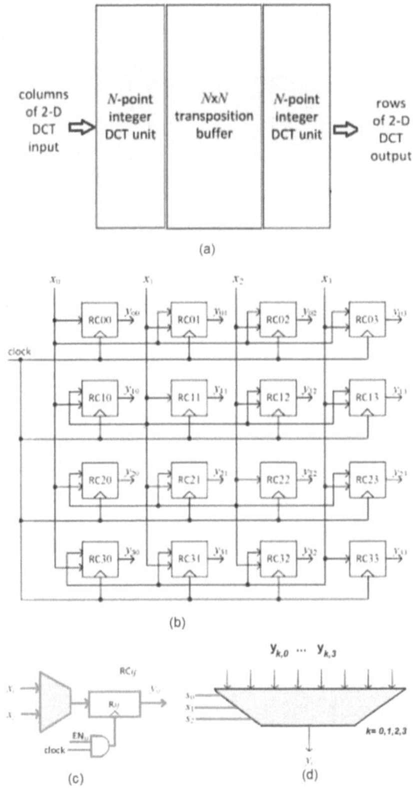

Figure 6.14: Full-parallel structure of $(N \times N)$-point 2D integer DCT. (a) Full parallel 2D DCT architecture. (b) Structure of the transposition buffer for input size 4×4. (c) Register cell RC_{ik}. (d) 7:1 MUX for 4×4 transposition buffer [322].

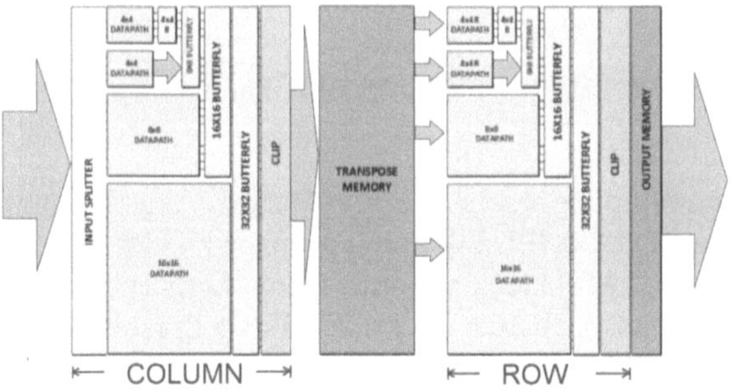

Figure 6.15: 2D IDCT and IDST [241].

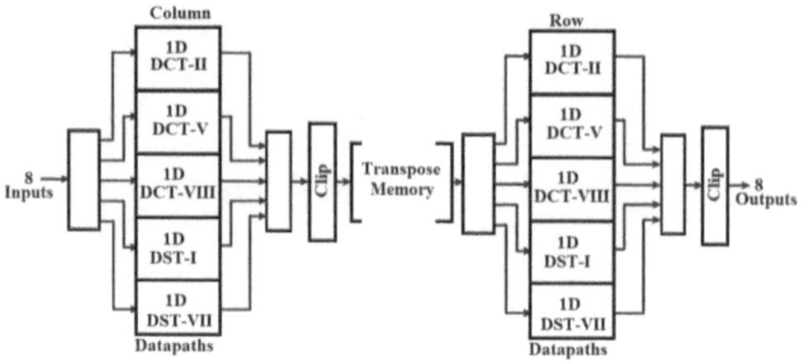

Figure 6.16: FVC baseline 2D transform hardware [324].

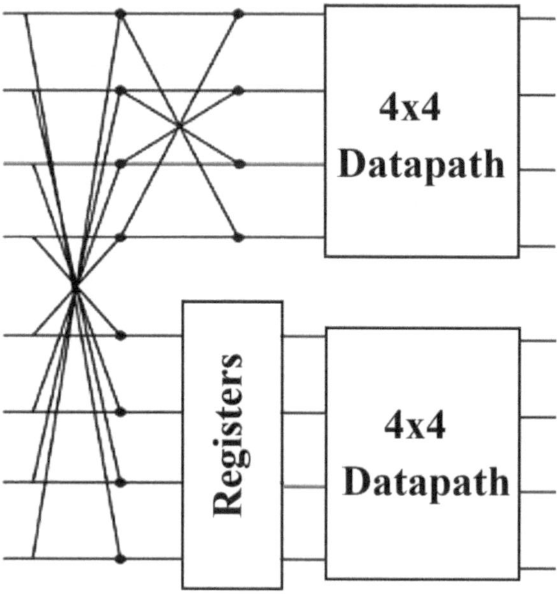

Figure 6.17: 1D DCT-II/DST-I column datapath [225].

There are four multiplier blocks in a 4×4 datapath and 8 multiplier blocks in an 8×8 datapath. The 1D DCT-II/DST-I column datapath [225] is shown in Figure 6.17. Each multiplier (M1 to M4) block performs the multiplications between 1 input and 8 transform coefficients. An output of each DCTs/DSTs multiplier blocks is selected and added or subtracted accordingly.

The transpose memory [225] of Figure 6.18 is implemented using 8 BRAMs; 4 and 8 BRAMs are used for 4×4 and 8×8 TU sizes respectively. Each BRAM

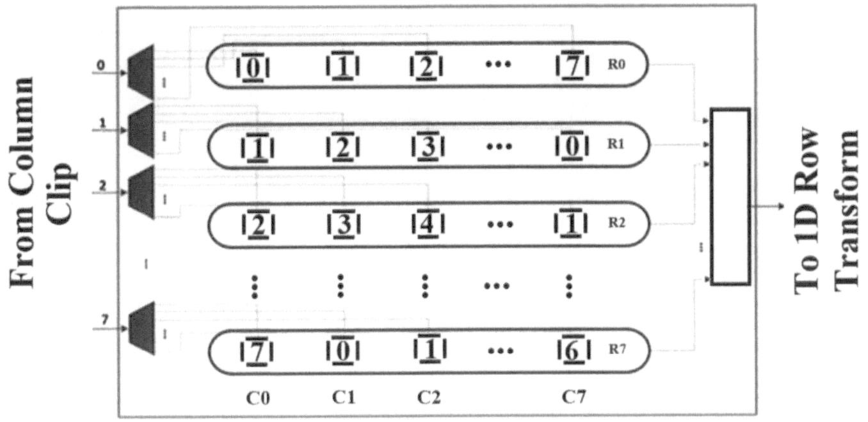

Figure 6.18: Transpose memory (8 BRAMs) [225].

address can store 32-bits and each transformed coefficient of 1D column DCT/DST is 16-bits, each BRAM address can store two transformed coefficients.

FVC reconfigurable 2D transform hardware for 4×4 and 8×8 TU sizes is similar to the proposed baseline hardware, it performs 2D DCT/DST by processing along columns of a TU, and then along rows. The column clip hardware, row clip hardware, and transpose memory in the reconfigurable hardware are the same as in the baseline hardware. The 2D hardware calculates eight transformed coefficients per clock cycle for both 4×4 and 8×8 TU sizes. The baseline hardware uses separate datapaths (see Figure 6.16). However, as shown in the reconfigurable hardware, one reconfigurable datapath is used for implementing all 1D column DCT/DST types and one reconfigurable datapath for implementing all 1D row DCT/DST types. Therefore, N-point DCT-II and DST-I are also performed by carrying out one N-point DCT-II and DST-I the same as DCT-IV, DCT-VIII and DST-VII.

Jridi, Alfalou and Meher [238] developed a reconfigurable architecture to approximate the DCT of lengths $N = 8, 16$ and 32. The hardware performs the calculation of one 32-point or two 16-point in parallel or four 8-point DCTs in parallel. There is one 32-point input adder unit to combine two 16-point DCTs, two 16-point input adder units and four 8-point DCT units. There are three control blocks composed of 64 2:1 MUXes and 30 3:1 MUXes to achieve reconfigurability. The control lines *sel32* and *sel16* select the length of the DCT to be computed. The output control block reorders the output depending on the selected length of the DCT, this is controlled by the 3:1 MUXes and the lines *sel16* and *sel32*. The throughput is of 32 coefficients per cycle [238]. Figure 6.19 shows the reconfigurable architecture.

The architecture is orthogonal and performance close to the DCT, it has low complexity and supports high lengths to work with modern video coders, tracking, surveillance, etc.

Budagavi and Sze [55] proposed a unified forward and inverse transform architecture for HEVC. The architecture takes advantage of the symmetry properties of

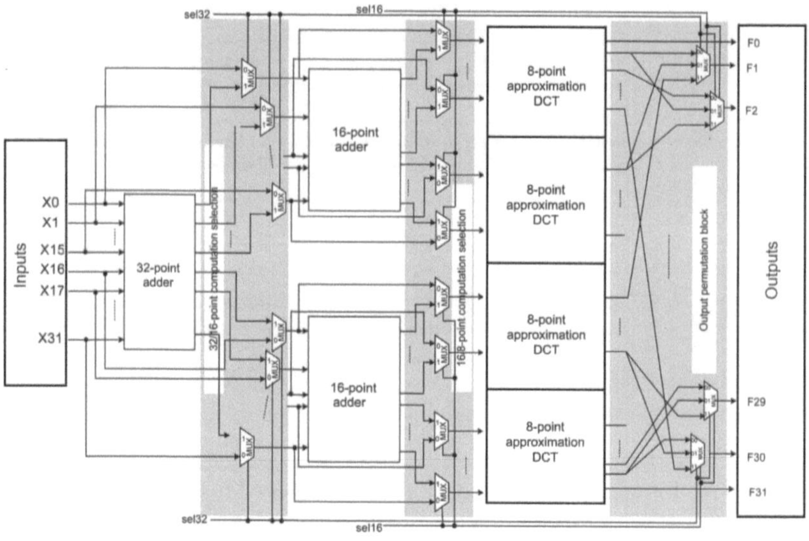

Figure 6.19: Reconfigurable architecture to approximate DCT of lengths $N = 8, 16$ and 32 [238].

the forward and inverse transform matrices in HEVC to achieve hardware sharing across different transform sizes and also between forward and inverse transforms to reduce implementation cost. Also, this architecture takes advantage of the property that transforms of size 16, 8 and 4 are sub-matrices of the transform matrix of size 32 [54] (see (6.21) to (6.24)). The forward/inverse transform was implemented for a throughput of a 32-pt 1D transform per cycle. A 32×32 2D transform requires 64 cycles. The implementations were synthesized in 45-nm CMOS (for more information refer to [55]).

Kalali, Mert and Hamzaoglu [242] proposed two implementations. The first implementation is a computation of an energy reduction technique for HEVC DCT for all TUs. The proposed implementation reduces the computational complexity significantly at the expense of a slight decrease in peak signal-to-noise ratio (PSNR) and slight increase in bit rate by only calculating several pre-determined low frequency coefficients of TUs and assuming that the remaining coefficients are zero. The hardware was described and synthesized using Verilog HDL. The transform calculates 4, 8, 16 and 32 DCT coefficients per clock cycle for 4×4, 8×8, 16×16 and 32×32 TUs sizes, respectively. The authors report that, in the worst case, the implementation can process 48 Quad Full HD (3840×2160) video frames per second. In the second implementation of the HEVC DCT, the hardware is able to process four 4×4 TUs or two 8×8 TUs in parallel to calculate 16 DCT coefficients per clock cycle for 4×4, 8×8, and 16×16 TU sizes, and 32 DCT coefficients per clock cycle for 32×32 TU size. It, in the worst case, can process 53 Ultra HD (7680×4320) video frames per second.

Clock gating is used to reduce the energy consumptions of both hardware. Hcub Multiplierless Constant Multiplication (MCM) algorithm [500] is used to obtain a reduction in the number and size of the adders of the lower and higher utilization hardware of up to 17.9% and 18.9%, respectively. The implementation discards the several pre-determined high-frequency coefficients of each TU and keeps the rest of the low frequency coefficients. To reduce the computational complexity three sets of coefficient are defined. When the first set is used, only $1/4$ of the low-frequency coefficients are calculated for all TU sizes. When the second set is used, $9/16$ of the low-frequency coefficients are calculated for 4×4 and 16×16 TUs and $9/64$ of the low-frequency coefficients for 8×8 and 32×32 TU sizes. When the third coefficient set is used, the number of coefficients to calculate are adaptively changed during the DCT operations. In this set, three levels and for each TU are used as well as the no computation option for the coefficients that are not calculated. In level-0 DCT, all DCT coefficients are calculated for all TUs, this is the initial level used. In level-1, $1/4$ of the low frequency coefficients are calculated for 4×4 TUs, $35/64$ of the low-frequency coefficients are calculated for 8×8, 16×16 and 32×32 TUs. In level-2 DCT, $9/64$ of the low-frequency coefficients are calculated for 8×8, 16×16 and 32×32 TUs. If the distortion of the current TU obtained by the current DCT operation in any level is smaller than 90% of the previous distortion value for the same TU or same size TU obtained by the previous DCT operation, the DCT level for this TU size is increased. If the distortion value for current TU obtained by the current DCT operation is larger than 110% of the previous distortion value for the same TU or same size TU obtained by the previous DCT operation, the level for this TU size is decreased.

Figure 6.20 shows the HEVC 2D DCT lower utilization hardware for all TU sizes including the input splitter and the output multiplexer, clock gating, Hcub MCM algorithm, and the technique with coefficient set 3.

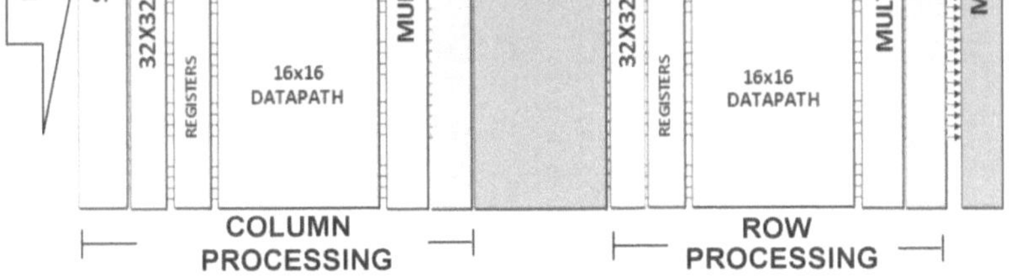

Figure 6.20: HEVC 2D DCT lower utilization hardware [242].

Figure 6.21 shows the detail of the forward transform input splitter. The input splitter selects the proper DCT inputs for each TU size. Output multiplexers select the proper DCT outputs for each TU size. Column and row clip modules scale the outputs of 1D column DCT and 1D row DCT to 16 bits, respectively. One 4×4 datapath is used for 4×4 TU size. Two 4×4 datapaths are used for 8×8 TU size. Two 4×4 datapaths and one 8×8 datapath are used for 16×16 TU size. All datapaths (two 4×4, one 8×8 and one 16×16) are used for 32×32 TU size. The transpose memory is shown in Figure 6.21 and is implemented using 32 BRAMs. The 4, 8, 16 and 32 BRAMs are used for 4×4, 8×8, 16×16 and 32×32 TU sizes.

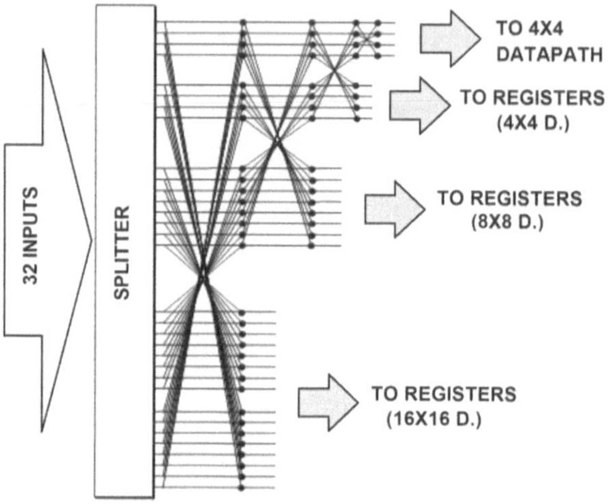

Figure 6.21: Column butterfly structure. Detail of the forward transform input splitter [242].

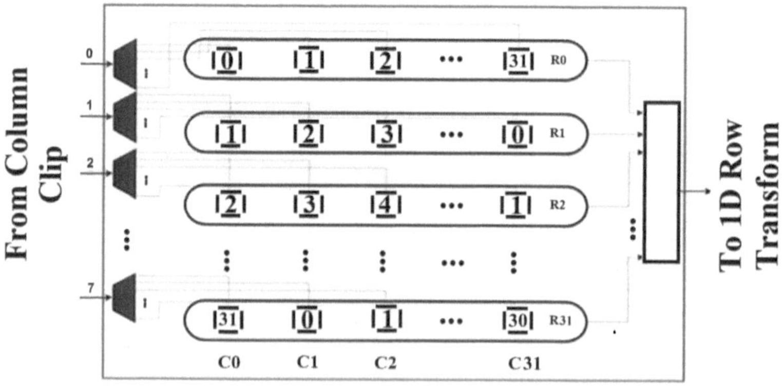

Figure 6.22: Transpose memory (32 BRAMs) [242].

Garrido, Pescador and Chavarrías [150] described in VHDL language a deeply pipelined high performance architecture to implement 4×4, 8×8, 16×16 and 32×32 size 1D horizontal or vertical transforms in 4, 32, 256, and 2048 clocks cycles, respectively and it supports HD and UHD resolutions. The design was synthesized for different FPGA chips, being able to process up to 182 frames per second (fps) @3840×2160 for 4×4 transform sizes. The adaptive multiple core transform (AMT) was used for the FVC standard for coding both intra and inter coded blocks. Figure 6.23 shows the top of the 1D-AMT processor.

Figure 6.24 shows the internal architecture of the datapaths of Figure 6.23 to compute an element of the 1D transform output matrix. The processor consists of

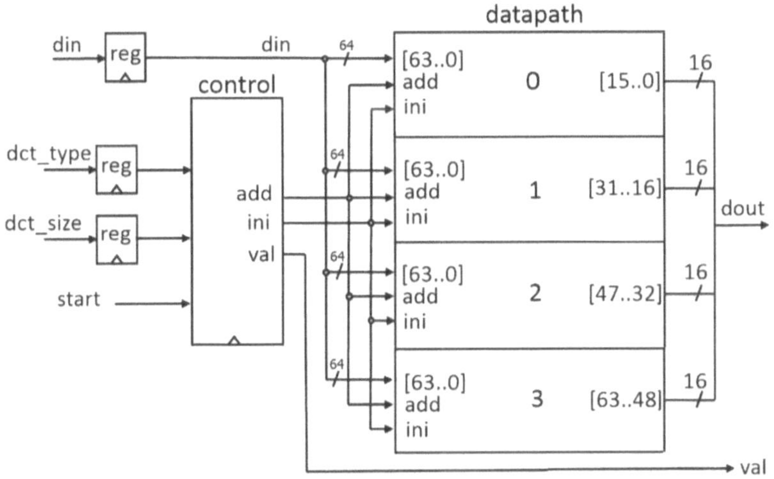

Figure 6.23: Architecture of the 1D-AMT processor [150].

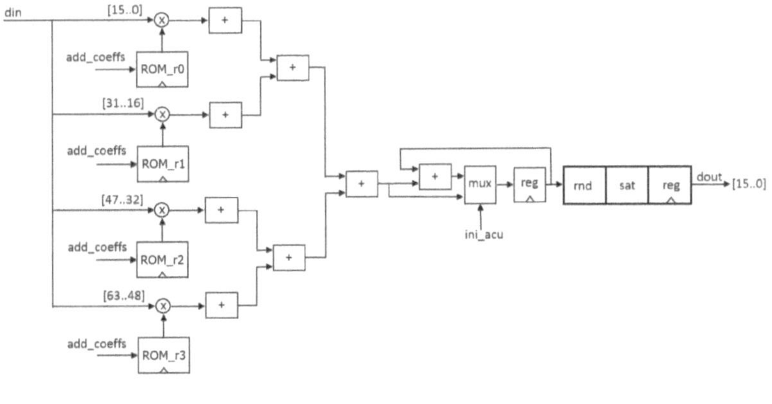

Figure 6.24: 1D Datapath [150].

four data paths to process data, computing one 16-bit output (dout) after a latency (val_out). A positive pulse starts a new transform with size and type defined by t_size and tt_type inputs with are entered at a column basis through the 16-bits input (din), starting in the clock cycle following the input start signal.

Four synchronous read only memory ROMs modules (ROM_rX) store one transform coefficient for each transform type. Four 16-bit multipliers and a 3-level add chain compute the sum of products of the inputs (din) times the coefficients. Results are rounded (rnd), saturated (sat) to 16 bits and stored in an output register (reg). Each datapath calculates an output. The 4×4 size transform is the basic transform in this case. In order to increase the processor performance the architecture is fully pipelined. The registers of the pipeline are shown in Figure 6.25 highlighted with a gray shadow. Notice that in the output of the ROMs, inside the multipliers there are 3 registers.

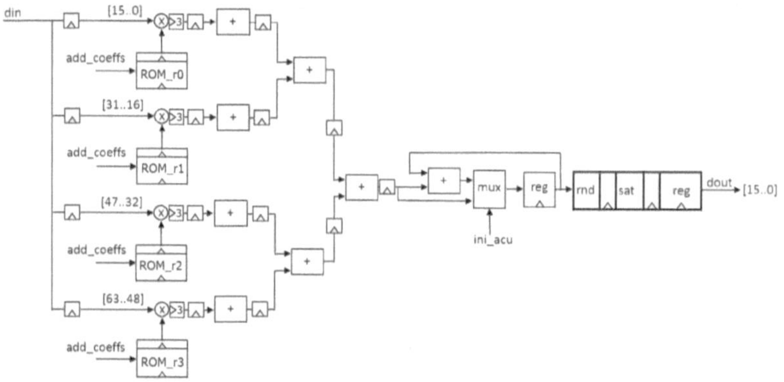

Figure 6.25: Fully pipelined datapath to increase the processor performance [150].

For each $N \times N$ block, the 2-D forward transform computation can be carried out by first applying N 1D horizontal transforms to its N rows followed by N 1D vertical transforms to its N columns followed by N 1D vertical transforms to its N columns to take advantage of the separability property of the transform,

$$Y_{int} = C \cdot X^T \tag{6.122}$$

$$Y = C \cdot Y_{int}^T = C \cdot \left(C \cdot X^T\right)^T = C \cdot X \cdot C^T. \tag{6.123}$$

where X, Y and C are $N \times N$ matrices containing the input block, the output block and the transform coefficients, Y_{int} is an intermediate $N \times N$ matrix to store the first N 1D horizontal transforms. For example, the output lines 0, 1, 2 and 3 $(Y_{int,0,i}, Y_{int,1,i}, Y_{int,2,i}, Y_{int,3,i})$ for a 4×4 input block are computed using the following sum of products,

$$Y_{int,0,i} = \sum_{k=0}^{3} X_{k,0} \cdot C_{k,i} \quad i = 0 \cdots 3 \tag{6.124a}$$

$$Y_{int,1,i} = \sum_{k=0}^{3} X_{k,1} \cdot C_{k,i} \quad i = 0 \cdots 3 \tag{6.124b}$$

$$Y_{int,2,i} = \sum_{k=0}^{3} X_{k,2} \cdot C_{k,i} \quad i = 0 \cdots 3 \tag{6.124c}$$

$$Y_{int,3,i} = \sum_{k=0}^{3} X_{k,3} \cdot C_{k,i} \quad i = 0 \cdots 3 \tag{6.124d}$$

Figure 6.26 shows the computation of the 1D transforms for a 4×4 input block applying (6.124). Four 1D transforms are needed to transform a 4×4 input block.

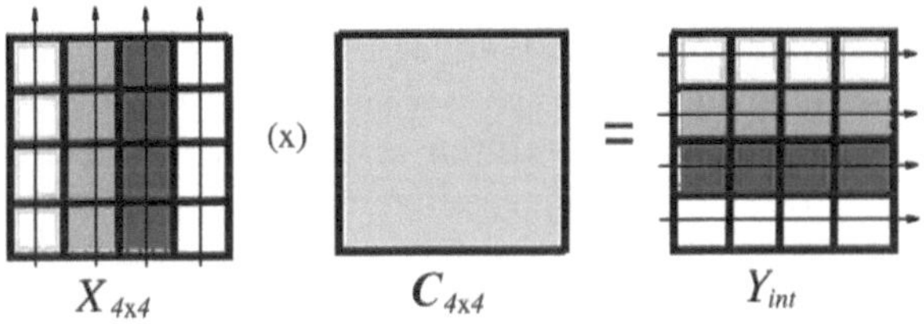

Figure 6.26: 1D transforms for a 4×4 input block [150].

Computing of the 1D transforms for an 8×8 input block is carried out with 4-size 1D transforms. Four 4-size 1D-transforms are needed to compute a line. The transform block is divided into four quadrants each containing a set of 4×4 coefficients to carry out the sum of products as in (6.124). The transformation of the 8×8 input block requires 32 4-size transforms.

APPLICATIONS

The evolution of ultra high definition video or images such as 4K or 8K and the current need for its transmission, reproduction or storage has forced academia and industry to propose new standards, seeing the need to include other methodologies to deal with these signals. Especially, in applications where time is critical or a low level of energy consumption is needed. One of these methodologies is the IntDCT which is suitable for applications such as image and video coding [47], [163], [196], [35], [398], [309], [389], [409], [462], [468], [28], [319], audio coding [153], [154], [192], [280], [564], [581], [567], [510], mobile applications [248], [397], [487], data hidding and encryption [155], [195], [568], [425], [446], watermarking [148], [492], [291], shape representation [429], luminance adaption effects [582], etc.

A recent application of the IntDCT is in the new codec released by the Alliance for Open Media (AOM) called AV1 (royalty free, open source software) [136]. The codec includes several sizes of the IntDCT. AV1 is mainly intended for video transmissions over the Internet and able to complete with HEVC. AV1 is meant to succeed its predecessor VP9 [325]. AOM charter members are Amazon, Cisco, Google, Intel Corporation, Microsoft, Mozilla, and Netflix among others. The formation consolidated the development of three potentially competitive open source codecs; Cisco's Thor, Google's VP10 and Mozilla's Daala. In April 2016, ARM, AMD, Apple and NVIDIA joined the alliance to help ensure that the codec is hardware friendly and to facilitate and accelerate AV1 hardware support. Hulu joined AOM as a member on 29 June 2017. NGCodec has joined as a member on 11 July 2017. Facebook has joined AOM as a founding member (Nov. 2017). There are now 25+ members (March 2018).

Generally, HDR images have much greater bit depth of pixel values and much wider color gamut. Conventional standards can not efficiently compress HDR images due to its greater bit depth and uncommon pixel format including a floating point based pixel encoding [313]. Nevertheless, in recent years a new standard for HDR image coding has been released by the ISO/IEC called JPEG eXTension (JPEG XT) [399] and it is backward-compatible with legacy JPEG with the ability to use lossy and lossless modes in images of higher precision and higher dynamic range. JPEG XT uses the IntDCT implemented in lifting steps where the butterflies are replaced by lifting rotations [372]. The implementation requires approximately 191 shifts per row and column or 39 multiplications per row and column and 230 adds per shift and column [19], [20], [127], [313], [312], [400], [401], [402], [407], [403], [406],[404], [405], [424], [528].

SUMMARY

The recent transforms released, especially the IntDCT, have made the video and image coding more efficient because these transforms are comparable to the performance of their original transform and inherit its properties such as linearity, orthogonality, energy compaction, efficiency, etc. The advantage is that the integer transforms are faster and can be implemented in a reconfigurable hardware using fixed point arithmetic.

Several methods to construct the integer DCT matrix have been described. At first, examples of construction of several matrices utilized in different video coders such as H.264, WMV9, AVS1, H265 and AVS2. Then, we described important strategies such as kernel replacement, lifting steps and lifting matrices, square wave transform and round-off approximation to approximate the integer DCT. We described the recent Tchebishef method that is a natural strategy to construct integer matrices close to the DCT with the disadvantage that method yields a high dynamical range. Finally, we presented the metrics to evaluate a transform and described the recent architectures to implement the IntDCT, some of them are useful for FVC.

PROBLEMS

6.1. Write a MATLAB code to compare the coding gain C_g of the T_4 transform matrices (6.2), (6.6), (6.17) and (6.21) also compare the coding gain of the transform matrices (6.20) and (6.24) for a first-order Gaussian–Markov process with zero-mean, unit variance and correlation coefficient $\rho = 0.90$ and 0.95.

6.2. Write a MATLAB code to compare the MSE of the T_4 transform matrices (6.2), (6.6), (6.17) and (6.21) for a first-order Gaussian–Markov process with zero-mean, unit variance and correlation coefficient $\rho = 0.90$ and 0.95.

6.3. Write a MATLAB code to compare the efficiency η of the T_4 transform matrices (6.2), (6.6), (6.17) and (6.21) for a first-order Gaussian-Markov process with zero-mean, unit variance and correlation coefficient $\rho = 0.90$ and 0.95.

6.4. Evaluate the transform T_4 defined by (6.2) and calculate the coefficients of the following 4×4 block of pixels,

$$\begin{bmatrix} 166 & 60 & 57 & 52 \\ 177 & 128 & 65 & 54 \\ 198 & 156 & 84 & 54 \\ 196 & 194 & 128 & 74 \end{bmatrix}.$$

6.5. Repeat problem 6.4. for the transform defined by (6.6).

6.6. Repeat problem 6.4. for the transform defined by (6.17).

6.7. Repeat problem 6.4. for the transform defined by (6.21).

6.8. Derive a 4×4 ICT matrix by kernel component replacement.

6.9. Plane rotations and permuted plane rotation matrices are shown in (6.125) and (6.126) respectively.

$$\begin{bmatrix} y_0 \\ y_1 \end{bmatrix} = \begin{bmatrix} \cos\theta & \sin\theta \\ -\sin\theta & \cos\theta \end{bmatrix} \begin{bmatrix} x_0 \\ x_1 \end{bmatrix}, \tag{6.125}$$

$$\begin{bmatrix} y_1 \\ y_0 \end{bmatrix} = \begin{bmatrix} -\sin\theta & \cos\theta \\ \cos\theta & \sin\theta \end{bmatrix} \begin{bmatrix} x_0 \\ x_1 \end{bmatrix}. \tag{6.126}$$

Derive the scaled lifting structure as in (6.32) and draw the flowgraphs for both matrices.

6.10. Derive lifting structures of problem 6.9. for a rotation angle of $\theta = \pi/4$, using plane rotations and approximate the matrices using dyadic-rational coefficients of the form $a/2^b$ where a and b are integers.

6.11. Derive lifting structures of problem 6.9. for rotation angles of $\theta = 3\pi/8$ and $7\pi/16$, using permuted plane rotations and approximate using dyadic-rational coefficients of the form $a/2^b$ where a and b are integers.

6.12. Chen, Smith and Fralick [73] proposed a fast algorithm for computation of the DCT. Replace the plane rotation in their flowgraph using plane rotation and permuted plane rotations where possible. Replace the resulting coefficients using dyadic-rational numbers and compare with the implementation proposed in [282].

6.13. Loeffler, Lightenberg and Moschytz [290] proposed a fast algorithm for the computation of the DCT with 11 multiplications. Replace the plane rotation in their flowgraph using plane rotation and permuted plane rotations where possible. Replace the resulting coefficients using dyadic-rational numbers and compare with the implementation proposed in [282].

6.14. Using factorization (6.49) develop a forward ICT matrix for $\left[C_4^{II}\right]$ for kernel (6.48).

6.15. Derive the inverse ICT matrix $\left[C_4^{II}\right]^{-1}$ of problem 6.14.

6.16. Repeat problem 6.14. to develop an ICT matrix $\left[C_8^{II}\right]$.

6.17. Derive the inverse ICT matrix $\left[C_8^{II}\right]^{-1}$ of problem 6.16.

6.18. We know that the inverse of DCT-II is DCT-III (see (3.34)). Using (6.57) develop a forward $\left[\bar{C}_4^{III}\right]$ ICT matrix and verify the orthogonality with respect to the $\left[\bar{C}_4^{II}\right]$ ICT matrix.

6.19. The kernel of DCT-IV (ignoring the constant factor) in (3.20) can be expressed in terms of the DCT-II by $\left[C_N^{IV}\right] = U_N C_N^{II} E_N F_N$ and approximated by $\left[\bar{C}_N^{IV}\right] = U_N \bar{C}_N^{II} \bar{E}_N \bar{F}_N$, compute the coefficients of a forward and inverse integer DCT-IV matrix DCT-IV of length $N = 4$ knowing that $\left[\bar{C}_N^{IV}\right]^{-1} = \bar{F}_N^{-1} \bar{E}_N^{-1} \left[\bar{C}_N^{II}\right]^{-1} U_N^{-1}$.

6.20. Using the MATLAB code of §6 and (6.1) generate a DTT of length $N = 4$ for $\alpha = 2$ and compare the parameters C_g and η with those obtained in (6.110).

6.21. Using the MATLAB code of §6 and (6.1) generate a DTT of length $N = 8$ for $\alpha = 4$ and compare the parameters C_g and η with those obtained in (6.111).

6.22. Repeat problem 6.21. for $\alpha = 2$ and $\alpha = 1.5$.

6.23. Draw the flowgraphs of matrices generated in problems 6.20., 6.21. and 6.22.

6.24. Verify the deviation from orthogonality of matrices generated in problems 6.20., 6.21. and 6.22. using Frobenius norm (6.109).

6.25. Verify the deviation from orthogonality of matrices 6.110 and 6.111 using Frobenius norm (6.109).

6.26. In Chapter 6 of [471], Madhukar, Fuldseth and Bjontegaard [296] describe the 4×4, 8×8, 16×16 and 32×32 IntDCTs including the embedded process (small size transforms are embedded in large size transforms). Review the references listed at the end of this chapter. The inverse integer DCTs are transposes of their corresponding integer DCTs. Compute the orthogonal property of these integer DCTs. (Hint: Matrix multiply the integer DCTs with their corresponding transposes).

PROJECTS

P.6.1. In [533] Section 7.3.3, the DCT-based interpolation basic steps for a one-dimensional interpolation operation are explained. Derive the interpolation filter (IF) coefficients independently and confirm that they match with the coefficients listed in Tables 5.4 and 5.5. Plot the corresponding transfer functions (see Figs. shown in 7.16 and 7.17, pages 201–202 of this paper).

P.6.2. Sjövall [440] et al. have developed a high level synthesis implementation of HEVC 2D DCT/DST (8, 16 and 32-point DCT units and 4-point DCT/DST unit) on FPGA. Develop this architecture and verify the results shown in Tables 1 and 2.

P.6.3. Extend project P.6.2. to extend the high level synthesis to 2D 64-point DCT.

P.6.4. In [360] the authors have developed low-complexity integer approximation of DTT for sizes 4 and 8. Extend this to 16 and 32.

P.6.5. In [582] a measure was developed for focus and blurriness using reorganized DCT coefficients with application for autofocus. Use this metric for edge image enhancement as the authors suggest in the conclusion section of the paper.

P.6.6. Develop a hardware architecture for 4-point DTT using FPGAs.

P.6.7. Repeat project P.6.6. for 8-point DTT using FPGAs.

P.6.8. Explore the possibility to develop a reconfigurable architecture for 4 and 8-point DTT using simulation and synthesis tools for FPGAs.

P.6.9. Wang et al [515] have applied multi-scale DCT in visual recovery. They also state that the proposed DCT regularizer can be independently embedded into most image processing tasks, e.g., image in painting and image denoising. Implement this DCT regularizer and confirm the results shown in this paper.

P.6.10. Repeat project P.6.9. using IntDCT.

P.6.11. Jridi, Alfalou and Meher [238] developed a reconfigurable architecture to approximate the DCT of lengths $N = 8, 16$ and 32. Develop a C/C++ code to implement this architecture using graphics processing units (GPUs) and compare performance with those of sections 6 and 6.

P.6.12. Meher, Park and Mohanty [322] developed a reusable architecture of integer DCT (lengths $N = 8, 16$ and 32). Develop a C/C++ code to implement this architecture using GPUs.

P.6.13. Park and Meher [368] proposed three flexible architectures for the implementation of IntDCT of different lengths to be used in HEVC. Develop a C/C++ code to implement the architectures using GPUs.

P.6.14. A method to derive a recursive order 2^N ICT was developed in [135]. In this work, transforms up to order-32 are derived and implemented in HEVC. Represent the flowgraph of Figure 1 by lifting matrices where possible and compare performance in time with respect to the original proposal.

P.6.15. Repeat project P.6.14. by using Givens rotations and lifting matrices where possible and compare performance in time with respect to the original proposal.

P.6.16. The LCICT was developed in [78]. Use Givens rotations and lifting scheme to represent the signal flowgraph of Figure 3 and compare the performance in time with the original proposal.

P.6.17. Use a hardware description language to describe the structure of implementation in project P.6.14.

P.6.18. Use a hardware description language (HDL) to describe the structure of implementation in project P.6.16.

P.6.19. The JPEG XT suite of standards described in the paper [407] can lead to several projects. Review this paper in detail and implement these standards. Please access the reference software (see [401], [405]).

P.6.20. JPEG Privacy & Security: With social media having a huge impact on every individual, it disrupts their privacy. Hence, protecting the privacy and security is becoming very important, not only because of the social media, but also because images/ meta-data use the cloud for storage purposes over private repositories. Encrypting the images and providing access to only the authorized person can help in maintaining the privacy of the image/ meta-data. Implement a privacy and security scheme for the JPEG XT and the JPEG XR formats.

P.6.21. Recording Image Editing Operations: While editing any image or text, only the original data and the final output after editing is usually saved. But instead, if every intermediate step is also recorded, it gives a possibility to revert the data to the original image or any of the intermediate steps. Saving every step while developing a special effect (sharpen, emboss, smooth) can help to develop new, faster and easier effects. Develop and image editing process for the JPEG XT and the JPEG XR formats.

P.6.22. In the conclusion section the authors [19] state: The encoding or decoding hardware can be in fact designed based on a pair of existing JPEG coding chips, as shown in Figure 2, resulting in a minimal hardware change in the existing hardware infrastructure without influencing its real-time performances. Implement the encoding and decoding hardware.

P.6.23. See [424]. As part of JPEG XT future extensions, JPEG privacy and security, Shah has developed an algorithm to encrypt and decrypt the images using C++ and OpenCV libraries. This algorithm performs well for protecting an individual's identity by blurring the image. By resolving the compatibility issues between the OpenCV libraries and JPEG XT code this algorithm can be incorporated in the JPEG XT code to provide the security feature automatically. Explore this.

P.6.24. See P.6.23. Shah [424] suggests that the amount of blurring can be changed according to the user preference by changing the kernel size. This gives the user full control over how the image should be seen by others. Explore this in detail and draw some conclusions.

P.6.25. See P.6.23. Several other tasks also are proposed as future work in this thesis. Investigate and implement these tasks.

P.6.26. See [403], [1] and [404]. Richter has proposed some optimization techniques to improve the legacy JPEG in terms of rate distortion performance close to JPEG XT (See Fig. 5 in [403]). Using the software described in [1] and [404] confirm the enhanced JPEG rate distortion performance shown in [403].

P.6.27. JPEG XT standard defines a normative decoding procedure to reconstruct an HDR image from two JPEG regular code streams named the base layer (visible to legacy decoders) and an extension layer and how to merge them together to form one single image. See also [407].The standard does not however, define the encoding procedure and leaves large freedoms to the encoder for defining the necessary decoder configuration. Mantiuk, Richter and Artusi [312], explored the whole space of possible configurations to achieve the best possible R-D performance. Review this paper in detail and verify the performance results described in the figures [312].

P.6.28. See [407], [401] and [402]. The JPEG XT encoder (review the encoder/decoder block diagrams) is both forward and backward compatible with legacy JPEG encoder. See references on JPEG. Verify the forward compatible characteristic of JPEG XT. This implies that the JPEG XT decoder should be able to decode and display the images based on the JPEG coded bit streams.

P.6.29. See [403]. Verify the backward compatible characteristic of JPEG XT. This implies that legacy JPEG decoder should be able to extract part of the JPEG XT coded bit stream, decode and display the JPEG images.

P.6.30. Review the papers related to JPEG XT. Implement JPEG XT and evaluate its performance.

P.6.31. See [528]. Watanabe, Kobayashi and Kiya have proposed a two-layer coding method using the histogram packing technique with the backward compatibility to the legacy JJPEG. The authors claim that this method has a better lossless compression performance than that of the JPEG XT. Using the test images verify this claim?

P.6.32. See [528]. The authors claim improved performance of lossless compression using histogram packing technique. Go thru the various references listed and verify this claim.

P.6.33. In Figure 2 of [73] a flowgraph for the fast DCT is presented for up to $N = 32$. The algorithm may be extended to any desired value $2^m \geq 2$ where m is an integer. Expand the flowgraph for $N = 64$.

P.6.34. In Figure 2 of [73] a flowgraph for the fast DCT is presented for up to $N = 32$. The algorithm may be extended to any desired value $2^m \geq 2$ where m is an integer. Expand the flowgraph for $N = 128$.

P.6.35. Use lifting matrices to change the flowgraphs of P.6.33. and P.6.34. into an integer representation.

P.6.36. Using a HDL describe and synthesize the flowgraphs of P.6.35.

P.6.37. In [246] a hybrid video codec superior to H.264/AVC codec was proposed by adding additional features such as extended block sizes (up to 64×64), mode dependent directional transforms (MDDT) for intra coding, luma and chroma high precision filtering, geometric motion partitions, adaptive coefficient scanning, extended block size partition, adaptive loop filtering, large size integer transform, etc. By using several test sequences at different spatial resolutions, they have shown that the proposed codec outperforms the traditional H.264/AVC codec in terms of both subjective quality and objective metrics. Also, the proposed decoder requires only moderate increase in complexity of the encoder and decoder. Implement this new codec to obtain similar results to those described in the paper, include the SSIM in all simulations. Use the latest HM software for H.264/AVC.

P.6.38. In [236] the authors developed numerically stable, recursive, orthogonal, scaled integer transforms. They also embedded these transforms in HEVC/JMKTA framework. Develop 16-point, 32-point and 64-point scaled transforms and test them in the HEVC JMKTA software.

P.6.39. In [109] a new design for fast MDDT through integer lifting steps was developed. The scheme reduces the complexity with negligible loss of coding performance. Develop the fast MDDT using integer lifting steps for 4×4 and 8×8 and compare the performance with the DCT.

P.6.40. In [418] the authors compared TMuC 0.7 between the proposed adaptive DCT/DST as the primary transform and the DCT in intra prediction for 16×16, 32×32 and 64×64 block sizes for two cases, i.e., secondary transform (ROT) is off or on. Implement this scheme and verify the result shown in Tables 2 and 3 of this document. Use TMuC 0.7.

P.6.41. In [294] the authors carried out a detailed analysis of fractional pel interpolation filters both in H.264/AVC and in HEVC. Also they conclude that the interpolation filters contribute 10 % performance gain to HEVC compared to H.264/AVC at the cost of increased complexity. They also demonstrated that the frequency response of quarter pel filters in HEVC are superior to those in H.264/AVC. Their conclusions are based on reference software HM 5.2 using low resolution test sequences (416×240). Carry out this analysis using HM 15.0 software and high resolution test sequences.

P.6.42. In project P.6.41. go thru the analysis of how the discrete cosine transform interpolation filters (DCTIF) have been developed [294]. By plotting the frequency responses of filters in H.264/AVC and in HEVC, they showed that the pass band filters in HEVC are much flatter and have smaller ripples than those in H.264/AVC (see Figure 5 in [377]). Confirm these conclusions.

P.6.43. In [129] authors developed an efficient hardware design for implementing multiple forward and inverse transforms for various video coding standards. Explore the possibility to extend this to HEVC. Investigate this in detail. Note that HEVC 16×16 and 32×32 IntDCT are also valid besides 4×4 and 8×8 IntDCTs.

P.6.44. In [366] authors developed a 2D 16×16 and 32×32 inverse IntDCT architecture that can process 4K@30 fps video. Extend this architecture for 2D inverse 64×64 IntDCT that can process 8K@30 fps video.s. They discuss the advantages of this DCT in terms of complexity, energy efficiency, compression performance, etc. Confirm the results as shown in Figures 8 and 9 and develop the approximate DCT for length 64.

P.6.45. In [239] authors derived an approximate kernel for the DCT of length 4 from that defined in HEVC and used this for computation of DCT and its inverse of power-of-2 length. Confirm these results.

P.6.46. Performance in joint exploration test model (JEM) has high coding efficiency [267]. However, there is still industry need to further reduce the video bandwidth. In [267] authors proposed a variable block-sized signal-dependent transform (SDT) for HEVC. The proposed model has been adopted into the JEM in 2015 for the exploration of potential future video coding standard. Authors claim that their fast algorithm achieved a Bjøntegaard delta rate (BD-rate) savings of 2.2%, 2.4%, 3.3%, and 7.1% under AI-Main10, RA-Main10, RA-Main10, and LP-Main10 configurations, respectively, compared with the test model HM-12 of HEVC. They also claim that the fast algorithm makes it feasible to utilize large block size SDTs. But the further exploration is left as future work. Explore this future work.

P.6.47. Repeat project P.6.46. using JEM-4.0.

P.6.48. In Figure 1 of [267] structure of modes in the proposed scheme at the TU stage for sizes of 4×4, 8×8, 16×16 and 32×32 is shown. Extend this to 64×64 (proposed in JEM. See reference 4 cited in this paper.)

P.6.49. In Section V of [267] Experimental results for SDT modes for various test sequences are based on luma only. Extend these results for SDT modes for all the test sequences based on all color components.

P.6.50. In [578] the residual characteristics of intra prediction and the deficiency of DCT/DST for residual blocks are investigated. The authors state that "In the future, fast algorithms, such as early skip for SDT-SVD transform and fast template searching method, can be exploited to reduce the complexity of the proposed method." Explore this.

P.6.51. See P.6.50. The authors further state "To further improve the efficiency of the SDT-SVD transform, more work can be done to design an efficient block with more similar structures as the residual block for deriving transforms." Explore this.

P.6.52. In [360] the authors have implemented a video coding based on a low-complexity approximation of DTT and compared with H.264/AVC. Replace the IntDCT in HEVC [462] by the integer DTT (IntDTT) and compare its performance with the standard HEVC in video coding. Develop graphs similar to Figures 1 and 2 in [358].

P.6.53. The rate-distortion graphs shown in Figures 1 and 2 in [358] are for 1280×720 60 Hz and 1920×1080 24 video test sequences and class E and B (see Table VIII in [358]). Extend the performance comparison (IntDTT and IntDCT) based on video test sequences - classes A, C, D, and E listed in Table VIII.

P.6.54. In [264] authors provided an overview of intra coding techniques in HEVC and compared its performance with H.264/AVC intra coding. Replace IntDTT by IntDCT in HEVC intra coding and compare its performance with

H.264/AVC intra coding. Develop tables similar to those in [264] using test sequences of Table IX.

P.6.55. In [548] authors provided an overview of the tools and coding performance associated with the HEVC screen content coding (SCC) extension. In this article, Figure 1 shows the encoder block diagram. Replace the blocks "Transform & Quant" and "De-quant & Inv. Transform" by the blocks "IntDTT & Quant" and "De-quant & Inverse IntDTT" respectively and compare results with [548]. Based on these results what conclusions can be drawn?.

P.6.56. In [277] a content-adaptive joint image compression and encryption scheme is proposed. The encryption operations include three parts: new orthogonal 8×8 transforms transformation, DC coefficients encryption, and AC encryption. The authors generate new transforms by introducing rotation angles into order-8 DCTs flow-graph structure, and apply them alternatively for 8×8 blocks transformation, controlled by the encryption key. DC coefficients are encrypted after quantization by 8×8 blocks' permutation and XOR operation. For AC coefficients encryption, we embed the 256-bit encryption key into the bitstream of some certain AC coefficients, and the data embedding/extracting operation is controlled by another secret key, called the embedding key. In the conclusion section several extensions to this approach are suggested. Explore these.

7 The Directional Discrete Cosine Transform

INTRODUCTION

The 2D-DCT of a square or a rectangular block shape is used for almost all block-based transform schemes for image and video coding [20], [76], [389], [398], [462], [463]. The 2D-DCT can be separately implemented through two 1D transforms, one to process the block along the vertical direction and another along the horizontal direction. Conventional DCT seems to be the best choice for image blocks in which vertical and/or horizontal edges are dominating. However, this is not always the best scenario. When the conventional DCT is applied to an image block in which other directional edges dominate, the non-zero coefficients are not well aligned across different columns or rows and the 1D DCT may produce more non-zero coefficients. Therefore, the transform becomes inefficient especially for compression applications. The following example depicts the problem,

Example 7.1. *7.1. and 7.2. show two blocks of pixels of size* 4x4 *with a vertical and a horizontal dominant edge respectively. When the DCT is applied the edges are well represented by the transform coefficients producing many zero-coefficients. 7.3. shows a block of highly correlated pixels. In the DCT domain the block is represented with only one coefficient. The block of pixels of 7.4. form a diagonal dominant edge with direction down-left. The pixels are not well aligned across the rows neither the columns. Therefore, after DCT the information is not well represented by the transform coefficients.*

$$7.1.\quad \begin{bmatrix} 10 & 10 & 0 & 0 \\ 10 & 10 & 0 & 0 \\ 10 & 10 & 0 & 0 \\ 10 & 10 & 0 & 0 \end{bmatrix} \xrightarrow{[DCT]} \begin{bmatrix} 20.00 & 18.47 & 0.00 & -7.65 \\ 0 & 0 & 0 & 0 \\ 0 & 0 & 0 & 0 \\ 0 & 0 & 0 & 0 \end{bmatrix}$$

$$7.2.\quad \begin{bmatrix} 10 & 10 & 10 & 10 \\ 10 & 10 & 10 & 10 \\ 0 & 0 & 0 & 0 \\ 0 & 0 & 0 & 0 \end{bmatrix} \xrightarrow{[DCT]} \begin{bmatrix} 20.00 & 0 & 0 & 0 \\ 18.47 & 0 & 0 & 0 \\ 0.00 & 0 & 0 & 0 \\ -7.65 & 0 & 0 & 0 \end{bmatrix}$$

$$7.3.\quad \begin{bmatrix} 10 & 10 & 10 & 10 \\ 10 & 10 & 10 & 10 \\ 10 & 10 & 10 & 10 \\ 10 & 10 & 10 & 10 \end{bmatrix} \xrightarrow{[DCT]} \begin{bmatrix} 40.00 & 0 & 0 & 0 \\ 0 & 0 & 0 & 0 \\ 0 & 0 & 0 & 0 \\ 0 & 0 & 0 & 0 \end{bmatrix}$$

7.4. $$\begin{bmatrix} 10 & 10 & 10 & 10 \\ 10 & 10 & 10 & 0 \\ 10 & 10 & 0 & 0 \\ 10 & 0 & 0 & 0 \end{bmatrix} \xrightarrow{[DCT]} \begin{bmatrix} 25.00 & 11.15 & 0.00 & 0.79 \\ 11.15 & -5.00 & -4.61 & 0.00 \\ 0.00 & -4.61 & 5.00 & 1.91 \\ 0.79 & 0.00 & 1.91 & -5.00 \end{bmatrix}$$

This chapter explains the directional DCT (DDCT) [570] and the steerable DCT [138]. Both transforms can be oriented according to the characteristics of the block. The orientation of the DDCT is limited by a number of modes while the steerable DCT can be oriented in almost any desired angle. Nevertheless, the steerable DCT is more complex computationally.

THE DIRECTIONAL DCT

In Example 7.1, the conventional DCT is not very efficient to represent blocks of samples with dominant edges which are not vertical or horizontal. However, a new block-based DCT framework, called directional discrete cosine transform, in which the first transform may choose to follow a direction other than the vertical or horizontal one is utilized to deal with this problem [143], [165], [198], [398], [569], [570].

The direction of transform is chosen according to the dominating edge(s) within each individual image block. Different directional transforms can be obtained. Coefficients produced by all directional transforms are arranged appropriately; a second transformation can be applied to the coefficients that are best aligned with one another [66], [339], [547], [569], [570].

Example 7.2. *Suppose that we want to represent the block of samples using the DCT, with a diagonal dominant edge with direction down-left shown next,*

$$x = \begin{bmatrix} 10 & 10 & 10 & 10 \\ 10 & 10 & 10 & 2 \\ 10 & 10 & 2 & 2 \\ 10 & 2 & 2 & 2 \end{bmatrix}$$

Notice that there are two areas (triangles) with similar values (10 and 2). Therefore, each side of the edge is built of similar pixels. This can be used to transform separately into two similar regions. Still, we must modify the way we apply DCT a bit, since the kernel only allows rectangular regions. Fortunately, the 2D DCT is separable. So that, we can use 1D-transforms of different sizes in two simple steps. For example, using the following arrangement for the 4×4 block,

n →

m ↓				
	$x_{0,0}$	$x_{0,1}$	$x_{0,2}$	$x_{0,3}$
	$x_{1,0}$	$x_{1,1}$	$x_{1,2}$	$x_{1,3}$
	$x_{2,0}$	$x_{2,1}$	$x_{2,2}$	$x_{2,3}$
	$x_{3,0}$	$x_{3,1}$	$x_{3,2}$	$x_{3,3}$

In a first step, we can transform the block x by applying the 1D DCT in the direction of the dominant edge (down-left) as follows.

Sample positions	Samples in the direction of the dominant edge	1D DCT			
DCT($\{x_{0,0}\}$; 1)	DCT(10;1)	10.00			
DCT($\{x_{0,1}, x_{1,0}\}$; 2)	DCT(10,10;2)	14.14	0.00		
DCT($\{x_{0,2}, x_{1,1}, x_{2,0}\}$;3)	DCT(10,10 10;3)	17.32	0.00	0.00	
DCT($\{x_{0,3}, x_{1,2}, x_{2,1}, x_{3,0}\}$;4)	DCT(10,10,10,10;4)	20.00	0.00	0.00	0.00
DCT($\{x_{1,3}, x_{2,2}, x_{3,2}\}$;3)	DCT(2,2,2;3)	3.46	0.00	0.00	
DCT($\{x_{2,3}, x_{3,2}\}$;2)	DCT(2,2;2)	2.82	0.00		
DCT($\{x_{3,3}\}$;1)	DCT(2;1)	2.00			
		DC	*AC*1	*AC*2	*AC*3

The number after semicolon means the size of the DCT. The results are column vectors with the DC component placed at left. In a second step, we apply the 1D DCT to the columns of the 1D DCT from previous results, because the columns contain all DC components and each other column contains all AC components with the same index.

DCT(10.00, 14.14, 17.32, 20.00, 3.46, 2.82, 2.00 ;7)	26.36	12.11	-9.81	-6.50	4.04	1.75	-4.90
DCT(0.00, 0.00, 0.00, 0.00, 0.00; 5)	0.00	0.00	0.00	0.00	0.00		
DCT(0.00, 0.00, 0.00; 3)	0.00	0.00	0.00				
DCT(0.00; 1)	0.00						

Notice that the final representation of the dominant edge, in the transform domain, contains fewer non-zero coefficients than the diagonal dominant edge of Example 7.1 7.4. and the original block can be recovered by inverting the process. Also, an important issue is to find the dominant edge in a block.

Figure 7.1 shows several directional modes for a 2D (8×8) DDCT [570], for different dominant edges. In 7.2 we used the directional mode 3.

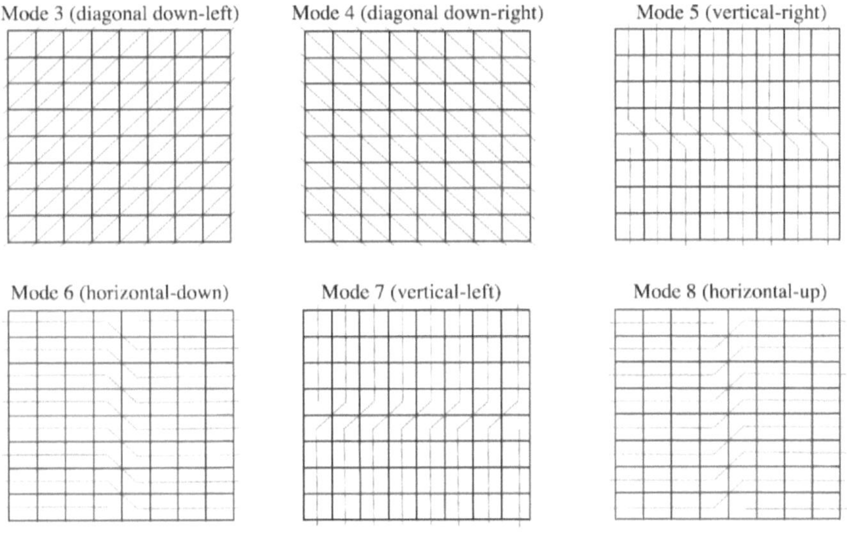

Figure 7.1: Six directional modes for the block size 8×8. (The vertical and horizontal modes are not included here) [570]

Procedure 1. We can summarize the process to calculate the DDCT for Mode-3, using Example 7.2, in the following steps [547],

Step 1: $\{x_{0,0}, x_{0,1}, \cdots, x_{3,3}\}$ are the pixels in the 2D spatial domain.

Step 2: Perform 1D-DCT for the $N \times N$ block in diagonal down-left position (Mode 3) with lengths $L = 1, 2, \cdots N, N-1, N-2, \cdots 2, 1$ in our example $N = 4$ then, $L = 1, 2, 3, 4, 3, 2, 1$ and we obtain the coefficients $\{A\}$, $\{B,C\}$, $\{D,E,F\}$, $\{G,H,I,J\}$, $\{K,L,M\}$, $\{N,O\}$, $\{P\}$ for each 1D-transform.

Step 3: Rearrange the coefficients after 1D DCT as

$$\begin{array}{ccccccccl} A & B & C & D & H & L & P & & \rightarrow 7 \text{ coefficients} \\ & E & F & G & K & O & & & \rightarrow 5 \text{ coefficients} \\ & & I & J & N & & & & \rightarrow 3 \text{ coefficients} \\ & & & M & & & & & \rightarrow 1 \text{ coefficient} \end{array}$$

Step 4: Apply horizontal 1D-DCT for lengths $L = N*2-1, N*2-3, \cdots 3, 1$. In our example the lengths were $L = 7, 5, 3, 1$. Then, arrange coefficients in the same pattern to obtain the following coefficient arrangement.

$$\begin{array}{ccccccc} \hat{A} & \hat{B} & \hat{C} & \hat{D} & \hat{H} & \hat{L} & \hat{P} \\ & \hat{E} & \hat{F} & \hat{G} & \hat{K} & \hat{O} & \\ & & \hat{I} & \hat{J} & \hat{N} & & \\ & & & \hat{M} & & & \end{array}$$

Step 5: Move all 2D (4×4) DDCT coefficients to the left.

$$\begin{array}{ccccccc} \hat{A} & \hat{B} & \hat{C} & \hat{D} & \hat{H} & \hat{L} & \hat{P} \\ \hat{E} & \hat{F} & \hat{G} & \hat{K} & \hat{O} & & \\ \hat{I} & \hat{J} & \hat{N} & & & & \\ \hat{M} & & & & & & \end{array}$$

Therefore, we can implement quantization followed by 2D variable length coding (VLC) for compression/coding applications along the zigzag scan. This scanning helps to increase the runlength of zero (transform) coefficients leading to reduced bit rate in the 2D-VLC coding (similar to JPEG [501]).

The next two MATLAB functions are examples on how to calculate an $N \times N$ forward and inverse DDCT Mode-3 respectively, where $N = 4, 8, 16, 32, 64$.

```
%%% Forward DDCT Mode 3
function Y = DDCTmode3(Im)
% Im is the MxM (M = 4, 8, 16, 32) block od pixels
% Y is the M x 2*N-1 block of 2D forward DDCT Mode-3
[M,N] = size(Im);
X = zeros(M, 2*N-1);
k = 0;
for i = 1:M
```

```
    X(i:i,i:N) = Im(i:i, 1:M-i+1);
    if(i < M)
        X(i:i,N+1:end-k) = Im(i+1:M, M-k:M-k)';
    end
    k = k+1;
end
% 1D DCT along the cols of X
k=1;
for i = 2: 2*M-2
    if(i > M)
        X(1:M-k,i:i) = dct(X(1:M-k,i:i),M-k);
        k = k+1;
    else
        X(1:i, i:i) = dct(X(1:i, i:i),i);
    end
end
Y = zeros(size(X));
% 1D DCT along the rows of X
for i = 1:M
    Y(i:i, 1:2*(M-i)+1)=dct(X(i:i,i:2*M-i), (2*(M-i)+1));
end

%%% Inverse DDCT Mode 3
function x = InvDDCTmode3(Y)

% Y is the M x N block of 2D-DDCT coefficients obtained by
  DDCT Mode-3
% x is the M x M block of pixels after 2D-Inverse DDCT

[M,N] = size(Y);
X = zeros(M,N);
x = zeros(M,M);

% 1D Inverse DCT on each row of Y
for i = 1:M
    X(i:i,i:end-i+1)= idct(Y(i:i, 1:end-2*(i-1)), (2*(M-i)+1));
end

% 1D Inverse DCT on each colum of X to obtain the pixel samples
k = 1;
for i= 2 : N-1
    if (i <= (N+1)/2)
        X(1:i, i:i) = idct( X(1:i, i:i), i);
    else
        X(1:M-k, i:i) = idct( X(1:M-k, i:i), M-k);
        k = k+1;
    end
end
```

```
% Align pixels
k = (N+1)/2 + 1;
j=1;
for i = k: N-1
    X(j+1:M, i:i) = X(1:M-j, i:i)
    j = j+1;
end

% Recover the M x M block of samples
for i =1:M
   x(i:i,1:M) = X(i:i, i:M+i -1);
end
x(M,M) = X(1,2*M-1);
```

Example 7.3. *Suppose that we want to represent the following block of samples using the DCT*

$$x = \begin{bmatrix} 10 & 12 & 14 & 6 \\ 10 & 12 & 14 & 6 \\ 8 & 10 & 12 & 14 \\ 8 & 10 & 12 & 14 \end{bmatrix} \xrightarrow{[DCT]} \begin{bmatrix} 43.00 & -2.38 & -5.00 & \\ -0.92 & 6.03 & -4.61 & 2.50 \\ 0.00 & 0.00 & 0.00 & 0.00 \\ 0.38 & -2.50 & 1.91 & -1.03 \end{bmatrix}$$

When we apply the forward DCT to the block of samples, the result is not sparse. However, according to samples distribution inside the block, if we use Mode-5 of Figure 7.1 we can take advantage of the pattern the samples describe. First, we apply 1D-DCT in the direction of the dominant edge (vertical-right). For the visual clarity, the samples can be rearranged as,

$$\begin{bmatrix} 8 & 10 & 12 & 14 & 6 \\ 8 & 10 & 12 & 14 & 6 \\ & 10 & 12 & 14 & \\ & 10 & 12 & 14 & \end{bmatrix}.$$

The samples' first and last columns can be transformed using 1D-DCT of length $N = 2$ and the central columns using 1D-DCT of length $N = 4$ to obtain the following result,

	1D-DCT			
DCT(8,8;2)	11.31	0.00		
DCT(10,10,10,10;4)	20.00	0.00	0.00	0.00
DCT(12,12,12,12;4)	24.00	0.00	0.00	0.00
DCT(14,14,14,14;4)	28.00	0.00	0.00	0.00
DCT(6,6;2)	8.48	0.00		
	DC	*AC1*	*AC2*	*AC3*

The number after semicolon means the size of the DCT. The result are column vectors with the DC component placed at left, followed by the first AC component and so on. In a second step, we apply the 1D DCT to the columns of the 1D DCT from previous result, because the columns contain all DC components and each of other column contains all AC components with the same index.

	1D-DCT				
DCT(11.31, 20, 24, 28,8.48;5)	41.05	-1.27	-14.42	5.86	-5.51
DCT(0, 0, 0, 0, 0; 5)	0.00	0.00	0.00	0.00	0.00
DCT(0, 0, 0; 3)	0.00	0.00	0.00		
DCT(0, 0, 0; 3)	0.00	0.00	0.00		

Procedure 2. We can summarize the process to calculate the DDCT for Mode-5, using Example 7.3, in the following steps [547],

Step 1: $\{x_{0,0}, x_{0,1}, \cdots, x_{3,3}\}$ are the pixels or samples in the 2D spatial domain.

Step 2: Perform 1D-DCT for the $N \times N$ block in vertical-right position, as shown in Figure 7.1 Mode-5, with lengths $L = 2$ to the set of samples $\{x_{2,0}, x_{3,0}\}$ to yield the DCT coefficients I, M, $L = 4$ to the set of samples $\{x_{0,0}, x_{1,0}, x_{2,1}, x_{3,1}\}$ to produce the DCT coefficients A, E, J, N, $L = 4$ to the set of samples $\{x_{0,1}, x_{1,1}, x_{2,2}, x_{3,2}\}$ to yield the DCT coefficients B, F, K, O, $L = 4$ to the set of samples $\{x_{0,2}, x_{1,2}, x_{2,3}, x_{3,3}\}$ to obtain the DCT coefficients C, G, L, P and $L = 2$ to the set of samples $\{x_{0,3}, x_{1,3}\}$ to obtain the DCT coefficients D, H as shown next,

$$\begin{array}{cccc} A & B & C & D \\ E & F & G & H \\ I & J & K & L \\ M & N & O & P \end{array}$$

Step 3: Arrange the coefficients vertically in the same pattern as following,

$$\begin{array}{cccccl} I & A & B & C & D & \rightarrow 5 \text{ coefficients} \\ M & E & F & G & H & \rightarrow 5 \text{ coefficients} \\ & J & K & L & & \rightarrow 3 \text{ coefficients} \\ & N & O & P & & \rightarrow 3 \text{ coefficients} \end{array}$$

Step 4: Apply horizontal 1D DCT for lengths $L = 5, 5, 3$ and 3 and arrange in the same pattern to obtain,

$$\begin{array}{ccccc} \hat{I} & \hat{A} & \hat{B} & \hat{C} & \hat{D} \\ \hat{M} & \hat{E} & \hat{F} & \hat{G} & \hat{H} \\ & \hat{J} & \hat{K} & \hat{L} & \\ & \hat{N} & \hat{O} & \hat{P} & \end{array}$$

Step 5: Move all 2D (4×4) DDCT coefficients to the left. Here, we can implement quantization followed by 2D VLC for compression/coding zigzag scan. This scanning helps to increase the runlength of zero (transform) coefficients leading to reduced bit rate in the 2D-VLC coding (similar to JPEG [501]).

$$\begin{array}{ccccc} \hat{I} & \hat{A} & \hat{B} & \hat{C} & \hat{D} \\ \hat{M} & \hat{E} & \hat{F} & \hat{G} & \hat{H} \\ \hat{J} & \hat{K} & \hat{L} & & \\ \hat{N} & \hat{O} & \hat{P} & & \end{array}$$

The next two MATLAB functions are examples on how to calculate an $N \times N$ forward and inverse DDCT Mode-5 respectively, where $N = 4, 8, 16, 32, 64$.

```
%%% Forward DDCT Mode 5
function  Y = DDCTmode5(x)
%%%  x is the MxM (M = 4, 8, 16, 32) block of pixels
%%%  Y is the M x M+1 block of 2D forward DDCT Mode-3

[M, N] = size(x);

X = zeros(M,N+1);
X(1:M/2, 2:N+1) = x(1:M/2,1:N);
X(M/2+1:M,2:N) = x(M/2+1:M,2:N);
X(1:M/2,1:1) = x(M/2+1:M,1:1);

for i = 1 : N+1
    if ((i > 1) & (i < N+1))
            X(1:M, i:i) = dct(X(1:M, i:i));
    else
        X(1:M/2, i:i) = dct(X(1:M/2, i:i));
    end

end

Y = zeros(size(X));
for i=1:M
        if i > M/2
            Y(i,1:N-1) = dct(X(i,2:N));
        else
            Y(i,:) = dct(X(i,:));
        end
end

%%% Inverse DDCT Mode 5
function x = InvDDCTmode5(Y)
%%% Y is the M x N block of 2D-DDCT coefficients obtained by
%%% DDCT Mode-5
%%% x is the M x M block of pixels after 2D-Inverse DDCT

[M,N] = size(Y);

X = zeros(size(Y));
x = zeros(M,M);

for i = 1:M
    if(i > M/2)
        X(i:i,2:N-1) = idct(Y(i:i,1:N-2));
    else
        X(i:i,1:N) = idct(Y(i:i,1:N));
    end
end
```

```
for i = 1:N
    if((i > 1) & (i < N))
        X(1:M, i:i) = idct(X(1:M,i:i));
    else
        X(1:M/2,i:i) = idct(X(1:M/2,i:i));
    end
end
X(M/2+1:M,1:1)=X(1:M/2,1:1);
X(1:M/2,1:N-1)=X(1:M/2,2:N);
x= X(1:M, 1:M);
```

Extension to other modes is straightforward. For instance, for the diagonal down-right mode (Mode-4), we can simply flip it (either vertically or horizontally), and the flipped image block falls into the case of Mode-3. For Modes 6–8, we can simply flip or transpose the image block first and then the manipulated block falls into the case of Mode-5.

THE BASIS IMAGES FOR DDCT MODE 3

Figures 7.2a and 7.2b show the basis images for blocks of 4×4 and 8×8 respectively.

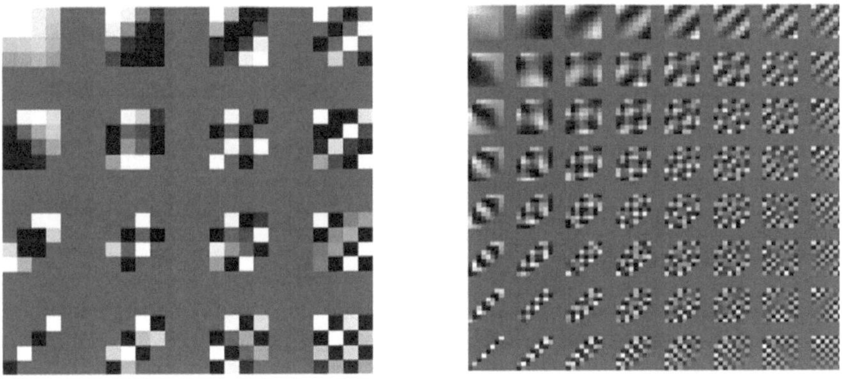

(a) Basis images for mode 3 and N=4. (b) Basis images for mode 3 and N=8.

Figure 7.2: Basis images of mode 3 (diagonal down-left) for block sizes (a) 4×4 and (b) 8×8.

For a 4×4 block and mode 3, the basis images are obtained as follows,

1. Suppose we have the following 4×4 block as:

$$\begin{matrix} A & B & C & D \\ E & F & G & H \\ I & J & K & L \\ M & N & O & P \end{matrix}$$

Here $N = 4$.

2. Set $A = 1$ and the rest to zero. Then rearrange the block as in the step 3 of process 1:

$$\begin{array}{ccccccccl}
A & 0 & 0 & 0 & 0 & 0 & 0 & & \rightarrow 7 \text{ coefficients} \\
 & 0 & 0 & 0 & 0 & 0 & & & \rightarrow 5 \text{ coefficients} \\
 & & 0 & 0 & 0 & & & & \rightarrow 3 \text{ coefficients} \\
 & & & 0 & & & & & \rightarrow 1 \text{ coefficient}
\end{array}$$

3. Apply horizontal 1D-DCT of lengths $L = 7,5,3,1$ to obtain:

$$\begin{array}{ccccccc}
\hat{A} & \hat{B} & \hat{C} & \hat{D} & \hat{H} & \hat{L} & \hat{P} \\
 & 0 & 0 & 0 & 0 & 0 & \\
 & & 0 & 0 & 0 & & \\
 & & & 0 & & &
\end{array}$$

4. Apply vertical 1D-DCT of lengths $L = 1,2,3,4,3,2,1$.

$$\begin{array}{ccccccc}
\bar{A} & \bar{B} & \bar{C} & \bar{D} & \bar{H} & \bar{L} & \bar{P} \\
 & \bar{E} & \bar{F} & \bar{G} & \bar{K} & \bar{O} & \\
 & & \bar{I} & \bar{J} & \bar{N} & & \\
 & & & \bar{M} & & &
\end{array}$$

5. Rearrange the block to obtain the basis image:

$$\begin{array}{cccc}
\bar{A} & \bar{B} & \bar{C} & \bar{D} \\
\bar{E} & \bar{F} & \bar{G} & \bar{H} \\
\bar{I} & \bar{J} & \bar{K} & \bar{L} \\
\bar{M} & \bar{N} & \bar{O} & \bar{P}
\end{array}$$

6. Go to step 1., set the next coefficient to one and the rest to zero.
7. Repeat steps 2. to 6. until obtaining the 16 basis images.

THE STEERABLE DCT

When block-based separable transforms are used, the resulting coefficients tend to represent inefficiently the block when it contains arbitrarily shaped discontinuities. DDCTs incorporate directional information but in a limited number of directions. In Section 7 we introduced the DDCT that consists in a separable 2D DCT. In a first step the 1D DCT may follow a direction other than the vertical or horizontal one. The resulting coefficients are rearranged and a second 1D DCT can be applied. However, one drawback of this method is that it requires 1D DCTs of various lengths and are not always a power of two. In some cases, the second DCT may not always be applied to coefficients of similar AC frequencies [557].

The steerable DCT (SDCT) [137], [138], [139] is a transform that can be steered in any choosen direction by rotating the 2D DCT basis by a single given angle for each image block. The transform allows to rotate in a flexible way pairs of basis vectors and enables precise matching of directionality in each image block. The SDCT relies on the graph theory and more particularly on the graph Fourier transform and the connection between this transform and the DCT.

The aim of using a transform matrix whose basis functions adapt to the orientation of the input signal is to obtain a more compact and accurate representation of the signal in the transform domain. The total energy remains unchanged, but one obtains a more sparse signal representation in each eigenspace. In the next subsections we present the basics of the graph theory and how the SDCT is obtained.

BASIC GRAPH THEORY

A graph (G) is formed by a set of vertices (V) and a set of edges (E). The edges ($e_i, \quad i \in \{0, 1 \cdots n\}$) are links that connect vertices ($v_k, \quad k \in \{0, 1 \cdots m\}$) as shown in Figure 7.3, this graphs follows a line topology. Notice that $V = \{(v_1, v_2)\}$ and $E = e_1 = \{(v_1, v_2) = (v_2, v_1)\}$. In this case, the edge connecting (v_1, v_2) is the same as the edge connecting (v_2, v_1). In other words, the pair is not ordered. Also, we say that vertex v_1 is adjacent of vertex v_2 and viceversa because there is an edge between both vertices.

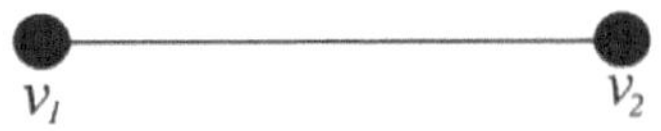

Figure 7.3: Example of a graph with line topology with two vertices (v_1), (v_2) connected by one edge e_1.

Figure 7.4 shows two graphs with four edges and four vertices each. The set of vertices is $V = \{v_1, v_2, v_3, v_4\}$. In the first graph the set of edges is $E = \{(v_1, v_2), (v_1, v_3), (v_2, v_4), (v_3, v_4)\}$ and in the second one is $E = \{(v_1, v_2), (v_1, v_3), (v_1, v_4), (v_3, v_4)\}$

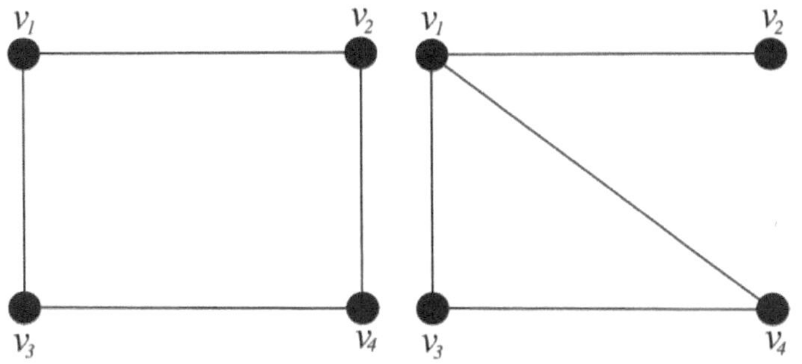

Figure 7.4: Two graphs with the same number of edges and vertices with different configurations.

If an edge is drawn from vertex to itself, it is called a loop. If there is only one edge connected to the vertex, the vertex is called pendent vertex and if no edges are formed at a vertex, the vertex is an isolated vertex. Figure 7.5 shows the loop at vertex v_2 with edge (v_2, v_2), the isolated vertex v_6 and the pendent vertex v_5.

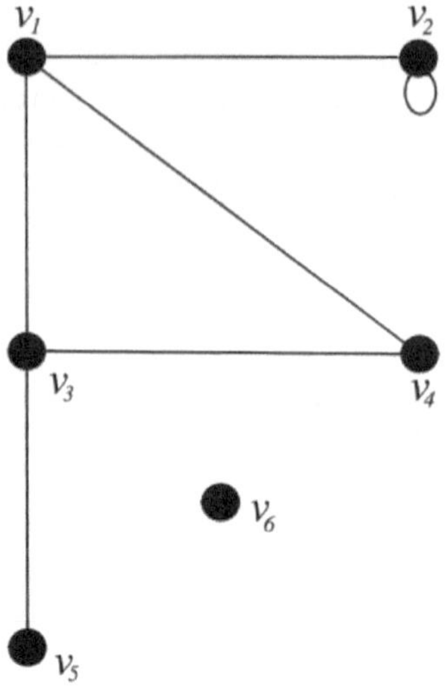

Figure 7.5: Graph with a loop in v_2, an isolated vertex v_6 and a pendent vertex v_5.

The graphs can be undirected and directed. An undirected graph has no directed edges. Figures 7.3 thru 7.5 are examples of undirected graphs. In directed graphs, edges have a specific direction. Figure 7.6 shows an example of a directed graph.

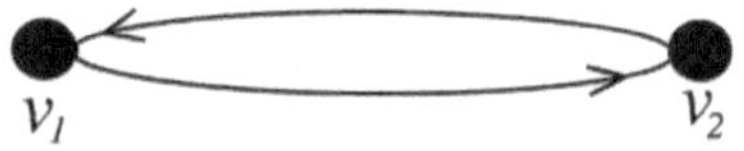

Figure 7.6: Example of a directed graph with two vertices v_1 and v_2 connected by two directed edges.

Another directed graph is shown in Figure 7.7 with four vertices. The arrows show the relation of a vertex with its adjacent. For example, v_1 is adjacent to v_2 while v_1 is not adjacent to v_3. There is however no edge from node v_2 to node v_1. Notice that there can be no more than two edges between any two vertices.

In a simple graph (finite graphs without loops or parallel edges) with n number of vertices, the degree of any vertex is $\deg(v) = n - 1 \quad \forall v \in G$. For example, in Figure 7.5, $\deg(v_1) = 3$, $\deg(v_3) = 3$, $\deg(v_5) = 1$ and $\deg(v_6) = 0$. In a directed graph, each vertex has an indegree (number of edges which are coming into the vertex) and an outdegree (number of edges which are going out from the vertex). Observe that in Figure 7.3 both vertices v_1 and v_2 are pendent vertices because their degree is one. If a pair of vertices are connected by more than one edge, edges are called parallel edges and the graph is known as multigraph. When the degrees of all vertices in a graph are arranged in descending or ascending order, then the sequence obtained is known as the degree sequence of the graph. A graph having no edges is called a null graph. A graph with only one vertex is called a trivial graph.

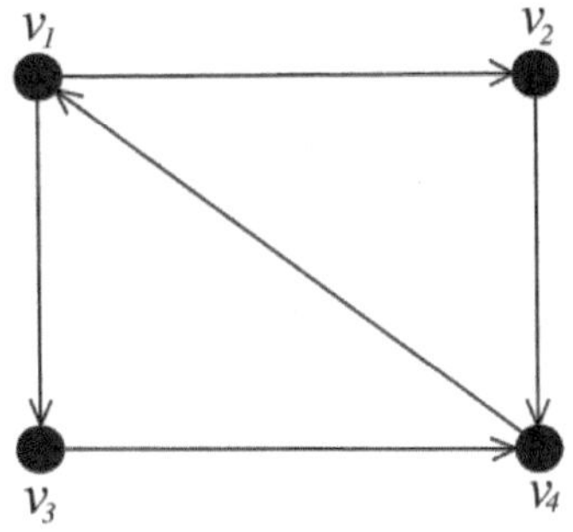

Figure 7.7: Example of a directed graph with four vertices v_1, v_2, v_3 and v_4.

ADJACENCY MATRIX

There is a strong relation between graphs and matrices. Suppose we are given a graph with n vertices. Then we construct an $n \times n$ adjacency matrix A associated to it with $|V| = N$. The adjacency matrix of a graph is a squared matrix that indicates whether pairs of vertices are adjacent or not in the graph. In other words, for an undirected graph, we define the adjacency matrix $A \in \mathbb{R}^{N \times N}$ where $[A]_{m,n} = 1$ if there is an edge between vertex m and vertex n and $[A]_{m,n} = 0$ otherwise. For example, the adjacency matrix of the graph represented in Figure 7.3 is shown in Table 7.1,

Table 7.1
Adjacency matrix corresponding to graph of Figure 7.3

	v_1	v_2
v_1	0	1
v_2	1	0

Zeros in the main diagonal mean that v_1 and v_2 do not have an edge drawn from itself (loops). Tables 7.2 and 7.3 show the adjacency matrices of graphs in Figure 7.3. In case of a directed graph we need to observe the direction of the arrow from one edge to another to determine their adjacency. Table 7.4 shows the adjacency matrices of the directed graph in Figure 7.7.

The following properties can be immediately seen from the adjacency matrices for undirected graphs

1. The adjacency matrix is a $n \times n$ real-symmetric (i.e., equal to its transpose) if the graph is undirected.

Table 7.2
Adjacency matrix of the left graph of Figure 7.4

	v_1	v_2	v_3	v_4
v_1	0	1	1	0
v_2	1	0	0	1
v_3	1	0	0	1
v_4	0	1	1	0

Table 7.3
Adjacency matrix of the right graph of Figure 7.4

	v_1	v_2	v_3	v_4
v_1	0	1	1	1
v_2	1	0	0	0
v_3	1	0	0	1
v_4	1	0	1	0

2. The spectrum is the set of n real eigenvalues. Each eigenvalue is associated with an eigenvector. The set of eigenvectors defines an orthonormal basis.
3. The adjacency matrix has n real eigenvalues and its n real eigenvectors form an orthonormal basis.
4. The adjacency matrix has zeros on its main diagonal (unless the graph has loops) and ones off-diagonal to indicate the presence of connections.
5. A graph can easily be reconstructed from its adjacency matrix.
6. The adjacency matrix of a graph G depends on a choice of ordering of the vertices. However, two adjacency matrices derived from the same graph G with different orderings are related by permutation similarity.

LAPLACIAN MATRIX

Given a simple graph G, with n vertices, and the adjacency matrix A, the Laplacian matrix $L \in \mathbb{R}^{N \times N}$ of G is defined as

$$L = D - A, \tag{7.1}$$

Matrix D is a diagonal matrix that contains information about the degree of each vertex. In other words, $[D]_{m,m}$ is equal to the number of edges (degree) incident to

Table 7.4
Adjacency matrix of the directed graph of Figure 7.7

	v_1	v_2	v_3	v_4
v_1	0	1	1	0
v_2	0	0	0	1
v_3	0	0	0	1
v_4	1	0	0	0

vertex m. When G is a simple graph, A is constructed of only ones and zeros then,

$$[L]_{m,n} = \begin{cases} \deg(v_m), & \text{if } m = n \\ -1, & \text{if } m \neq n \text{ and } v_m \text{ is adjacent to } v_n \\ 0, & \text{otherwise} \end{cases} \tag{7.2}$$

L is a symmetric and positive semi-definite matrix. Therefore, it has n non-negative, real-valued eigenvalues. In case of directed graphs, D is a diagonal matrix with $\deg(v_m)$ equal to the outdegree of vertex m .

Example 7.4. *The Laplacian matrix of adjacency matrix of Table 7.2 is defined as $L = D - A$ where*

$$L = \begin{bmatrix} 2 & 0 & 0 & 0 \\ 0 & 2 & 0 & 0 \\ 0 & 0 & 2 & 0 \\ 0 & 0 & 0 & 2 \end{bmatrix} - \begin{bmatrix} 0 & 1 & 1 & 0 \\ 1 & 0 & 0 & 1 \\ 1 & 0 & 0 & 1 \\ 0 & 1 & 1 & 0 \end{bmatrix} = \begin{bmatrix} 2 & -1 & -1 & 0 \\ -1 & 2 & 0 & -1 \\ -1 & 0 & 2 & 1 \\ 0 & -1 & -1 & 2 \end{bmatrix}$$

Example 7.5. *The Laplacian matrix of adjacency matrix of Table 7.3 is defined as $L = D - A$ where*

$$L = \begin{bmatrix} 3 & 0 & 0 & 0 \\ 0 & 1 & 0 & 0 \\ 0 & 0 & 2 & 0 \\ 0 & 0 & 0 & 2 \end{bmatrix} - \begin{bmatrix} 0 & 1 & 1 & 1 \\ 1 & 0 & 0 & 0 \\ 1 & 0 & 0 & 1 \\ 1 & 0 & 1 & 0 \end{bmatrix} = \begin{bmatrix} 3 & -1 & -1 & -1 \\ -1 & 1 & 0 & 0 \\ -1 & 0 & 2 & -1 \\ -1 & 0 & -1 & 2 \end{bmatrix}$$

We also can see that the diagonal numbers are the absolute sum of each row.

THE NORMALIZED ADJACENCY AND NORMALIZED LAPLACIAN MATRICES

The normalized adjacency matrix is defined as,

$$\mathscr{A} \equiv D^{-1/2} \cdot A \cdot D^{-1/2}, \tag{7.3}$$

where

$$D^{-1/2} = \begin{bmatrix} \frac{1}{\sqrt{\deg(v_1)}} & 0 & \cdots & 0 \\ 0 & \frac{1}{\sqrt{\deg(v_2)}} & \cdots & 0 \\ \vdots & \vdots & \ddots & \vdots \\ 0 & 0 & 0 & \frac{1}{\sqrt{\deg(v_n)}} \end{bmatrix} \tag{7.4}$$

and the normalized Laplacian matrix defined as,

$$\mathscr{L} \equiv I - \mathscr{A}, \tag{7.5}$$

where I is the identity matrix, then

$$\mathscr{L} = I - \mathscr{A} = D^{-1/2} \cdot (D - A) \cdot D^{-1/2} = D^{-1/2} \cdot L \cdot D^{-1/2}, \tag{7.6}$$

where L is the unnormalized Laplacian matrix.

Example 7.6. *The normalized Laplacian matrix of Example 7.5 is*

$$\mathscr{L} = \begin{bmatrix} \frac{1}{\sqrt{3}} & 0 & 0 & 0 \\ 0 & 1 & 0 & 0 \\ 0 & 0 & \frac{1}{\sqrt{2}} & 0 \\ 0 & 0 & 0 & \frac{1}{\sqrt{2}} \end{bmatrix} \begin{bmatrix} 3 & -1 & -1 & -1 \\ -1 & 1 & 0 & 0 \\ -1 & 0 & 2 & -1 \\ -1 & 0 & -1 & 2 \end{bmatrix} \begin{bmatrix} \frac{1}{\sqrt{3}} & 0 & 0 & 0 \\ 0 & 1 & 0 & 0 \\ 0 & 0 & \frac{1}{\sqrt{2}} & 0 \\ 0 & 0 & 0 & \frac{1}{\sqrt{2}} \end{bmatrix}$$

$$= \begin{bmatrix} 1 & -\frac{1}{\sqrt{3}} & -\frac{1}{\sqrt{6}} & -\frac{1}{\sqrt{6}} \\ -\frac{1}{\sqrt{3}} & 1 & 0 & 0 \\ -\frac{1}{\sqrt{6}} & 0 & 1 & -\frac{1}{2} \\ -\frac{1}{\sqrt{6}} & 0 & -\frac{1}{\sqrt{2}} & 1 \end{bmatrix}$$

SPECTRUM OF THE LAPLACIAN MATRIX

The $n \times n$ symmetric matrix has n real eigenvalues $\lambda_1 \leq \lambda_2 \cdots \leq \lambda_n$ and eigenvectors $x_1, x_2, \cdots, x_n$ such that

$$\mathscr{L} x_i = \lambda_i x_i \tag{7.7}$$

Then, $\lambda_1 = 0$ and $\lambda_n \leq 2$. The sum of eigenvalues is bounded by the number of vertices n , with equality holding if and only if there are no isolated vertices. The spectrum is the graph eigenvalues and is the union of the spectrum of its connected components. For example, the eigenvalues of $\mathscr{L}$ of Example 7.6 are $\lambda_1 = 0$, $\lambda_2 = 0.7713$, $\lambda_3 = 1.5$ and $\lambda_4 = 1.7287$ is the spectrum of the graph and the sum is equal to 4.

Example 7.7. *Consider the following undirected path graph G, with four vertices and three edges, with line topology. Using MATLAB, determine the matrices of eigenvectors U and eigenvalues Σ of the Laplacian matrix.*

$$L = \left(\begin{bmatrix} 1 & 0 & 0 & 0 \\ 0 & 2 & 0 & 0 \\ 0 & 0 & 2 & 0 \\ 0 & 0 & 0 & 1 \end{bmatrix} - \begin{bmatrix} 0 & 1 & 0 & 0 \\ 1 & 0 & 1 & 0 \\ 0 & 1 & 0 & 1 \\ 0 & 0 & 1 & 0 \end{bmatrix} \right) = \begin{bmatrix} 1 & -1 & 0 & 0 \\ -1 & 2 & -1 & 0 \\ 0 & -1 & 2 & -1 \\ 0 & 0 & -1 & 1 \end{bmatrix}$$

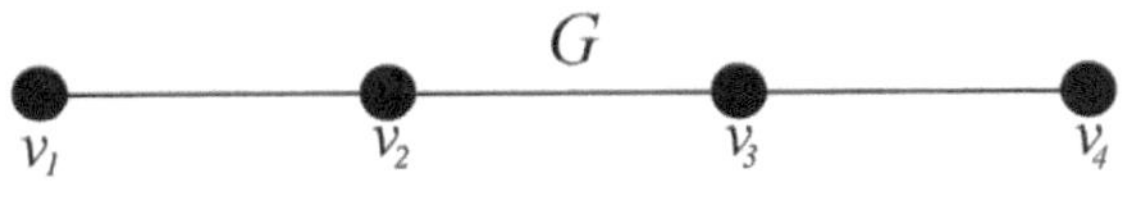

Then,

$$U = \begin{bmatrix} 0.5 & 0.6533 & 0.5 & 0.2706 \\ 0.5 & 0.2706 & -0.5 & -0.6533 \\ 0.5 & -0.2706 & -0.5 & 0.6533 \\ 0.5 & -0.6533 & 0.5 & -0.2706 \end{bmatrix}, \quad \Sigma = \begin{bmatrix} 0 & 0 & 0 & 0 \\ 0 & 0.5858 & 0 & 0 \\ 0 & 0 & 2 & 0 \\ 0 & 0 & 0 & 3.4142 \end{bmatrix}.$$

Notice that $L = U \cdot \Sigma \cdot U^T$.

The Cartesian product of two graphs G_1 and G_2 with line topology and n vertices is a square grid topology with $N = n^2$ vertices (see §7). The spectrum of the Laplacian of this new graph depends on the spectrum of two generator graphs as stated in the following theorem.

Theorem 7.1

Let G_1 and G_2 be graphs on N_1 and N_2 vertices respectively. Then, the eigenvalues (λ's) of the Laplacian matrix of the resulting product graph (see §7) $G_1 \times G_2$ are all possible sums of $\lambda_i(G_1) + \lambda_k(G_2)$, with $0 \le i \le N_1 - 1$ and $0 \le k \le N_2 - 1$. Moreover, if v^i is an eigenvector of G_1 corresponding to the eigenvalue $\lambda_i(G_1)$ and v^k an eigenvector of G_2 corresponding to the eigenvalue $\lambda_k(G_2)$, then, the Kronecker product $v^i \otimes v^k$ is an eigenvector of the graph G corresponding to $\lambda_i(G_1) + \lambda_k(G_2)$ [323]. ■

KRONECKER PRODUCT

If A is an $m \times n$ matrix and B is a $p \times q$ matrix, the Kronecker product $A \otimes B$ is the $mp \times nq$ matrix,

$$\begin{bmatrix} a_{1,1}B & \cdots & a_{1,n}B \\ \vdots & \ddots & \vdots \\ a_{m,1}B & \cdots & a_{m,n}B \end{bmatrix} \tag{7.8}$$

If u is a $m \times 1$ vector and v is a $n \times 1$ vector the Kronecker product $u \otimes v$ is the $m \times n$ matrix,

$$u \otimes v = \begin{bmatrix} u_1 \\ u_2 \\ \cdots \\ u_m \end{bmatrix} \begin{bmatrix} v_1 v_2 \cdots v_n \end{bmatrix} = \begin{bmatrix} u_1 v_1 & u_1 v_2 & \cdots & u_1 v_n \\ u_2 v_1 & u_2 v_2 & \cdots & u_2 v_n \\ \vdots & & \ddots & \vdots \\ u_m v_1 & u_m v_2 & \cdots & u_m v_n \end{bmatrix} \tag{7.9}$$

For complex vectors

$$u \otimes v = uv^*, \tag{7.10}$$

where v^* is the conjugate transpose of v also denoted by v^H. Then we can cite the important theorem in [323],

CARTESIAN PRODUCT OF TWO GRAPHS

Recall that the Cartesian product of two sets X and Y is $X \times Y = \{(x,y) | x \in X, y \in Y\}$. The Euclidean plane is an example of a Cartesian product of two sets of real numbers where $\mathbb{R} \times \mathbb{R}$ is the Cartesian product of the two sets (i.e., $X \in \mathbb{R}$ and $Y \in \mathbb{R}$). Another simple example is $A = \{x, y\}$ and $B = \{-1, 0, 1\}$ the $A \times B = \{(x,-1),(x,0),(x,1),(y,-1),(y,0),(y,1)\}$.

The Cartesian product of two graphs G_1, G_2, or simply called the product of two graphs, with disjoint vertex sets $V_1 = \{u_1, u_2, \cdots, u_m\}$ and $V_2 = \{v_1, v_2, \cdots, v_n\}$ respectively, and disjoint edge sets E_1 and E_2 respectively, is the graph with vertex set $V_1 \times V_2 = \{(u_i, v_k) | u_i \in V_1, v_k \in V_2\}$.

Then, we determine where the edges must be placed in the product graph. We see that e is an edge of the product graph if and only if $e = (u_i, v_k)(u_m, v_n)$, where either $i = m$ and $v_k, v_n \in E_2$ or $k = n$ and $u_i, u_m \in E_1$.

Example 7.8. *Find the Cartesian product $G = G_1 \times G_2$ of the following two graphs*

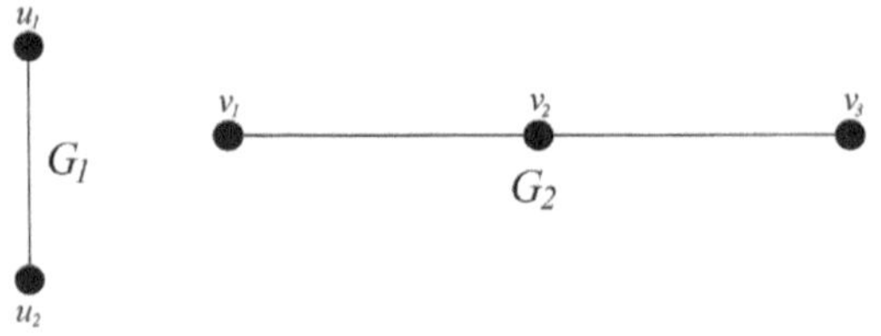

First, we write all the possible vertices of the new graph.

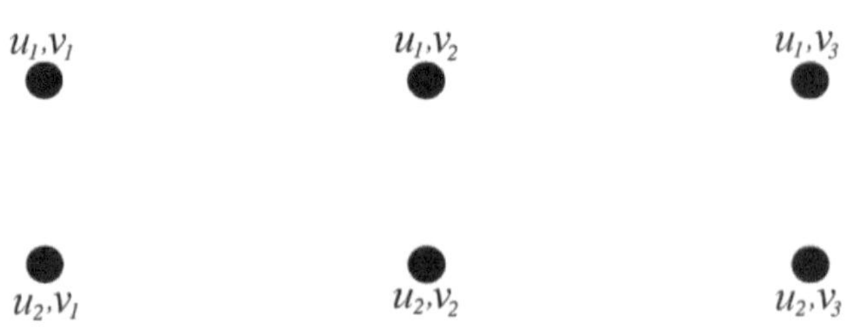

Now, we determine where the edges must be placed, observe that the product vertices (u_1, v_1) and (u_2, v_1) are adjacent in the product because u_1 and u_2 are

adjacent in the graph G_1, the same happens in the product vertices (u_1, v_2) and (u_2, v_2), (u_1, v_3) and (u_2, v_3).

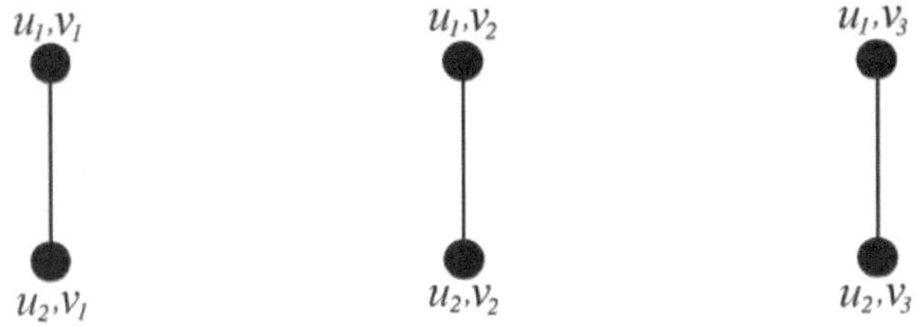

Then, we determine the edges produced in the product graph by the graph G_2. Observe vertices (u_1, v_1) and (u_1, v_2), since v_1 and v_2 are adjacent in graph G_2 we draw an edge between these two product vertices. We see that the same reasoning can be applied to the product edges (u_2, v_1) and (u_2, v_2).

Vertices v_2 and v_3 are adjacent in graph G_2, the product vertices (u_1, v_2) and (u_1, v_3) share a common edge as well as edges (u_2, v_2) and (u_2, v_3). We can continue to analyze and we will end up with the following product graph.

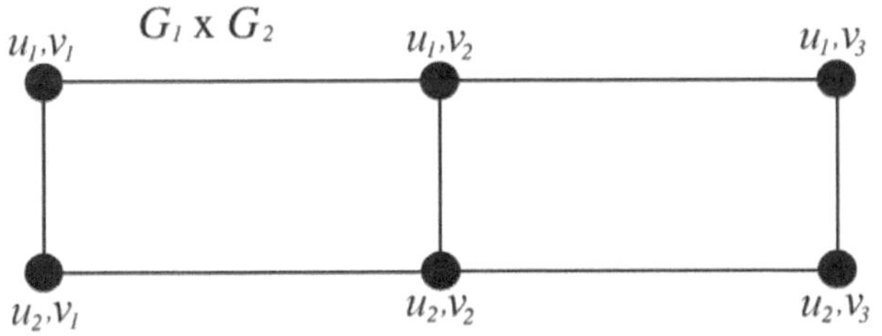

Example 7.9. *Find the Cartesian product of $G = G_1 \times G_2$ of the following two graphs,*

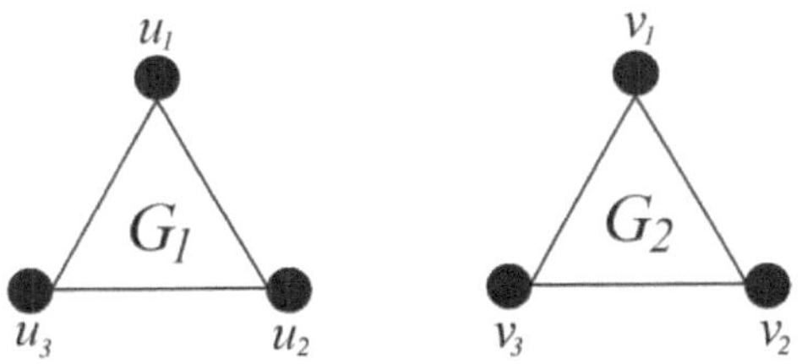

We write all the possible vertices of the new graph as shown below. Edges that happen due to G_1 occur where we have the second component of our vertex label fix. For example, we have (u_1, v_1) adjacent to (u_2, v_1) adjacent to (u_1, v_3) adjacent to (u_1, v_1) and so on. Notice that we have three copies of G_1.

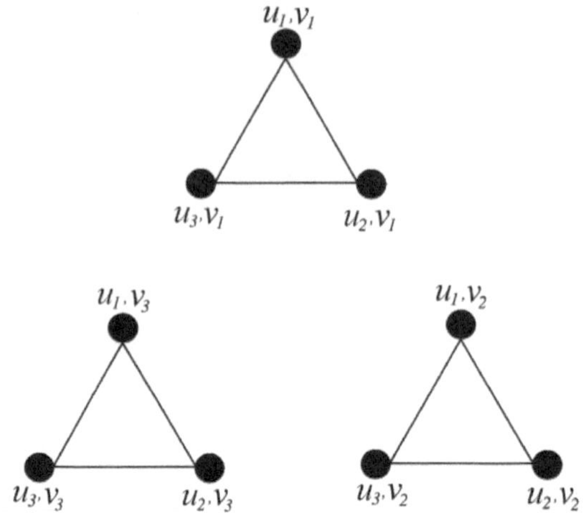

Following, we investigate the edges that happen due to G_2. These edges occur where the first component of our vertex label is fixed. For example, if we take vertex (u_1, v_1), we see that the first component exists also in (u_1, v_2) and (u_1, v_3). all making a triangle. This is because v_1, v_2 and v_3 are adjacent in G_2. We can continue to analyze the remaining edges and end up with the following product graph (the dashed and dotted lines correspond to edges due to G_2).

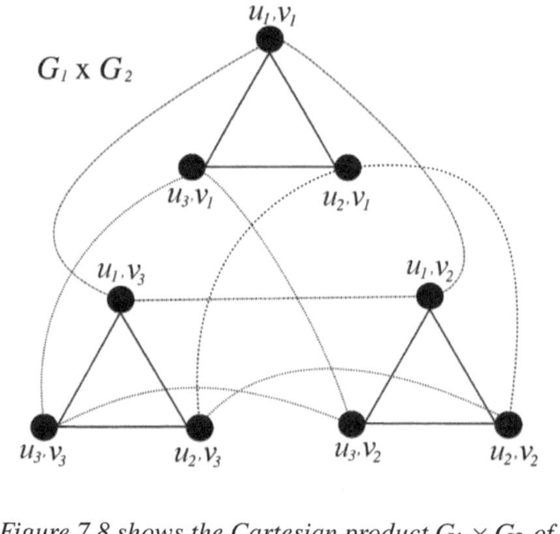

Example 7.10. *Figure 7.8 shows the Cartesian product $G_1 \times G_2$ of two graphs with line topology.*

Any signal $x \in \mathbb{R}^N$ can be associated with a graph G_x with $|V| = N$; each component x_n, $n = 1, \cdots, N$ is associated with vertex $v_i \in V$ through the so-called graph Fourier transform [432] as $X = U^T \cdot x$, where U is the matrix whose columns are

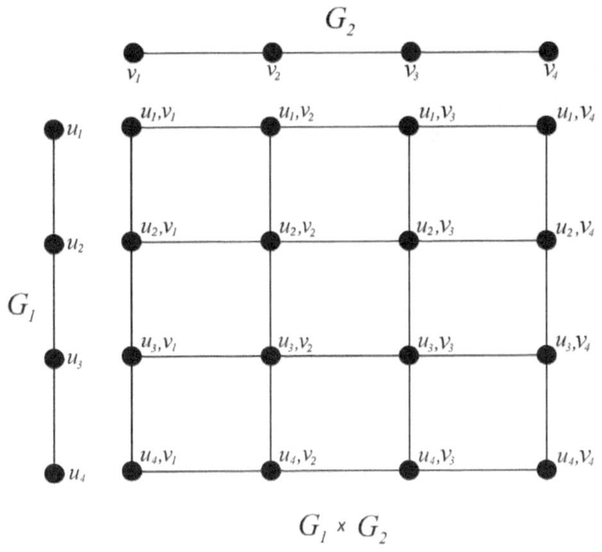

Figure 7.8: Cartesian product $G_1 \times G_2$ of two graphs with line topology and four vertices each one.

the eigenvectors of the Laplacian matrix of G_x (see Example 7.7). This operation can be inverted by $x = U \cdot X$. It is known that the eigenvectors of a graph G with line topology, as the one defined in example 7.7, and more general, a path graph of N vertices, are equal to the N basis vectors (eigenvectors of the Laplacian of the path graph) $\{v^k\}_{k=0}^{N-1}$ of the 1D DCT-II defined by,

$$v_l^k = \cos\left[\frac{\pi k}{N}(l+0.5)\right], \quad l,k = 0,1,\cdots,N-1, \tag{7.11}$$

associated with the corresponding eigenvalue

$$\lambda_k = 4\sin^2\left(\frac{\pi k}{2N}\right) = 2 - 2\cos\left(\frac{\pi k}{N}\right) \tag{7.12}$$

Given that the multiplicity of the eigenvalues in (7.12) is always one, the 1D DCT basis is the unique eigenbasis for the Laplacian of the path graph. Therefore, the graph Fourier transform for a signal represented by a path graph is equivalent to the 1D DCT [137]. Using Theorem 7.1 we can compute the eigenvalues and eigenvectors

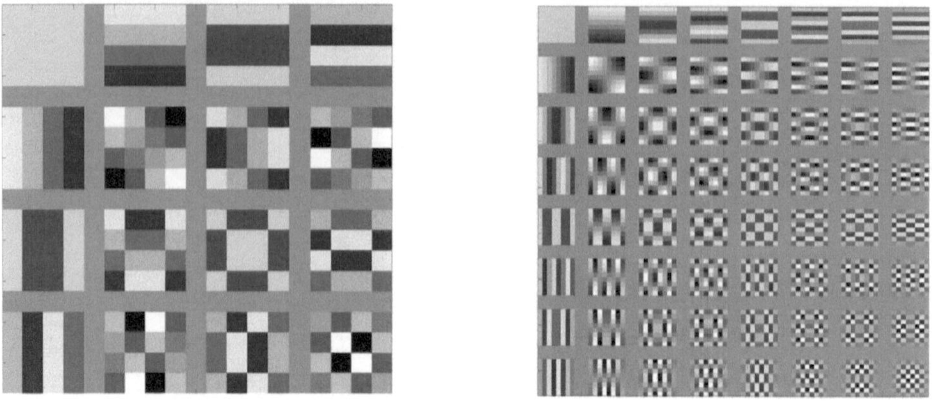

(a) 2D-DCT Basis images for N=4.

(b) 2D-DCT Basis images for N=8.

Figure 7.9: 2D-DCT Basis images for (a) N=4 and (b) N = 8.

of the grid graph of Figure 7.8 or an $N \times N$ grid graph by,

$$\begin{aligned} \lambda_{k,l} &= \lambda_k + \lambda_l = 4\sin^2\left(\frac{\pi k}{2N}\right) + 4\sin^2\left(\frac{\pi l}{2N}\right) \\ &= 4 - 2\left(\cos\left(\frac{\pi k}{N}\right) + \cos\left(\frac{\pi l}{N}\right)\right) \\ v^{k,l} &= v^k \otimes v^l, \quad k,l = 0,1,\cdots,N-1, \end{aligned} \tag{7.13}$$

where v^k and v^l are the kth and lth eigenvectors of the Laplacian matrix of $G1$ (or $G1$) corresponding to the eigenvectors λ_k and λ_l respectively. This extends the concept to the 2D DCT. Figure 7.9 shows the 2D DCT basis computed by the Kronecker product of eigenvectors (7.11).

The resulting eigenvalues for $N = 8$ are,

$$\lambda_{k,l} = \begin{bmatrix} \mathbf{0} & 0.1522 & 0.5858 & 1.2346 & 2 & 2.7654 & 3.4142 & 3.8478 \\ 0.1522 & \mathbf{0.3045} & 0.7380 & 1.3869 & 2.1522 & 2.9176 & 3.5665 & 4 \\ 0.5858 & 0.7380 & \mathbf{1.1716} & 1.8204 & 2.5858 & 3.3512 & 4 & 4.4335 \\ 1.2346 & 1.3869 & 1.8204 & \mathbf{2.4693} & 3.2346 & 4 & 4.6488 & 5.0824 \\ 2 & 2.1522 & 2.5858 & 3.2346 & 4 & 4.7654 & 5.4142 & 5.8478 \\ 2.7654 & 2.9176 & 3.3512 & 4 & 4.7654 & \mathbf{5.5307} & 6.1796 & 6.6131 \\ 3.4142 & 3.5665 & 4 & 4.6488 & 5.4142 & 6.1796 & \mathbf{6.8284} & 7.2620 \\ 3.8478 & 4 & 4.4335 & 5.0824 & 5.8478 & 6.6131 & 7.2620 & \mathbf{7.6955} \end{bmatrix}$$

We can see two eigenvalues with multiplicity 2 (they represent eigenvectors at the same frequency, one in the horizontal direction and the other in the vertical direction), eigenvectors of eigenvalues $\lambda = 4$ with multiplicity $N-1$ and $N-1$ eigenvectors of eigenvalues with algebraic multiplicity 1 (there are only $N-1$ eigenvalues $\lambda_{k,k}$, $k \neq N/2$ with algebraic multiplicity equal to 1). Some repeated eigenvalues are present due to symmetry $\lambda_{k,l} = \lambda_{l,k}$ and in the case $\lambda_{k,k}$ with $k \neq N/2$, the algebraic multiplicity is equal to 1, the rest have algebraic multiplicity 2 except the eigenvectors of eigenvalues $\lambda_{k,N-k}$.

The following MATLAB code calculates the 2D DCT basis images using (7.11) and (7.13) for $N = 8$ from a 1D DCT of $N = 4$.

```
%    BasisIM   is the image of 2D basis
%    Basis2D   is the array of 2D basis

clear all;
N       = 4;   % Row (column) length
U       = zeros(N,N); % 1D eigenbasis matrix
DCT2     = zeros(N*N,N*N);

% Eigenvalues
lambda1D =zeros(1,N);

for k = 0:N-1
    if (k==0)
        Ck = sqrt(1/N);
    else
        Ck = sqrt(2/N);
    end

    lambda1D(k+1) = 2-2*cos(pi*k/N);

    for l = 0:N-1
        U(k+1,l+1) =Ck*cos(pi*k*(2*l+1)/(2*N));
    end
end

lambda2D = zeros(1,N*N);
V = U;
U = U';
BasisIM = zeros(N*N,N*N);
Basis2D = zeros(N,N,N,N);
Gap = N/2;
for k = 0:N-1
   for l = 0:N-1

       BasisIM((N+Gap-1)*l+1:(N+Gap-1)*l+1+N-1, ...
           (N+Gap-1)*k+1:(N+Gap-1)*k+1+N-1) = ...
           U(:,k+1:k+1)*V(l+1:l+1, :);
       Basis2D(:,:,k+1,l+1)= BasisIM((N+Gap-1)*l+1: ...
           (N+Gap-1)*l+1+N-1, (N+Gap-1)*k+1: ...
           (N+Gap-1)*k+1+N-1);
       lambda2D(N*k+l+1) =lambda1D(k+1)+...
           lambda1D(l+1);
   end;
end
```

```
for k =0:N-1
    for l = 0:N-1;
        DCT2(N*k+l+1:N*k+l+1,1:end) = ...
            reshape(Basis2D(:,:,l+1,k+1), 1, N*N);
    end
end

lambda2D=reshape(lambda2D,N,N);
```

STEERABLE DCT DEFINITION

In Example 7.7 we calculated the eigenvectors matrix U for the Laplacian matrix L of a graph with line topology, this eigenvectors matrix is equal to the 4×4 1D DCT matrix. However, the matrix U is not unique. For example, another selection of U can be,

$$U = \begin{bmatrix} 0.5 & -0.2706 & -0.5 & -0.6533 \\ 0.5 & 0.6533 & 0.5 & -0.2706 \\ 0.5 & -0.6533 & 0.5 & 0.2706 \\ 0.5 & 0.2706 & -0.5 & 0.6533 \end{bmatrix},$$

Then, the 2D eigenbasis are the 2D DCT basis images and can be obtained by the Kronecker product of the 1D eigenbasis by using (7.13). The sum of their corresponding eigenvalues are useful to determine the multiplicity. In case that the resulting $\lambda_{k,l} = \lambda_{l,k}$ their corresponding eigenvectors are still linearly independent, because the Kronecker product is not commutative. Therefore, the geometric multiplicity is equal to the algebraic multiplicity. In this sense, Fracastoro *et al.* [137] give the following proposition,

Proposition 1: The 2D-DCT is not the unique eigenbasis for the Laplacian of a square grid graph [137].

The proposition is useful to find the possible eigenbases that better fits the properties of the specific image block and choose it as transform matrix. From the proposition we can see that the two eigenvectors with multiplicity two ($v^{k,l}$ and $v^{l,k}$) with eigenvalue $\lambda_{k,l}$ of the L, corresponding to the graph $G_1 \times G_2$ can be written in terms of rotations with any angle $\theta_{k,l}$ as a Givens rotation in the 2D space described by the pair of eigenvectors as,

$$V(\theta) = \begin{bmatrix} v^{(k,l)'} \\ v^{(l,k)'} \end{bmatrix} = \begin{bmatrix} \cos\theta_{k,l} & \sin\theta_{k,l} \\ -\sin\theta_{k,l} & \cos\theta_{k,l} \end{bmatrix} \begin{bmatrix} v^{k,l} \\ v^{l,k} \end{bmatrix} \tag{7.14}$$

Equation (7.14) is the steerable DCT basis. The eigenvectors corresponding to eigenvalues $\lambda_{k,l}$ with multiplicity two, rotate with its corresponding eigenvector $\lambda_{l,k}$. The $N-1$ eigenvectors of eigenvalues $\lambda = 4$ are rotated in pairs $v^{k,N-k}$ and $v^{N-k,k}$. When N is even the eigenvectors corresponding to eigenvalues with multiplicity one ($v^{N/2,N/2}$) are not rotated.

The rotated eigenvalues form the basis for the new transform $V(\theta) \in \mathbb{R}^{N^2 \times N^2}$ in terms of the rotation angle $\theta \in \mathbb{R}^p$ and the number of pairs of basis rotated is $p = N(N-1)/2$. From Figure 7.9 we see that $V(0) = V$ is the original 2D DCT (no rotation applied). Therefore, the steerable DCT is defined in terms of the rotation matrix $R(\theta) \in \mathbb{R}^{N^2 \times N^2}$ and the original DCT matrix [137] as,

$$V(\theta) = VR(\theta). \tag{7.15}$$

Figure 7.10 shows the 2D basis images of the steerable DCT for angles of (a) 20, (b) 45 and (c) 75 degrees respectively. Notice that the basis images can be rotated by any angle.

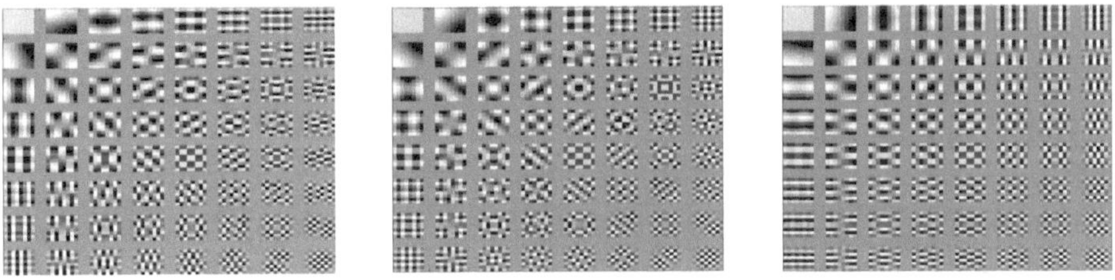

(a) $\theta = 20$ degrees. (b) $\theta = 45$ degrees. (c) $\theta = 75$ degrees.

Figure 7.10: 2D Basis images of the steerable DCT for $N = 8$ at (a) $\theta = 20$, (b) $\theta = 45$ and (c) $\theta = 75$ degrees.

The following MATLAB code rotates all the $N \times (N-1)/2$ pairs of eigenvectors generated by the previous MATLAB code by one specific angle in degrees according to (7.14).

```
function [BasisIM Basis2D] = RotateBasis(Basis2D, Angle)

% Generates the rotated eigenbasis for a specified angle
% Angle in degrees
% Basis2D is the array of 2D basis
% BasisIm is the image of 2D Basis
% N is the row (column) length

[N,M] = size(Basis2D(:,:,1,1));
Gap = N/2;
Theta = Angle*pi/180;
Rot = [cos(Theta) sin(Theta); ...
       -sin(Theta) cos(Theta)];

BasisIM = zeros(N*N,N*N);
Gap = N/2;
for k = 0:N
    for l = 1:k-1
       Vlk= Basis2D(:,:,k,l);
```

```
        Vkl= Basis2D(:,:,l,k);
        v1 = reshape(Vlk,1,N*N);
        v2 = reshape(Vkl,1,N*N);
        C = Rot*[v1;v2];
        v1 = C(1:1, :);
        v2 = C(2:2, :);
        Basis2D(:,:,k,l) = reshape(v1,N,N);
        Basis2D(:,:,l,k) = reshape(v2,N,N);
    end
end

for k = 0: N-1
  for l = 0: N-1
      BasisIM((N+Gap-1)*l+1:(N+Gap-1)*l+1+N-1,...
          (N+Gap-1)*k+1:(N+Gap-1)*k+1+N-1)= ...
          Basis2D(:,:,k+1,l+1);
  end
end
```

However, the code can be modified defining the variable θ as a vector of p components, where each component represent a different angle. Nevertheless, the computational cost increases and in case of compression applications the cost required to transmit all the rotation angles may be high. Notice that $V(\theta)$ is a transform matrix of L with a orientation different from that of the DCT.

ROTATION ANGLE OPTIMIZATION

Until now, we have studied the form to rotate the eigenspace of Laplacian matrix L from the cartesian product $G_1 \times G_2$ of two undirected graphs. We learned that the eigenspace of L is the same as DCT-II. Also, we designed the appropriate MATLAB functions to implement the eigenspace and rotate it manually according to the desired angle. Still, the rotation angle should be the optimum according to the input block.

To choose the rotation angle we need to use an automatic method that finds among all the possible angles one that better fits the characteristics of the input block. First, we need to find a function to optimize, it can be defined in terms of the energy of transform coefficients after applying a rotation angle and we want this function to be maximum. Therefore, we can define an objective function $J(\theta)$ as the total energy of a transform block of the $M < (N \times N)$ largest coefficients. Also, from the total number of coefficients we can select the $M < (N \times N)$ largest coefficients to calculate the energy. The objective function is the energy of the block containing the M largest coefficients and it is defined as,

$$J(\theta_k) = \sum_{k=1}^{M} c_{i,k}^2, \tag{7.16}$$

where $c_{i,k}$ are the transform coefficients at the kth rotation angle. The algorithm 1 was proposed by Fracastoro el al. [137] taking as the stopping criteria the number of rotation angles to analyze. Notice that using this stopping criteria the convergence can be slow if the optimum angle is far from the initial angle. The algorithm finds maximum energy of the transform coefficients among all the angles between 0° and 90° with stepsize of 1°.

Algorithm 1 Steerable DCT using the same rotation angle for all the eigenspace [137]

Initialization:
θ : set of R rotation angles between 0° and 90°
$J_{opt} \leftarrow 0$
$V \leftarrow$ 2D DCT-II matrix
Set the optimal angle to 0
for $k := 1$ to R **do**
  Get samples x from image
  Rotate each pair of vectors in V by θ_k
  Build a new transform matrix V' with rotated vectors
  Compute $J(\theta_k) \leftarrow E_n[V'x]$ using (7.16)
  if $J(\theta_k) > J_{opt}$ **then**
    Set the optimal angle to θ_k
    $J_{opt} \leftarrow J(\theta_k)$
  end if
end for

Example 7.11. *Equation* (7.17) *shows a block of* 4×4 *pixels with* 45° *of orientation.*

$$\begin{bmatrix} 100 & 100 & 100 & 100 \\ 0 & 100 & 100 & 100 \\ 0 & 0 & 100 & 100 \\ 0 & 0 & 0 & 100 \end{bmatrix}. \tag{7.17}$$

Matrices (7.18a) *and* (7.18b) *show the transform coefficients after applying the steerable DCT and the DCT respectively. Notice the sparseness of the transform coefficients yielded by the steerable DCT. In this case there are twice as many number of zeros.*

$$\begin{bmatrix} 250 & -157.71 & 0 & -11.21 \\ 0 & 50.00 & -65.32 & 0 \\ 0 & 0 & 50.00 & -27.06 \\ 0 & 0 & 0 & 50.00 \end{bmatrix} \tag{7.18a}$$

$$\begin{bmatrix} 250.00 & -111.52 & 0 & -7.92 \\ 111.52 & 50.00 & -46.19 & 0 \\ 0 & 46.19 & 50.00 & -19.13 \\ 7.92 & 0 & 19.13 & 50.00 \end{bmatrix} \tag{7.18b}$$

Several issues can be observed. There is no stopping criteria, the eigenspace is rotated with the same angle, this means that the algorithm will find the direction of the strongest dominant edge only. However, if two or more dominant edges exist the algorithm decides only one of them.

APPLICATIONS

Most of the schemes for image and video coding have chosen the 2D-DCT of a square block shape. In order to cope with the disadvantages of conventional DCT to handle the diagonally oriented edges, a new set of directional DCTs was introduced in [570] and tested on Advanced Video Coding (AVC). The idea was motivated by the works done in shape adaptive DCT (SA-DCT) [134], [158], [247], [433], [434] which was originally proposed in MPEG-4 in region (object) coding but was never adopted. Also, to represent images using multi-directional information some filter-banks have been proposed [33], [111], [292], [494]. However, for video coding, these directional filter banks are limited by the size of the block to decorrelate.

Biometric attributes are distinctive to each individual. They are used to detect and recognize human physical characteristics. Some of the attributes are fingerprints, face, DNA, palm print, hand geometry, iris, and retina. Feature extraction is one of the most important steps in computer vision and pattern recognition and where DCT has been extensively applied [9], [16], [23], [25], [26], [38], [71], [80], [114], [120], [126], [147], [181], [218], [230], [260], [286], [288], [392], [396], [451], [465], [502], [514], [574].

The digital media is growing fast and new technology is emerging. Copying, tampering and distributing of digital content has become an important issue [98], [22]. The wide distribution of such media have led to the development of copyright protection and forgery detection tools. As technology evolves, new robust watermarking systems are proposed for the proof of ownership and media authentication [97]. Watermarks, are signatures embedded in the signal in some domain to protect the original media. Some watermarking works proposed in the DCT domain that can find applications in the DDCT or SDCT are: [10] [21], [68], [74], [125], [124], [128], [146], [161], [175], [194], [293] [295], [311], [321], [331], [332], [357], [410], [438], [437], [456], [461], [479], [503], [539], [542], [543], [550], [577], [574], [576], [580].

Image denoising, compression and segmentation showed tremendous applications in medical imaging systems. However, segmentation of medical images is one of the challenging tasks in computer vision. It includes coding of region of interest (ROI) delineation of anatomical structures and identification and marking of other (ROI), [77], [112], [167], [186], [310], [328], [411], [448], [458], [466], [473], [529], [536], [540], [541].

In many applications only partial data in an image or a video sequence are available because of tampering, impulsive noise, detector limitations, transmission in noisy channels, etc. Recovery of missing data is called inpainting. Inpainting is a very active area and the DCT has been extensively used in this area [67], [99], [141], [149], [244], [265], [281], [299], [363], [486], [572], [585].

DCT has been extensively used for block analysis, scanning area analysis for surveillance applications. Also, these areas are niches of opportunities for DDCT and SDCT [86], [131], [169], [171], [180], [224], [229], [243], [252], [329], [355], [364], [512].

SUMMARY

Nearly all block-based transform techniques for image and video coding applications have chosen the 2D-DCT of a square block shape. This conventional 2D-DCT is always applied separately through two 1D transforms, one along the vertical and one along horizontal directions, respectively. Block-based separable transforms tend to be inefficient to represent image blocks when blocks contain arbitrarily shaped discontinuities because only horizontal and vertical image orientation information is taken into account. The DDCT may follow a direction other than the vertical or horizontal one, making it suitable to represent blocks of images with different dominant edges. The DDCT is also a separable transform. However, before transformation the blocks of pixels are rearranged according to the dominant edge of the blocks. The coefficients produced by all the directional transforms in the first step are arranged appropriately so that the second transform can be applied to the coefficients that are best aligned with one another. Compared with the conventional DCT, the resulting DDCT provides a better coding performance scenario for image blocks with directional edges and the rate distortion coding performance can be improved remarkably [457]. The new steerable DCT can be developed using graph theory or the DCT-II. This transform provides more flexibility finding directional information inside a block of pixels because the basis can be rotated and adapted to find the optimal directions inside a block. Facastoro et al. [137] showed that the steerable DCT outperforms both DCT and state-of-the-art directional transforms.

PROBLEMS

7.1. Develop a strategy to select among the different modes of DDCT.

7.2. Use MATLAB to calculate and display the 4×4 basis images for modes $4,5,6,7,8$ of Figure 7.1.

7.3. Use MATLAB to calculate and display the 8×8 basis images for modes $4,5,6,7,8$ of Figure 7.1.

7.4. In Algorithm 1 use the sparseness of the transform coefficients as stopping criteria and compare the performance in time.

7.5. Algorithm 1 finds only one optimal rotation angle. However, the number of pairs of basis can be $N(N-1)/2$ different angles [138]. Propose an algorithm similar to 1 to find the optimal vector of rotation angles.

7.6. In Algorithm 1, rotation of each pair of vectors is carried out in steps of 1 degree until a total of 90 degrees are completed. This is called the brute force. Modify the algorithm to find the optimal rotation angle faster. For example, take a pair of rotation angles and compare the energy of the truncated coefficients (see (7.17)) after transformation, then advance in the direction where the energy increases (like a logarithmic search).

7.7. Propose a different stopping criteria for Algorithm 1 using the sparseness of transform coefficients (i.e., by counting the total number of non-zero elements in the transform matrix).

7.8. In [179], Hernandez developed a 4×4 integer SDCT to be implemented in 16 bits. Use a similar strategy to extend transform sizes to 8×8, 16×16 and 32×32.

PROJECTS

P.7.1. Develop a strategy to select among different modes of DDCT.

P.7.2. Develop a strategy to encode blocks of 4×4 pixels for high resolution images using the DDCT including modes shown in Figure 7.1.

P.7.3. Repeat project P.7.2. for blocks of size 8×8, 16×16 and 32×32 and 64×64.

P.7.4. Use the IntDCT to develop different modes of DDCT.

P.7.5. In HEVC [387], [389], [462] the input frame is divided into blocks of 64×64 samples called coding units (or CTUs) consisting of three blocks, namely luma (Y) and chroma samples (Cb and Cr), and associated syntax elements. Each block is divided into other coding units CUs or encoded depending on the characteristics of the block. Use the same quad-tree division and integrate projects P.7.1. to P.7.4. to develop an encoder based on the DDCT.

P.7.6. Perng and Chen [370] developed a global image restoration scheme using the DCT. Propose a similar scheme using DDCT and SDCT and compare the three schemes.

P.7.7. Turgay, Huseyin and Hasan [488] proposed a method for facial feature extraction using complex wavelet transform. Use the SDCT to propose a similar method to extract facial features.

P.7.8. Bhatnagar and Wu [38] proposed a biometrics inspired watermarking technique. Develop a similar watermarking method inspired in the SDCT.

P.7.9. Use a hardware description language (HDL) to implement the SDCT of problem 7.8.

P.7.10. Image analysis has proven to be an efficient tool for artist identification using the discrete wavelet transform (DWT). For example, Johnson *et al.* [233] made a computational analysis of the brushstrokes orientation in paintings mostly by van Gogh with the DWT. Explore the DDCT and the SDCT to develop a similar system.

P.7.11. Alahmadi Amani et al. [15] proposed a passive image forgery detection method based on local binary pattern and the DCT to detect copy−move and splicing forgeries. Propose a similar method using SDCT and explore the possibility to use it in project P.7.10.

P.7.12. In [23], a low memory, face recognition based on the DCT pyramid decomposition of the face images is proposed. The pyramid generates multi-resolution face images for face feature extraction. This methodology outperforms the DWT methods. Replace the DCT by SDCT in the pyramid and compare with the reported reported.

P.7.13. The dual-tree complex wavelet transform (DT-CWT) is a nearly shift invariant and directionally selective transform in two and higher dimensions [421]. In [188], the author takes advantage of the directionality of the DT-CWT to propose an illumination invariant face recognition. Propose a similar system based on the SDCT and compare results.

P.7.14. Maheshwari [298] proposed a robust watermarking scheme using DCT-based pyramid transform. Use a similar scheme to develop a watermarking method based on the SDCT.

P.7.15. Positron emission tomography (PET) scanners are used in medicine to help evaluate the organs and tissue functions. However, because of the separation between detectors the recovered sinograms present gaps or missing parts. Tuna et al. [486] proposed a DCT domain approach with a gap-dependent filter. Explore the possibility to use the SDCT instead the DCT to propose a similar inpainting algorithm for PET images.

P.7.16. Hangarge, Santosh and Pardeshi [172] proposed a DDCT-based word level handwritten script identification. Implement a similar method using SDCT.

P.7.17. In [106] extensions to H.264/AVC FRExt such as larger motion vector (MV) search range, larger macroblock, skipped block size and 1D DDCT were added. The performance with motion JPEG 2000 using HD video sequences (4096×2160) was compared and the former showed significant improvement in terms of PSNR at various bit rates. Implement the extended H.264/AVC and motion JPEG 2000 and confirm that the former has a superior performance using HD test sequences.

P.7.18. Access S. Subbarayappas's thesis from EE5359, MP Lab web site [261], "Implementation and Analysis of Directional Discrete Cosine Transform in Baseline Profile in H.264." Obtain the basis images for all the directional modes related to 4×4 and 8×8 DDCT. Modes 4, 6, 7 and 8 can be obtained from modes 3 and 5. Use this approach to obtain all the basis images.

P.7.19. In [137] authors developed the steerable DCT. Implement the 4×4 steerable DCT in H.264/AVC video coding standard and check the performance and complexity.

P.7.20. Repeat project P.7.19. for 8×8 steerable DCT.

P.7.21. In [137] authors developed the steerable DCT. Implement this transform for 4×4 in HEVC video coding standard and check the performance.

P.7.22. Repeat project P.7.21. for 8×8 steerable DCT.

P.7.23. In [138] the authors fixed 8 quantization levels for the angles distributed uniformly between [0 and π]. They state, "In order to improve the compression performance, as future work, we may consider a non-uniform angle quantization." Investigate this thoroughly. Consider all possible non-uniform angle quantizations, the main objective being improved compression performance.

P.7.24. In [138] the steerable DCT and its integer approximation have been proposed and applied to image coding and the authors suggest the efficient implementation of an integer steerable DCT in HEVC standard. As the transform has to be inserted in the rate-distortion optimization loop, auxiliary information may be signaled and this is not an easy task. Using the IntDCT of HEVC explore the possibility to rotate this transform using the technique developed in [137] for all profiles and compare with the HEVC HM software.

P.7.25. See P.7.24. Using the steerable DCT and its integer approximation in HEVC invariably results in increased implementation complexity. Investigate this thoroughly.

P.7.26. Performance comparison of steerable DCT versus DDCT is shown in Tables I and IV of [137] with block size limited to 32×32. For super high resolution videos it is suggested to use larger block sizes such as 64×64. Extend these tables to 64×64 block size.

P.7.27. In [316] the authors have applied directional transforms for video coding based on lifting on graphs. The proposed method is compared with JM 15.1 (H.264). Extend this comparison to HEVC.

8 Transform Mirroring and Rotation

There exist important operations carried out in image and video sequences such as scaling, overlaying, mirroring, rotation and filtering [223], [430]. There are software packages in the market that use these operations as part of their engines. Some examples are BeFunky, Fotor, Gimp, Paint, Paint.net, Photo Pro, PhotoScape, Photoshop and Pixlr among many others which make use of these operations to edit images. For example, these software packages can create a mirror effect by flipping the image on the right horizontally so it appears as a reflection of the one on the left. AVS Video Editor, DVDVideoSoft, Filmora, Movavi and VideoPad are some of the popular video editors in the market that utilize mirroring and rotation. Mobile devices also have applications to manipulate images and video such as Adobe photoshop Express, AfterFocus, ArtStudio, BeFunky, Color Splash FX, Image Combine, PicsArt PhotoStudio, Pixlr, Snapseed, ToonPAINT, VSCO Cam, Magistro, Vlogit, FilmoraGo, VivaVideo, Quik, KineMaster, Funimate, Magisto, WeVideo, AndroVid, Adobe Premiere Clip, and VidTrim Pro among others.

In a mirroring or reflection operation, a shape is "flipped over" the mirror line (2D) or plane (3D). The easiest mirror operation uses one of the primary coordinate system axes as the plane and simply multiplies all x coordinates (or all y-coordinates, or z-coordinates) by -1. However, it is computationally expensive to recover the image to apply any operation in the spatial domain and recompress it again. However, compressed-domain image manipulation can be carried out [442] and the computational cost is much lower [430] because visual information is generally stored and transmitted in compressed form.

Image or video coding standards like JPEG, MPEG-1,2,4, H.264/AVC, HEVC, AV1 by AOM and AVS China utilize DCT or IntDCT [389], [462] to compress image or video data and currently, the mirroring and rotation operations are performed in the compressed domain.

This chapter is devoted to the study of mirroring and rotation operations in the compressed domain specifically when using the DCT.

SEQUENCE MIRRORING PROPERTIES FOR DISCRETE ORTHOGONAL TRANSFORMS CONSISTING OF THE EVEN AND ODD SYMMETRIC ROW VECTORS

The sequence mirroring properties for discrete orthogonal transforms consisting of the even and odd symmetric row vectors were presented by Kim and Rao in [250]. These properties are shared by the DCT, DST (also in their integer versions as IntDCT [76], integer discrete sine transform (IntDST) [52], ICT, IST, etc.), Slant and Hadamard transforms and also the variants of ICT developed for H.264/AVC, WMV9 and AVS China. The properties for the DCT have been presented by Shen and Ishwar [430] and generalized for other transforms in [250] as follows,

Let $[S_4]$ be a 4×4 orthogonal matrix with even and odd row vectors (R) defined by,

$$[S_4] = \begin{bmatrix} a & b & b & a \\ c & d & -d & -c \\ e & -f & -f & e \\ g & -h & h & -g \end{bmatrix} \begin{matrix} \rightarrow \text{Row vector even } (S_{R,0}) \\ \rightarrow \text{Row vector odd } (S_{R,1}) \\ \rightarrow \text{Row vector even } (S_{R,2}) \\ \rightarrow \text{Row vector odd } (S_{R,3}) \end{matrix} \tag{8.1}$$

Therefore, we can define $[S_4]$ in terms of its row vectors,

$$[S_4] = \begin{bmatrix} s_{R,0} \\ s_{R,1} \\ s_{R,2} \\ s_{R,3} \end{bmatrix}, \tag{8.2}$$

or in terms of its column vectors (C)

$$[S_4] = \left[s_{C,0}, s_{C,1}, s_{C,2}, s_{C,3}\right]. \tag{8.3}$$

Notice that the column vectors are related as,

$$s_{C,0} = [K_4^I]s_{C,3}$$

and

$$s_{C,1} = [K_4^I]s_{C,2}$$

where

$$[K_4^I] = \text{diag}\,(+1, -1, +1, -1). \tag{8.4}$$

The diagonal element of $[K_4^I]$ is $+1$ if the column vector is of even symmetry and -1 if the column vector is of odd symmetry.

The transpose version of $[S_4]$ can be defined by,

$$[S_4]^T = \begin{bmatrix} a & c & e & g \\ b & d & -f & -h \\ b & -d & -f & h \\ a & -c & e & -g \end{bmatrix} = \left[s_{R,0}^T, s_{R,1}^T, s_{R,2}^T, s_{R,3}^T\right] \tag{8.5}$$

$$= \begin{bmatrix} s_{C,0}^T \\ s_{C,1}^T \\ s_{C,2}^T \\ s_{C,3}^T \end{bmatrix}.$$

Since $[S_4]$ is orthogonal with orthonormal rows it follows that $[S_4][S_4]^T = [I_4]$. Suppose that we set the input sequence x in reverse order and then we transform it. We will obtain the mirrored image transform of M coefficients X_M as

$$[X_4] = [S_4][\tilde{I}_4][x] \tag{8.6}$$

Remember that $[\tilde{I}_4]$ is a 4×4 opposite diagonal identity matrix (permutation matrix) defined by

$$[\tilde{I}_4] = \begin{bmatrix} 0 & 0 & 0 & 1 \\ 0 & 0 & 1 & 0 \\ 0 & 1 & 0 & 0 \\ 1 & 0 & 0 & 0 \end{bmatrix}. \tag{8.7}$$

Notice that

$$\begin{aligned} [S_4][\tilde{I}_4] &= \begin{bmatrix} a & b & b & a \\ c & d & -d & -c \\ e & -f & -f & e \\ g & -h & h & -g \end{bmatrix} \begin{bmatrix} 0 & 0 & 0 & 1 \\ 0 & 0 & 1 & 0 \\ 0 & 1 & 0 & 0 \\ 1 & 0 & 0 & 0 \end{bmatrix} \\ &= \begin{bmatrix} a & b & b & a \\ -c & -d & d & c \\ e & -f & -f & e \\ -g & h & -h & g \end{bmatrix} \end{aligned} \tag{8.8}$$

Observe that the signs of the elements of the odd vectors change, the signs of the elements of the even vectors are left unchanged as well as the absolute values of all the elements. Using (8.4)

$$\begin{aligned} [S_4][\tilde{I}_4] &= \left([K_4^I]s_{C,0}, [K_4^I]s_{C,1}, [K_4^I]s_{C,2}, [K_4^I]s_{C,3}\right) \\ &= [K_4^I]\left(s_{C,0}, s_{C,1}, s_{C,2}, s_{C,3}\right) \\ &= [K_4^I][S_4] \end{aligned} \tag{8.9}$$

Now, let the mirror of the input sequence x be x_M. From (8.9),

$$\begin{aligned} X_M &= [S_4]x_M \\ &= [S_4][J_4]x \\ &= [K_4^I][S_4]x \\ &= [K_4^I]X \end{aligned} \tag{8.10}$$

or

$$\begin{aligned} x_M &= [S_4]^T X_M \\ &= [S_4]^T [K_4^I]X \end{aligned} \tag{8.11}$$

Block transform of length N of the form (8.1) is referred to as type I orthogonal transform and includes DCT, DST and slant transform and can be generalized as

$$[S_N]^T = \left[s_{R,0}, s_{R,1}, \cdots, s_{R,N-2}, s_{R,N-1}\right].$$

In a more general form, if we consider the even (e) and odd (o) symmetric column vectors of $[S_N]^T$,

$$[S_N]^T = \begin{cases} [e_R, o_R, \cdots, e_R, o_R] & \text{for } N \text{ even} \\ [e_R, o_R, e_R, \cdots, o_R, e_R] & \text{for } N \text{ odd} \end{cases} \tag{8.12}$$

The Walsh-Hadamard transform (WHT) has a similar property. If we let $[S^{WH}]$ be a 8×8 Walsh-Hadamard transform. Then,

$$[S^{WH}]^T = [e_R, o_R, o_R, e_R, o_R, e_R, e_R, o_R] \tag{8.13}$$

The type of (8.13) is referred to as type II orthogonal transform. To change an input vector into its reverse order, the signs of elements of even symmetric row vectors are not changed. however, the signs of elements of odd symmetric row vectors must change. Then we can define a diagonal matrix similar to (8.4) as,

$$[K_8^{II}] = \text{diag}\,(+1, -1, -1, +1, -1, +1, +1, -1)\,. \tag{8.14}$$

Therefore, similar to (8.10), we can compute X_M^{WH} transform coefficients of the mirrored sequence of an input sequence x as

$$X_M^{WH} = [S^{WH}][\tilde{I}_8]x = [K_8^{II}]X_M^{WH} \tag{8.15}$$

where $[\tilde{I}_8]$ is the 8×8 opposite diagonal matrix. Thus we can mirror a sequence in the transformed domain just by changing signs of transform coefficients.

Example 8.1. *In the following example the mirroring or rotation of an image is ilustrated using an* 8×8 *block. Given the original block,*

$$[x] = \begin{bmatrix} x_{00} & x_{01} & x_{02} & \cdots & x_{07} \\ x_{10} & x_{11} & x_{12} & \cdots & x_{17} \\ \vdots & \vdots & \vdots & \ddots & \vdots \\ x_{70} & x_{71} & x_{72} & \cdots & x_{77}, \end{bmatrix}, \tag{8.16}$$

the horizontal mirrored block of $[x]$ *is*

$$[x_h] = \begin{bmatrix} x_{07} & x_{06} & x_{05} & \cdots & x_{00} \\ x_{17} & x_{16} & x_{15} & \cdots & x_{10} \\ \vdots & \vdots & \vdots & \ddots & \vdots \\ x_{77} & x_{76} & x_{75} & \cdots & x_{70} \end{bmatrix}, \tag{8.17}$$

and the vertical mirrored block of $[x]$ *is*

$$[x_v] = \begin{bmatrix} x_{70} & x_{60} & x_{50} & \cdots & x_{00} \\ x_{71} & x_{61} & x_{51} & \cdots & x_{01} \\ \vdots & \vdots & \vdots & \ddots & \vdots \\ x_{77} & x_{67} & x_{57} & \cdots & x_{07} \end{bmatrix}. \tag{8.18}$$

TWO-DIMENSIONAL EXTENSION

Images can be divided into non-overlapping blocks of size $N \times N$ (i.e., $N = 8$). Each block can be represented as a matrix $[x] = \{x(n_1, n_2)\}_{n_1,n_2=0}^{7}$ as shown in Example 8.1.

Let $[S_8]$ be an orthogonal transform of size 8×8. The two-dimensional transform coefficients $[X]$ of an input image block $[x]$ are computed as

$$[X] = [S_8][x][S_8]^T. \tag{8.19}$$

Let the 2D-transform of horizontally mirrored sequence be $[X_H]$ and the 2D-transform of vertically mirrored sequence be $[X_V]$, then $[X_H]$ and $[X_V]$ can be computed as,

$$[X_H] = [X][K_8] \tag{8.20}$$

$$[X_V] = [K_8][X] \tag{8.21}$$

where $[K_8]$ can be $[K_8^I]$ or $[K_8^{II}]$ depending on the assocciated orthogonal transform. In other words,

$$[K_8^I] = \mathrm{diag}(1, -1, 1, -1, 1, -1, 1, -1) \tag{8.22}$$

for type I orthogonal transform and $[K_8^{II}]$ is defined in (8.14) for type II.

Procedure 3. The following is the general procedure applicable to $N \times N$ 2D-transform having even and odd symmetric vectors by choosing $[K_8^I]$ appropriately.

1. Apply 8×8 2D-transform to the non-overlapping blocks of size 8×8 of the image.
2. Set the size of a rectangular block to be horizontally flipped. Horizontal and vertical sizes should be integer multiples of eight according to the transform size.
3. Compute transform-domain image flipping for each 2D-transform block by using (8.20).
4. Rotate transform blocks horizontally within the rectangular block. The most left transform block goes to the most right, and vice versa.

The following MATLAB code implements the horizontal and vertical rotation operations of the complete Lena image using (8.20) and (8.21).

```
clear;
Im = imread('lena.pgm');
ImH = zeros(size(Im));
ImV = zeros(size(Im));
S = size(Im);
N = 8;
K8_1 =diag([1 -1 1 -1 1 -1 1 -1]);
%% Horizontal rotation
```

```
for i =1:8:S(1)
    k=0;
    for j =1:8:S(2)
        ImH(i:i+7, S(2)-N+1-(8*k):S(2)-(8*k)) = ...
            idct2(dct2(Im(i:i+7, j:j+7))* K8_1);
        k = k+1;
    end
end
%% vertical rotation
for i =1:8:S(1)
    k=0;
    for j =1:8:S(2)
        ImV(S(1)-N+1-(8*k):S(1)-(8*k), i:i+7 ) = ...
            idct2(K8_1*dct2(Im(j:j+7, i:i+7)));
        k = k+1;
    end
end

figure; imagesc(Im); colormap(gray);
figure; imagesc(ImH); colormap(gray);
figure; imagesc(ImV); colormap(gray);
```

Figure 8.1 shows (a) the Lena image and the rotated (b) horizontally and (c) vertically using the MATLAB code.

(a) Original. (b) Horizontal. (c) Vertical.

Figure 8.1: Lena image (a) original, rotated (b) horizontally and (c) vertically.

ROTATION BY 90 DEGREES

We can also rotate an image block by 90^0 in the 2D-transform domain by using the proposed scheme. Let us denote the 2D-transform of the 90^0 rotated blocks in the spatial domain $[x_{90^0}]$ as $[X_{90^0}]$. Then, we can compute 2D-transform of rotated blocks in the spatial domain using (8.9) as

$$\begin{aligned}[X_{90^0}] &= [S][x]^T[\tilde{I}_8][S]^T \\ &= [S][x]^T[S]^T[K_8^I] \\ &= [X]^T[K_8^I] = [X_V]^T. \end{aligned} \tag{8.23}$$

ROTATION BY 180 DEGREES

In order to rotate by 180^0 we need to define the matrix $[W]$ which can be $[W^I]$ or $[W^{II}]$ depending on the associated orthogonal transform. Here the matrices are defined as,

$$[W^I] = \begin{bmatrix} 1 & -1 & 1 & -1 & 1 & -1 & 1 & -1 \\ -1 & 1 & -1 & 1 & -1 & 1 & -1 & 1 \\ 1 & -1 & 1 & -1 & 1 & -1 & 1 & -1 \\ -1 & 1 & -1 & 1 & -1 & 1 & -1 & 1 \\ 1 & -1 & 1 & -1 & 1 & -1 & 1 & -1 \\ -1 & 1 & -1 & 1 & -1 & 1 & -1 & 1 \\ 1 & -1 & 1 & -1 & 1 & -1 & 1 & -1 \\ -1 & 1 & -1 & 1 & -1 & 1 & -1 & 1 \end{bmatrix}. \tag{8.24}$$

$$[W^{II}] = \begin{bmatrix} 1 & -1 & -1 & 1 & -1 & 1 & 1 & -1 \\ -1 & 1 & 1 & -1 & 1 & -1 & -1 & 1 \\ -1 & 1 & 1 & -1 & 1 & -1 & -1 & 1 \\ 1 & -1 & -1 & 1 & -1 & 1 & 1 & -1 \\ -1 & 1 & 1 & -1 & 1 & -1 & -1 & 1 \\ 1 & -1 & -1 & 1 & -1 & 1 & 1 & -1 \\ 1 & -1 & -1 & 1 & -1 & 1 & 1 & -1 \\ -1 & 1 & 1 & -1 & 1 & -1 & -1 & 1 \end{bmatrix}. \tag{8.25}$$

Let us denote the 2D-transform of the 180^0 rotated blocks in the spatial domain $[x_{180^0}]$ as $[X_{180^0}]$. Then we can compute 2D-transform of rotated blocks in the spatial domain using (8.9) as,

$$\begin{aligned}[X_{180^0}] &= [K_8][X][K_8] \\ &= [W] \odot [X] \end{aligned} \tag{8.26}$$

The symbol $\odot$ represents element-by-element multiplication.

ROTATION BY 270 DEGREES

Let us denote the 2D-transform of the 270^0 rotated blocks in the spatial domain $[x_{270^0}]$ as $[X_{270^0}]$. Then, we can compute the 2D-transform of rotated blocks in the spatial domain using (8.9) as,

$$\begin{aligned}[X_{270^0}] &= [S][\tilde{I}_8][x]^T[S]^T \\ &= [K_8][S][x]^T[S]^T \\ &= [K_8^I][X]^T = [X_H]^T \end{aligned} \tag{8.27}$$

Also, we can select a region of consecutive DCT blocks in the image and flip horizontally according to the following steps,

1. Apply the 8×8 2D-DCT to the image.
2. Set the size of a rectangular block to be horizontally flipped. Horizontal and vertical sizes should be multiples of eight according to the DCT size.
3. Compute DCT-domain image flipping each DCT block by using (8.20).
4. Rotate horizontally DCT blocks within the rectangular block. The most left DCT block goes to the most right, and vice versa.

A selected region of consecutive DCT blocks of the Lena image is shown in Figure 8.2. For the 2D-DCT domain image rotation of 90°, DCT coefficients matrix is transposed and then the result is flipped horizontally in the DCT-domain. Similarly we can rotate a rectangular image by 270°. The DCT-domain image rotation of 270° on Boat image is shown in Figure 8.3.

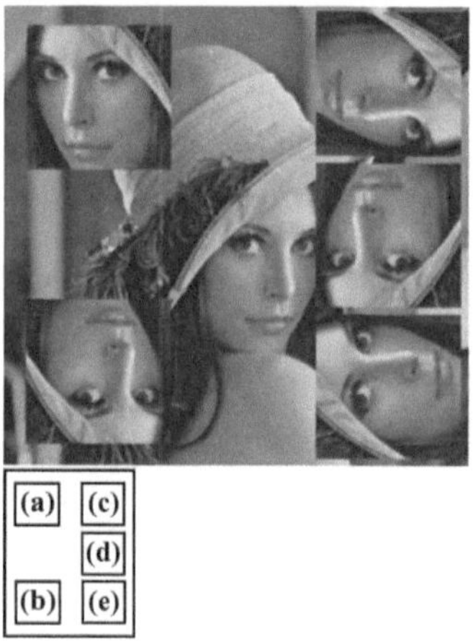

Figure 8.2: Mirroring or rotation of portion of the Lena image in the 2D-DCT domain: (a) horizontal mirroring, (b) vertical mirroring, (c) rotation by 90°, (d) rotation by 180° and (e) rotation by 270°.

Figure 8.3: Mirroring or rotation of portion of the Boat image in the 2D-DCT domain: (a) translation, (b) horizontal mirroring of (a), (c) vertical mirroring of (a), (d) translation, (e) rotation of (d) by 90°, (f) rotation of (d) by 180°, and (g) rotation of (d) by 270°.

IMAGE MIRRORING AND ROTATION IN THE WAVELET DOMAIN

We previously mentioned that Kim and Rao [250] generalized the rotation by 90, 180 and 270 degrees for any orthogonal transform. In this section, we introduce the mirroring and rotation of the image in the wavelet domain. The JPEG 2000 standard [87], [441] is a powerful tool with capabilities for designers and users of networked image applications and it is based on the discrete wavelet transform (DWT). At the input, the source image is decomposed and separated into components and each component divided into tiles which is the basic unit of the original or reconstructed image. Each tile is decomposed in frequency subbands using the DWT to describe the frequency characteristics of local areas of the tile components. The coefficients in the subbands are quantized and collected into rectangular arrays called code blocks. The bit planes of the code blocks are entropy coded and markers added for resilience error purposes.

The codestream has a main header at the beginning to describe the original image and the various decomposition and coding styles that are used to locate, extract, decode and reconstruct the image with the desired resolution, fidelity, region of interest and other characteristics [87]. The basic encoding engine of JPEG 2000 is based on the coder called embedded block coding with optimized truncation (EBCOT) algorithm which is one of the most time consuming parts of this standard. The filters used by JPEG 2000 for transformation are the 9/7 Daubechies and the 5/3 Le Gall filter banks. Several software implementations of the JPEG 2000 encoder/decoder can be obtained from [237].

To use operations of mirroring or rotation in the compressed images, one method can be to transform the image back into the spatial domain and proceed with the operation. However, mirroring and rotation operations in the transform domain for DCT and DST have already been implemented in [250], [383] and [430] and redefined in [250] to be used in the wavelet domain. This process reduces the complexity because it avoids the use of backward transformation.

PERFECT RECONSTRUCTION

Let us consider the 1D two channel filterbank shown in Figure 8.4. The filter performs one level of analysis and synthesis. At the analysis side there are one low pass and one high pass filters, $H_0(\omega)$ and $H_1(\omega)$, respectively. $F_0(\omega)$ and $F_1(\omega)$ are the low pass and the high pass filters on the synthesis side respectively [455].

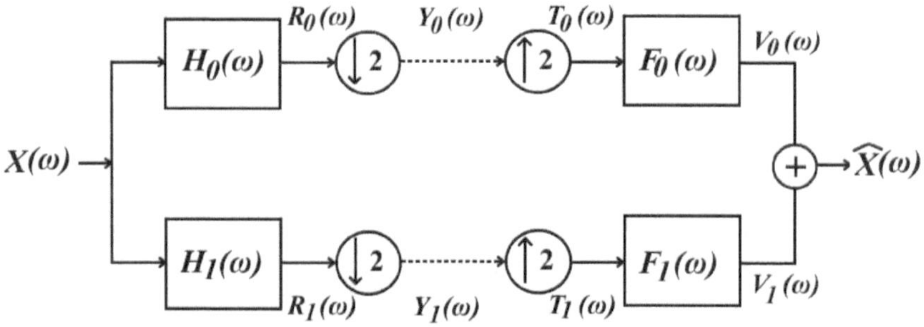

Figure 8.4: 1D-two channels filterbank.

At the analysis side the output of the filters are defined by,

$$R_0(\omega) = X(\omega)H_0(\omega) \tag{8.28}$$

$$R_1(\omega) = X(\omega)H_1(\omega). \tag{8.29}$$

The output of the analysis stage is

$$Y_0(\omega) = \frac{1}{2}\left[R_0\left(\frac{\omega}{2}\right) + R_0\left(\frac{\omega}{2} + \pi\right)\right] \tag{8.30}$$

$$Y_1(\omega) = \frac{1}{2}\left[R_1\left(\frac{\omega}{2}\right) + R_1\left(\frac{\omega}{2} + \pi\right)\right]. \tag{8.31}$$

Therefore,

$$Y_0(\omega) = \frac{1}{2}\left[X\left(\frac{\omega}{2}\right)H_0\left(\frac{\omega}{2}\right) + X\left(\frac{\omega}{2}+\pi\right)H_0\left(\frac{\omega}{2}+\pi\right)\right] \tag{8.32}$$

$$Y_1(\omega) = \frac{1}{2}\left[X\left(\frac{\omega}{2}\right)H_1\left(\frac{\omega}{2}\right) + X\left(\frac{\omega}{2}+\pi\right)H_1\left(\frac{\omega}{2}+\pi\right)\right]. \tag{8.33}$$

The outputs of the upsamplers are

$$T_0(\omega) = Y_0(2\omega) \tag{8.34}$$

$$T_1(\omega) = Y_1(2\omega). \tag{8.35}$$

Hence,

$$T_0(\omega) = \frac{1}{2}\left[X(\omega)H_0(\omega) + X(\omega+\pi)H_0(\omega+\pi)\right] \tag{8.36}$$

$$T_1(\omega) = \frac{1}{2}\left[X(\omega)H_1(\omega) + X(\omega+\pi)H_1(\omega+\pi)\right]. \tag{8.37}$$

The output of the synthesis filters are

$$V_0(\omega) = \frac{1}{2}\left[X(\omega)H_0(\omega) + X(\omega+\pi)H_0(\omega+\pi)\right]F_0(\omega) \tag{8.38}$$

$$V_1(\omega) = \frac{1}{2}\left[X(\omega)H_1(\omega) + X(\omega+\pi)H_1(\omega+\pi)\right]F_1(\omega). \tag{8.39}$$

Adding the output of the synthesis filters and reordering terms,

$$\hat{X}(\omega) = V_0(\omega) + V_1(\omega) = \frac{1}{2}\left[F_0(\omega)H_0(\omega) + F_1(\omega)H_1(\omega)\right]X(\omega) + \frac{1}{2}\left[F_0(\omega)H_0(\omega+\pi) + F_1(\omega)H_1(\omega+\pi)\right]X(\omega+\pi) \tag{8.40}$$

Notice that the second term is the aliasing term. Hence no distortion and aliasing cancellation conditions must be fulfilled, therefore,

$$F_0(\omega)H_0(\omega) + F_1(\omega)H_1(\omega) = 2e^{-j\omega l} \tag{8.41}$$

$$F_0(\omega)H_0(\omega+\pi) + F_1(\omega)H_1(\omega+\pi) = 0 \tag{8.42}$$

The non-distortion term (8.41) must be equal to a constant term multiplied only by l delays. For the aliasing cancellation we must select the filters configuration as [455],

$$F_0(\omega) = H_1(\omega+\pi) \tag{8.43}$$

$$F_1(\omega) = -H_0(\omega+\pi). \tag{8.44}$$

Substituting (8.43) and (8.44) into (8.41)

$$F_0(\omega)H_0(\omega) - F_0(\omega+\pi)H_0(\omega+\pi) = 2e^{-j\omega l} \tag{8.45}$$

Let the product filter be $P_0(\omega) = F_0(\omega)H_0(\omega)$ then (8.45) can be defined as,

$$P_0(\omega) - P_0(\omega+\pi) = 2e^{-j\omega l} \tag{8.46}$$

The left-hand side of (8.46) is an odd function then l in the right-hand side must be an odd number. Therefore, we can multiply both sides of (8.46) by $e^{j\omega l}$, incorporate these delays to the product filter and do $P_0(\omega+\pi) = -P_0(\omega+\pi)$ then,

$$P_0(\omega) + P_0(\omega+\pi) = 2, \tag{8.47}$$

which is the condition of the normalized product filter for perfect reconstruction. Notice that the product filter is a half band low pass filter. A half-band filter is a low-pass filter with same pass-band and stop-band ripples and the pass-band-edge and stop-band-edge frequencies are equidistant from the half-band frequency $\pi/2$ or $f_s/4$, where f_s is the sampling frequency. In other words, the frequency response of the filter is anti-symmetric around $\pi/2$ [455].

TWO-DIMENSIONAL EXTENSION

Our aim is to apply the mirroring and rotation operations in the wavelet domain, that is, at the output of the analysis side. Hence, both (8.32) and (8.33) can be expressed as,

$$Y_m(\omega) = \frac{1}{2}\left[X\left(\frac{\omega}{2}\right)H_m\left(\frac{\omega}{2}\right) + X\left(\frac{\omega}{2}+\pi\right)H_m\left(\frac{\omega}{2}+\pi\right)\right], \quad m = 0,1. \tag{8.48}$$

If we flip the image in time domain, after the analysis filter bank we get the expression,

$$Y_m(-\omega) = \frac{1}{2}\left[X\left(-\frac{\omega}{2}\right)H_m\left(-\frac{\omega}{2}\right) + X\left(-\frac{\omega}{2}+\pi\right)H_m\left(-\frac{\omega}{2}+\pi\right)\right], \quad m = 0,1. \tag{8.49}$$

Equation (8.49) can be seen as filtering of a reversed sequence by reversed analysis filters. In order to maintain conditions of perfect reconstruction and obtain the reverse sequence as our final output, we can reverse the synthesis filters. This is equivalent to say that reversing the input and the filter would produce the output reversed.

For image analysis, filtering along rows is followed by filtering along columns. The structure of the 2D analysis filter bank is shown in Figure 8.5 using the 1D analysis and synthesis filter banks.

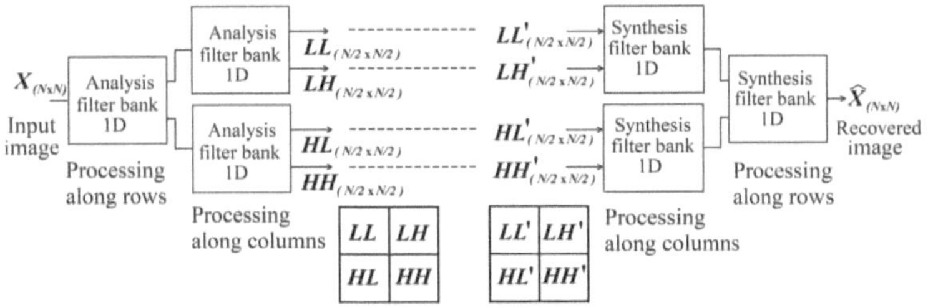

Figure 8.5: 2D filterbank showing one level of decomposition.

The input image X of size $N \times N$ is decomposed in four frequency subbands (LL, LH, HL and HH) of size $\frac{N}{2} \times \frac{N}{2}$. After processing the subbands, the input to the synthesis bank are the subbands LL', LH', HL' and HH'. If no processing is applied then we have perfect reconstruction, then we can say that input to the synthesis filterbank are the subbands at the output of the analysis filterbank multiplied by the identity matrix I of size $\frac{N}{2} \times \frac{N}{2}$. Considering the case where our analysis and synthesis filters are symmetric, as is the JPEG 2000 9/7 and 5/3 filters [441]. For symmetric filters, coefficients in forward and reverse order are the same, i.e., $H_m(\omega) = H_m(-\omega) \quad m = 0, 1$. Hence $H_m(-\omega)$ can be replaced by $H_m(\omega)$ and (8.49) can be seen as filtering of a reversed input by the analysis filter bank. Therefore, the synthesis filter bank can be applied without reversing the synthesis filters as well.

Horizontal and Vertical Mirroring

In the case of horizontal mirroring, we simply reverse each component and apply it to the synthesis filterbank. This is carried out by post-multiplying each subband by the reverse identity matrix $\tilde{I}$ of the same dimension as the subbands, then the horizontal mirrored image is recovered by applying the synthesis filterbank to the affected subbands,

$$LL' = LL \cdot \tilde{I} \tag{8.50}$$

$$LH' = LH \cdot \tilde{I} \tag{8.51}$$

$$HL' = HL \cdot \tilde{I} \tag{8.52}$$

$$HH' = HH \cdot \tilde{I} \tag{8.53}$$

For vertical mirroring, we pre-multiply each subband by the reverse identity matrix $\tilde{I}$ instead of post-multiplying as follows,

$$LL^{'} = \tilde{I} \cdot LL \tag{8.54}$$

$$LH^{'} = \tilde{I} \cdot LH \tag{8.55}$$

$$HL^{'} = \tilde{I} \cdot HL \tag{8.56}$$

$$HH^{'} = \tilde{I} \cdot HH \tag{8.57}$$

Rotation by 90 Degrees

In case of rotation by 90°, the subbands are post-multiplied by $\tilde{I}$, then transposed. Before applying the synthesis filter bank exchange $LH^{'}$ and $HL^{'}$,

$$LL^{'} = (LL \cdot \tilde{I})^T \tag{8.58}$$

$$LH^{'} = (HL \cdot \tilde{I})^T \tag{8.59}$$

$$HL^{'} = (LH \cdot \tilde{I})^T \tag{8.60}$$

$$HH^{'} = (HH \cdot \tilde{I})^T \tag{8.61}$$

Rotation by 180 Degrees

To rotate the image by 180° we simply have to pre-multiply and post-multiply by $\tilde{I}$, the subbands before the analysis the filterbank.

$$LL^{'} = \tilde{I} \cdot LL \cdot \tilde{I} \tag{8.62}$$

$$LH^{'} = \tilde{I} \cdot LH \cdot \tilde{I} \tag{8.63}$$

$$HL^{'} = \tilde{I} \cdot HL \cdot \tilde{I} \tag{8.64}$$

$$HH^{'} = \tilde{I} \cdot HH \cdot \tilde{I} \tag{8.65}$$

Rotation by 270 Degrees

In rotation by 90°, the subbands are pre-multiplied by $\tilde{I}$, then transposed. Before applying the synthesis filter bank exchange $LH^{'}$ and $HL^{'}$,

$$LL^{'} = (\tilde{I} \cdot LL)^T \tag{8.66}$$

$$LH^{'} = (\tilde{I} \cdot HL)^T \tag{8.67}$$

$$HL^{'} = (\tilde{I} \cdot LH)^T \tag{8.68}$$

$$HH^{'} = (\tilde{I} \cdot HH)^T \tag{8.69}$$

Figure 8.6 shows the Lena image and some portions of the same image mirrored and rotated after the Duablechies 9/7 analysis filter bank. Figure (a) is the horizontal mirroring, (b) rotation by 180°, (c) rotation by 270°, (d) vertical mirroring and (e) rotation by 90°.

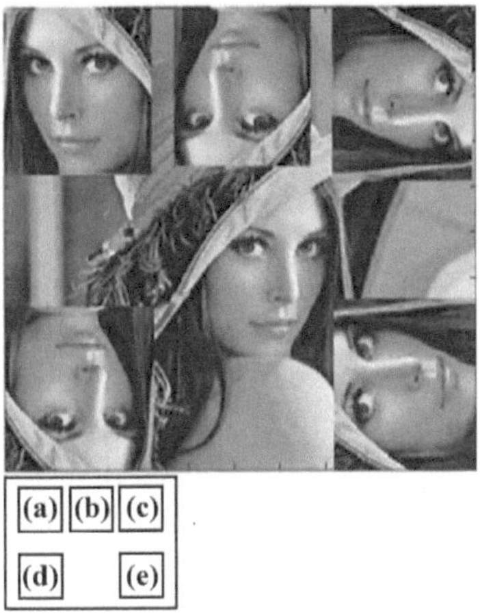

Figure 8.6: Lena image with a region of the same image mirrored and rotated after the Daubechies 9/7 analysis filter bank. (a) Horizontal mirroring, (b) rotation by 180°, (c) rotation by 270°, (d) vertical mirroring and (e) rotation by 90°.

Figure 8.7 shows the Girl image and some portions of the same image mirrored and rotated after the LeGal 5/3 analysis filter bank. Figure (a) is the horizontal mirroring, (b) rotation by 180°, (c) rotation by 270°, (d) vertical mirroring and (e) rotation by 90°.

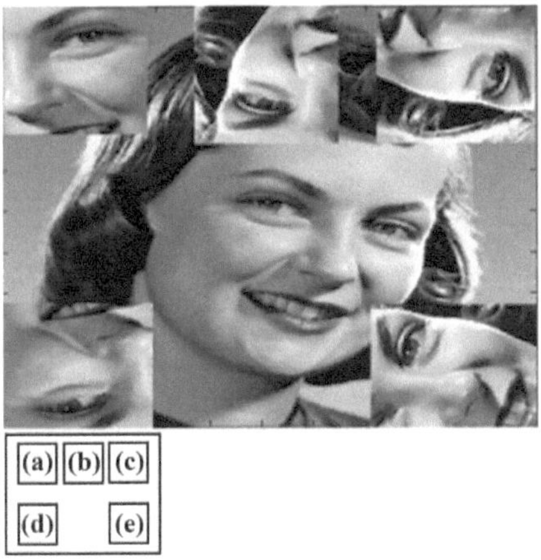

Figure 8.7: Girl image with a region of the same image mirrored and rotated after the Legal 5/3 analysis filter bank. (a) Horizontal mirroring, (b) rotation by 180°, (c) rotation by 270°, (d) vertical mirroring and (e) rotation by 90°.

The following MATLAB code executes the horizontal and vertical mirroring and the rotation by 90, 180 and 270 degrees operations on a gray scale image by using one level of DWT decomposition.

```
clear all;
% Read the image
X = imread('peppers.pgm');

wname = 'bior3.5';
[LL,LH,HL,HH] = dwt2(X,wname,'mode','sym');

S = size(LL);
Itilde = zeros(size(LL));
N = S(1,1);
for i = 1:S(1,2)
    Itilde(i,N) =1;
    N = N-1;
end

% Horizontal mirroring

LLh = LL*Itilde;
LHh = LH*Itilde;
HLh= HL*Itilde;
HHh = HH*Itilde;
Xhoriz = round(idwt2(LLh, LHh, HLh, HHh,wname ));

% Vertical mirroring

LLv = Itilde*LL;
LHv = Itilde*LH;
HLv = Itilde*HL;
HHv = Itilde*HH;
Xvert = round(idwt2(LLv, LHv, HLv, HHv,wname ));

% Rotation by 90 Degrees

LL90 = (LL*Itilde)';
HL90 = (LH*Itilde)';
LH90 = (HL*Itilde)';
HH90 = (HH*Itilde)';
X90 = round(idwt2(LL90, LH90, HL90, HH90,wname));

% Rotation by 180 Degrees

LL180=Itilde*LL*Itilde;
LH180=Itilde*LH*Itilde;
HL180=Itilde*HL*Itilde;
```

```
HH180=Itilde*HH*Itilde;
X180 = round(idwt2(LL180, LH180, HL180, HH180,wname));

% Rotation by 270 Degrees

LL270 = (Itilde*LL)';
LH270 = (Itilde*HL)';
HL270= (Itilde*LH)';
HH270 = (Itilde*HH)';
X270 = round(idwt2(LL270, LH270, HL270, HH270,wname));

figure(1);imagesc(X); colormap(gray);
figure(2);imagesc(Xhoriz); colormap(gray);
figure(3);imagesc(Xvert); colormap(gray);
figure(4);imagesc(X90); colormap(gray);
figure(5);imagesc(X180); colormap(gray);
figure(6);imagesc(X270); colormap(gray)
```

Notice that, after each processing the recovered image can be clipped by inserting the following code,

$$\text{Index} = \text{find(RecoveredImage} < 0\text{);}$$
$$\text{RecoveredImage(Index)} = 0;$$

APPLICATIONS

The set of applications of mirroring and rotation operations can be in the compressed domain as shown by Kim and Rao [250] and can be implemented in any place where compression is carried out using DCT or DWT and the decorrelation engine. Following this, there are some systems that use the above mentioned transforms as decorrelation engine and where mirroring and rotations can be applied when needed.

The use of ultra high definition (UHD), high dynamic rates (HDR) and high frame rate (HFR) video has increased the bandwidth requirement for transmission and storage of video content [107], [408], [535]. To ease such data intensive workflows a new standard called JPEG XS has been standardized. JPEG XS is a low energy format that defines compression algorithms with very low latency and very low complexity in hardware and software and target virtual reality (VR), high-resolution image formats like 4K and to work with 5G mobile networks and wireless systems [212]. The format offers mezzanine compression, visually lossless quality with compression ratios up to 6:1, frame rate from 24 to 120 fps and it can be implemented on different platforms such as CPU, GPU, FPGA and ASIC. [212]. JPEG XS is based on the 5/3 LeGall wavelet transform and it is designed to real-time process 4K 4:4:4 8 bit 60p content.

JPEG XT [206] defines a common framework for the lossy and lossless representation of HDR images [206] and specifies a series of backward compatible extensions to the legacy JPEG standard [202]. Modern display devices are appearing with the capability to display HDR images and video with a peak brightness of up to 4,000 nits and to support wide color gamut (WCG) [214]. JPEG XT has four profiles based on the DCT. This standard offers features such as lossy or lossless representation of WCG and HDR images as well as backward compatibility with the legacy JPEG standard [19], [20].

JPEG XR or JPEG-eXtended Range is a lossy and lossless still-image compression standard and file format [202] developed and patented by Microsoft under the name HD photo (Windows Media Photo). The format is preferred for Open XML Paper Specification documents. JPEG XR has image coding capabilities beyond the original baseline JPEG standard and it is intended for high/extended dynamic range imagery [439].

JPEG 2000 is an image coding standard that uses compression techniques based on the DWT [205]. It offers an extremely high level of scalability and accessibility. However the entropy encoder / decoder is far too complex because it uses EBCOT entropy coder [441].

The HD video and beyond HD formats (e.g., $4k \times 2k$ or $8k \times 4k$ resolution) have created the needs for higher coding efficiency. However, the new encoders significantly reduce bit rates compared to the preceding standards but at the expense of extremely high encoding complexity. The current and the future video coding standards are based on the IntDCT [24], [204], [208], [356], [359], [534] and the transform size tends to be larger. Consequently, the codecs are very complex especially at the encoder side where the complexity is exacerbated because the operations of filtering, motion estimation, motion compensation and transformation used to produce high quality video sequences.

Coding schemes have been analyzed using several objective and subjective metrics because the artifacts induced by the compression systems are very different. Test images and sequences utilized in the experiments must represent a broad gamut of content relevant to scheme under test. There are several metrics to evaluate the performance of codecs. The PSNR is one of the commonly used metrics to evaluate the quality of reconstruction of lossy compression codecs. However, most of the time PSNR does not represent the subjective image quality. Other perceptually driven objective metrics have been used in the past, for example, the structural similarity index (SSIM) [526] and multi-scale structural similarity index (MSSIM) [527] since both algorithms extract the structural information similar to the human visual system (HVS). Other important quality metrics are the visual information fidelity [427], [428], PSNR-HVS-M [374] and DC-Tune measurements [444]. Source codes can be obtained from [375]. Currently, the metrics used are the Bjøntegaard delta peak-signal-to-noise ratio (BD-PSNR), which corresponds to the average PSNR difference in dB for the same bit rate and the Bjøntegaard delta rate (BD-rate), which corresponds to the average bit rate difference in percent for the same PSNR. The Bjøntegaard model was proposed by Gisle Bjøntegaard [42] during the development of H.264/AVC (see Appendix B) and it is used by various experts to calculate the coding efficiency of compression standards and was used during the development of the H.264/AVC, multiview video coding (MVC) [204], HEVC [208], and multiview extensions of HEVC [497]. By default, this is an industry accepted metric.

SUMMARY

This chapter explains operations of mirroring and rotation in the compressed domain. Previously, when any of these operations had to be applied to an image or video sequence, the image had to be passed through a decoding/encoding process. However, these operations can be done in the compressed domain by using simple operations that do not add more complexity to the system and there is no need to decode the image.

PROBLEMS

8.1. Explore the possibility to develop a method to rotate an image by 45° in the DCT domain.
8.2. Rotate an image in the compress domain by 270°. Using the original and the rotated images, verify the angle of rotation (Hint: Use image registration techniques [53]).
8.3. Repeat problem 8.2. adding Gaussian noise with variance $\sigma^2 = 0.01$ to the image before compression.
8.4. Repeat problem 8.3. adding Gaussian noise with variance $\sigma^2 = 0.1$ to the image before compression.
8.5. Explore the possibility to develop a method to rotate an image by 45° in the DWT domain.
8.6. Compute the complexity of problem 8.1.
8.7. Compute the complexity of problem 8.5.

PROJECTS

P.8.1. JPEG XT is a backward-compatible extension of the JPEG graphics format [206]. Develop a C/C++ function to implement the mirroring operations in the compressed domain for this standard.
P.8.2. JPEG XT is a backward-compatible extension of the JPEG graphics format [206]. Develop a C/C++ function to implement the rotation operations in the compressed domain for this standard.
P.8.3. Compare the performance in time after adding to the JPEG XT reference software [217] the functions developed in P.8.1. and P.8.2.
P.8.4. In [3] the Jasper project was developed in C programming language by Michael Adams following the recommendation of JPEG 2000 [205], [3] and included in the JPEG-2000 Part-5 standard. Download the reference software and program a C function to implement horizontal mirroring in the DWT domain using the Daubechies 9/7 filters and the test images in [3].
P.8.5. Repeat project P.8.4. to program a C function to implement vertical mirroring in the DWT domain using the Daubechies 9/7 filters.
P.8.6. Repeat project P.8.4. to program a C function to implement rotation by 90° in the DWT domain using the Daubechies 9/7 filters.

P.8.7. Repeat project P.8.4. to program a C function to implement rotation by 180° in the DWT domain using the Daubechies 9/7 filters.

P.8.8. Repeat project P.8.4. to program a C function to implement rotation by 270° in the DWT domain using the Daubechies 9/7 filters.

P.8.9. Repeat projects P.8.4. to P.8.8. using LeGall 5/3 filters and compare the performance in time.

P.8.10. In [237] the project JJ200 to implement JPEG-2000 was developed in Java programming language. Develop the mirroring and rotation transformation operations for this language.

P.8.11. Implement mirroring and rotation in the compressed domain for H.264/AVC decoder.

P.8.12. Implement mirroring and rotation in the compressed domain for HEVC decoder.

P.8.13. Park and Ogunfunmi [367] developed a subjective measure for evaluating video quality based on 8×8 2D-DCT. They suggest that this metric is much simpler to implement compared to SSIM [526] and close in performance. Evaluate this based on various artifacts and also consider 4×4 and 16×16 besides 8×8 and explore the possibility to extend the concept to IntDCT.

P.8.14. In [13] the authors applied the rotational transform (ROT) ROT4 to 4×4 blocks and ROT8 to upper left sub matrices in all other cases (see Figures 2 and 3 in [13]). The results exhibit a BD-rate gain of 2.5 % on average for all test sequences (see Table 4 in [13]). Implement this technique using the test sequences and confirm the results.

P.8.15. In [350] the authors conducted a detailed performance comparison of HEVC main still picture (MSP) profile with other still image and video compression schemes (intra coding only) such as JPEG, JPEG 2000, JPEG-XR, H.264/AVC, VP8, VP9 and WebP. They used PSNR and BD-rate as the comparison metrics (see Figures 2–5 in [350]). Extend this comparison using test sequences with spatial resolution higher than 1280×1600. Another image compression standard to consider in this comparison is JPEG-LS. Consider this also regarding comparison with HEVC MSP profile. In project P.8.15. and [350]. Consider mean opinion score (MOS) as the metric for performance comparison of all these standards. Access reference 35 of [350] regarding HEVC subjective evaluation. Note that this project requires extensive man power, detailed testing lab and other resources.

P.8.16. In [391] a comparison of HEVC intra with JPEG, JPEG 2000 JPEG XR, JPEG LS and VP9 intra was carried out using standard definition (SD), HD, UHD and $4K$ test sequences. HEVC outperforms the other standards in terms of PSNR and SSIM at the cost of increased complexity. Consider BD-PSNR and BD-rate as comparison metrics.

P.8.17. Extend the performance comparison of project P.8.16. to $8K$ test sequences.

P.8.18. In [28] the authors developed a DCT-based local distortion detection probability (LDDP) model and proposed a HEVC complaint perceptual video coding (PVC). Using various test sequences (see V thru VII), they showed that the superiority of their PVC approach compared with the state-of-the-art PVC scheme for low delay (LD) and random access (RA) main profiles. Extend this technique to all intra (AI) main profile.

P.8.19. Extend the technique of [28] to 4*K* and 8*K* test sequences and develop tables similar to those shown in this paper. Based on these simulations, summarize the conclusions.

A Discrete Cosine Transform for Image Coding Standards

Standards ensure that products and systems are compatible and interoperable. Product manufactured according to standards are accepted worldwide. Therefore, international standardization requires collaboration between regions and countries.

The main purpose of the International Organization for Standardization (ISO) and the International Electrotechnical Commission (IEC) is to develop, maintain and promote standards in the fields of information technology (IT) and information and communications technology (ICT). The image and video coding standards impose only restrictions on the bitstream, syntax and decoder. For example, JPEG is a standard method of compressing photographic images and nowadays it is accepted around the world [202]. In the next sections we describe several image coding standards that have adopted the DCT as the main engine to decorrelate the data before compression.

JPEG

Features of the JPEG "baseline system" [202].

1. Represents a minimum of capabilities.
2. Compression of digital images with 8-bit representation.
3. Sequential processing of blocks, i.e., starting from upper left corner moving to right side.
4. Transform coding using 8x8 block DCT.
5. Scalar quantization of transform coefficients with different weighting adapted to human visual contrast sensitivity.
6. Zig-zag scan and successive entropy coding with prefix-free codes.

Features of extended DCT-based system [532]

1. Digital images with 8 and 12-bit representation.
2. Sequential as well as progressive block handling.
3. Prefix-free or arithmetic codes.

Lossless mode [532]

1. DPCM-based (no DCT).

Hierarchical mode [532]

1. Multiple pictures encoded differentially as well as non-differentially.
2. Extended DCT-based or lossless JPEG mode.

JPEG Process

1. Color conversion (RGB components $\rightarrow$ YCbCr components) to separate color from luminance.
2. Each component is divided into 8×8 blocks of samples.
3. Block shifting from [0-255] to [-128 to 127].
4. DCT is applied to each block from left to right and from top to bottom of each component.
5. Quantization is applied to each block to yield a quantized matrix.
6. Quantized matrix is entropy coded (two options for prefix free coding or three options for arithmetic coding).
7. The process is inverted to recover the image.

JPEG XR OR JPEG EXTENDED RANGE

JPEG XR [211] combines the benefits of optimized image quality and compression efficiency together with low-complexity encoding and decoding implementation requirements as well as additional features [94] compared to JPEG.

Features of JPEG XR

1. Better compression: Higher compression capability.
2. Less complexity: Lower computational and memory resource requirements.
3. Lossless and lossy compression: The signal processing steps in JPEG XR are the same for both lossless and lossy coding.
4. Image tile structure for random access and large image formats: The data for each region can be decoded separately enabling rapid access to parts of the image without decoding the entire image.
5. Support for low-complexity compressed-domain image manipulations: Full decoding is also unnecessary for certain editing operations such as cropping, horizontal or vertical flips, or cardinal rotations. Also, full decoding of the image is unnecessary for converting an image from a lossless to lossy encoding.
6. Support for embedded thumbnail images and progressive resolution refinement–embedded codestream scalability for both image resolution and fidelity.
7. Alpha plane support.
8. Bit-exact decoder results for fixed and floating point image formats.
9. Support for more color accuracy.
10. Transparency map support.
11. Metadata support.

A free copy of the JPEG XR reference software in C language can be downloaded from [216].

JPEG XT OR JPEG EXTENSION

JPEG XT [206] specifies a series of backwards compatible extensions to the legacy JPEG standard. The legacy JPEG format is often associated with a compressed image limited to a precision of 8 bits per component or 256 different values per color channel. JPEG images also use a conversion from an RGB space (Recommendation ITU-R 601) to YCbCr to decorrelate color components, and they use subsampling in the chroma signals. However, these features belong to the JPEG file interchange format (JFIF) [406], [203] and they do not limit JPEG because the original JPEG standard also defines a high-precision 12-bit mode. However, JPEG does not address current important applications that JPEG XT does in a completely backwards compatible way such as compression of images with higher bit depths (9 to 16 bits), high-dynamic-range imaging, lossless compression, and representation of alpha channels [406]. The features of JPEG XT are listed following,

1. Higher bit depth. The popular 8-bit mode still understands the JPEG XT codestream, although its output would have a precision of 8 bits per sample. JPEG XT encodes an 8-bit version of the high-precision input called the base layer and hides a second codestream, known as the enhancement layer (invisible to the legacy JPEG). Higher ranges of up to 16 bits per component or 48 bits in total can be obtained and sent to the decoder together with additional metadata to inform the decoder how to combine the base layer and the enhancement layer to form one single image of a higher precision [406].
2. High dynamic range. For natural outdoor scenes, the radiances in the image can vary over four to six orders of magnitudes, too large to represent the scene in 16-bit integer; in this case, we say that we are entering the high dynamic range (HDR). Then, the images are obtained acquiring multiple shots of the same scene at different exposure time. Here, one consistent image is computed from these inputs [406]. JPEG XT assembles the HDR image from the tone-mapped 8-bit image—the base layer—and the extension layer. Four profiles are defined to manage the complexity,

Profile A. The HDR image is described as the pixel-wise product of an inversely gamma-corrected 8-bit tone-mapped image multiplied by a common scale factor. The scale factor is encoded in logarithmic space. Profile A also includes an additive chroma residual that allows correction of color defects [19]. Figure A.1 shows the simplified encoding process of JPEG XT Profile A.

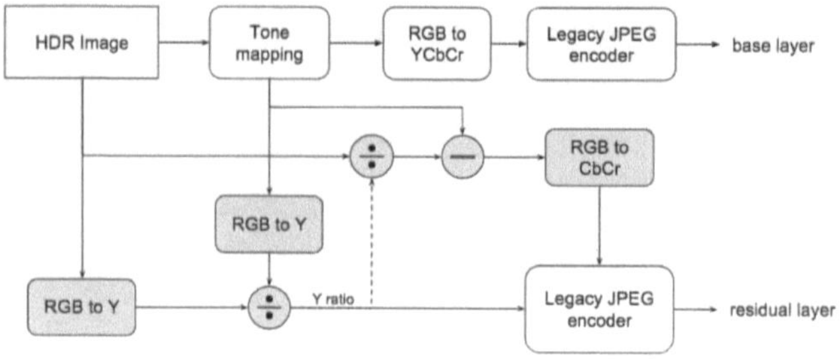

Figure A.1: Simplified encoding process of JPEG XT Profile A [85].

Profile B. The HDR image is described as the pixel-wise quotient of the inversely gamma-corrected tone-mapped image and a divisor image delivered by the extension layer. The output is scaled by one common factor, the "exposure value." Figure A.2 shows the simplified encoding process of JPEG XT Profile B.

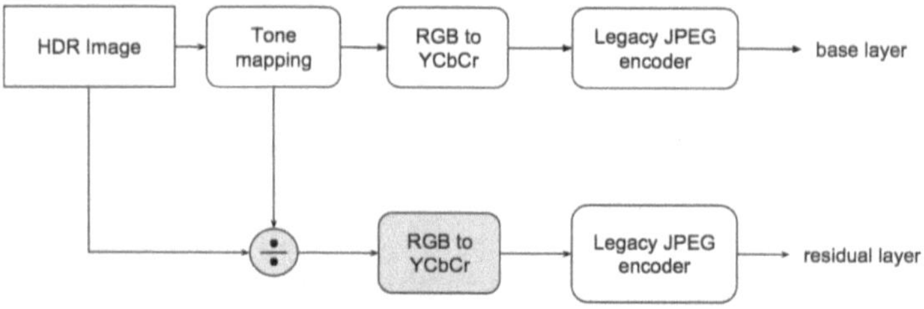

Figure A.2: Simplified encoding process of JPEG XT Profile B [85].

Profile C. Similar to profile A, in this profile the image is decomposed into an 8-bit image and scale factors. However, in this profile each component has its individual scale factor, whereas in Profile A there is only one profile factor [19]. Figure A.3 shows the simplified encoding process of JPEG XT Profile C.

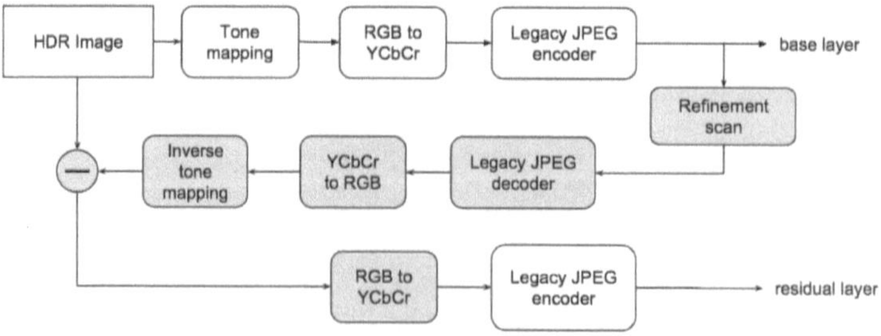

Figure A.3: Simplified encoding process of JPEG XT Profile C [85].

Profile D. Profiles A to C first reconstruct base and extension layers as independent JPEG streams and merge them pixel by pixel in the spatial domain. Profile D extends the resolution of the image precision by extending the precision of the DCT coefficients beyond the resolution allowed by legacy JPEG [399], [401], [406]. The reconstructed image data then undergoes an electro-optical transfer function (EOTF) signaled in the codestream, an idea related to the gamma transformation used in standard dynamic range.

3. Coding of opacity information. JPEG XT allows the possibility to include a transparency layer by the same means it includes extension information for higher bit depth. A second codestream, decoding to the opacity information, is hidden within the application markers of the actual foreground image, along with the meta-information to decode it [406].
4. Privacy protection and security. A portion of the image that is encrypted is encoded in the residual layer, and a decryption key is used to make it available to the end-user. First, the original image is segmented, manually or automatically, into public and protected regions that require privacy protection or other security mechanisms. Second, the identified image regions must be pixelated or protected. The differential image between the original and scrambled versions contains the pixel-wise difference between the two versions. Third, the data required for decoding the original image (the differential image) should be encrypted by a standard public/private key algorithm. Fourth, the encrypted data must be embedded into the codestream of the scrambled image [406].
5. JPEG XT Part 1 does not include lossless coding mode.

The JPEG XT consists of two layers: the base layer and the residual layer. The simplified decoder block diagram is shown in Figure A.4.

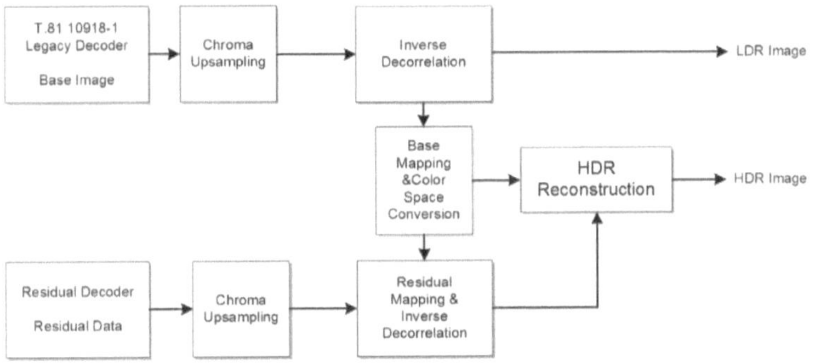

Figure A.4: JPEG XT simplified decoder block diagram [85].

A free copy of the JPEG XT reference software in C++ language can be downloaded from [217]. Also, JPEG XT Profile B reference software in C language can be downloaded from [480] as well as other third party contributions.

JPEG PLENO

Plenoptic is a term that is related to all the light, travelling in every direction in a given space. The light field is the complete spatial and angular distribution of light in certain space. In other words, light field are all the photons travelling in all directions throughout all points in space in a given area.

Plenoptic photography or light-field photography is a new imaging modality that makes possible to select part of the image to be in focus after it was acquired since all the information required to focus on any part of a scene is recorded and stored with the digital image. Therefore, a plenoptic camera captures the intensity of light in a scene, and the direction of light rays travelling in space [121].

JPEG Pleno provides a standard framework for representing and coding new imaging modalities, such as texture-plus-depth, light field, point cloud and holographic imaging [121] with the highest quality content representation with reasonable resource requirements [121]. JPEG Pleno intends to provide representation and compression mechanisms to this promising type of data [93].

In addition to the features already offered by existing image-coding formats, JPEG Pleno offers new features to enable new applications to be used in virtual, augmented and mixed-reality scenarios [121]. For example,

1. The change of field depth after capture.
2. The change of focus and refocus on objects of interest after capture.
3. The change of lighting in an already captured or synthesized scene.
4. The change of perspective and position of the observers within the scene, including free navigation.
5. Enhance the analysis and manipulation of objects within a scene, such as segmentation, modification, and even removal or replacement of specific objects.

The JPEG Pleno activities started in 2015 and will span beyond 2020.

JPEG AIC

The JPEG advanced image coding (AIC) [213], [210] is to provide better quality with reduced level of complexity while optimizing readability and clarity. It combines algorithms from the H.264/AVC [398] video coding standard and JPEG standard [202] by using H.264/AVC intra frame block prediction with a JPEG-style DCT. The entropy coder is a context adaptive binary arithmetic coding (CABAC) as in H.264/AVC. The compression scheme performs much better than JPEG and close to JPEG 2000 [3] and, sometimes outperforms JPEG 2000 for photographic images. One of the advantages is that JPEG AIC is much less complex and faster than the wavelet-based encoder JPEG 2000 but slower than JPEG.

For three channel images (RGB), the input image is converted into the YCbCr components. For grayscale the images have only one component. Each component is divided into 8×8 blocks of pixels. Each block is predicted from previously encoded and decoded blocks in the channel by using 9 prediction modes to choose the mode that minimizes the differences between the original and predicted block. In color images the block of chrominance uses the same prediction mode as the correspondent luminance block. Then, a residual is formed by subtracting the predicted block from the original block. Then DCT is applied to the residual block and the resulting coefficients are quantized using quantization levels that depends on the selected quality. The prediction mode and resulting quantized coefficients are entropy coded using CABAC. The DCT code is based on the code in the JPEG reference software from the Independent JPEG Group [200]. The features of JPEG AIC are listed following,

1. Much faster and less complex than JPEG 2000.
2. Flexible/Scalable complexity and power consumption at encoding and decoding.
3. Compressed domain manipulation.
4. Efficient 3D region of interest decoding/access.
5. For graphic images JPEG 2000 performs better.
6. Advanced error resilience .
7. For small images JPEG AIC performs better than JPEG 2000 for typical bitrates.
8. Floating point data compression to obtain the highest quality.
9. For photographic images and for some typical bit rates outperforms JPEG 2000.
10. Uses CABAC instead of Run-Length algorithm so that the quantized coefficients are not reordered.
11. Because of its high quality images, the applications of AIC include medical imaging.

A free copy of the JPEG AIC reference software in Delphi language can be downloaded from [123].

JPEG XS

The JPEG XS [212] standard provides a visually lossless, low-latency, lightweight codec for a wide range of applications including mezzanine compression in broadcast and Pro-AV markets and is also intended for the transmission of video signals between head mounted displays and the image generating source computer, requiring thus very low latencies [107].

For color images (RGB) the input color components are decorrelated by using lossless color transform as in [237] and an integer reversible DWT is applied. The wavelet coefficients are represented as sign-magnitude by a pre-quantizer and sliced the 8-least significant bit planes. The transform coefficients are analysed by a budget computation module that predicts the number of bits required for each possible quantization. A rate control algorithm calculates the smallest quantization factor that does not exceed the bit budget available. The resulting coefficients are entropy coded based on groups of four coefficients [408]. All data sections are combined into a packet structure and sent to the transmission channel via a smoothing buffer to ensure a constant bit rate at the output of the encoder. The features of JPEG XS [107], [408] are,

1. Wavelet-based codec.
2. A maximum of two vertical wavelet decompositions can be used.
3. A maximum of five horizontal wavelet decompositions can be used.
4. Targeted use cases are professional video links, IP transport, Ethernet transport, real-time video storage, video memory buffers, and omnidirectional video capture and rendering [107].
5. Ultra-low latency and ultra-low complexity while achieving visually lossless compression.
6. Permits multiple encoding and decoding cycles with minimum quality loss.
7. The standard can be implemented on different platforms such as CPU, GPU, FPGA and ASIC.
8. Visually lossless for compression ratios up to 6:1 for both natural and screen content and for at least seven encoding and decoding cycles.

SUMMARY

In the context of image coding, various aspects have to be taken into account, due to different characteristics in the natural camera captured content and hardware requirements. Therefore, new standards such as JPEG Pleno, JPEG XR, JPEG XT and JPEG XS have been released. These standards include several new tools such as the ability to deal with light field photography, backward compatibility to the legacy JPEG standard, higher bit depth, low latency, high dynamic-range imaging among others. The main core transform of these image coding standards is the DCT. This appendix presents the most important features of the most popular DCT-based image coding standards and useful projects that complements the explanations thru each chapter of the book.

PROJECTS

P.A.1. One of the three profiles in HEVC listed in final draft international standard (FDIS) (Jan. 2013) is intra frame (image) coding only. Implement this coding mode in HEVC and compare with other image coding standards such as JPEG, JPEG XR, JPEG XT, JPEG AIC and JPEG XS using MSE/PSNR, SSIM [SSIM1] and BD-rate as the metrics (see App. B). As before, perform this comparison using various test sequences at different spatial resolutions and bit rates.

P.A.2. Repeat P.A.1. for VVC reference software.

P.A.3. Implement the lossless coding of HEVC main profile proposed in [589] and validate the results. Also compare with current lossless coding methods such as JPEG 2000, etc., based on several test sequences at different resolutions and bit rates. Comparison metrics are MSE/PSNR, SSIM, BD-rates, etc. Consider the implementation complexity also in the comparison. In [57] the authors have also compared the performance of HEVC, H.264/AVC, JPEG 2000 and JPEG LS for both lossy and lossless modes. For lossy mode their comparison is based on $PSNR_{avg} = (6xPSNR_y + PSNR_u + PSNR_v)/8$ only. This is for 4:2:0 format and is by default.

P.A.4. Repeat project P.A.3. Extend this comparison based on SSIM, BD-rate and implementation complexity. Include also JPEG XR which is based on HD-Photo of Microsoft in this comparison. They have provided an extensive list of references related to performance comparison of intra coding of several standards.

P.A.5. In [589] the authors have implemented HEVC lossless coding for main profile by simply bypassing transform, quantization and in-loop filters and compared with other lossless coding methods such as JPEG 2000, ZIP, 7-Zip, WinRAR, etc. Implement this and also compare with JPEG, JPEG XR, JPEG XS, PNG, etc. Consider implementation complexity as another metric.

P.A.6. Several lossless image compression techniques have been developed– HEVC, H.264/AVC, JPEG, JPEG LS, JPEG 2000, 7-Zip, WinRAR, etc. [57], [575], [589]. Compare these techniques in terms of implementation complexity using several test images at different resolutions.

P.A.7. In [391] the author compared the performance of HEVC intra with JPEG, JPEG 2000, JPEG XR, JPEG LS and VP9 intra using SD, HD, UHD and 4K test sequences. HEVC outperforms the other standards in terms of PSNR and SSIM at the cost of increased implementation complexity. Consider BD-PSNR and BD-rate as comparison metrics.

P.A.8. See P.A.7. Extend the performance comparison to 8K test sequences.

P.A.9. Repeat P.A.6. Compare these lossless techniques in terms of implementation complexity using several test images at different resolutions for VVC technologies as in [351].

P.A.10. JPEG Privacy and Security: With social media having a huge impact on every individual, it disrupts their privacy. Hence, protecting the privacy and security is very important, not only because of the social media, but also because images/ meta-data use the cloud for storage purposes over private repositories. Encrypting the images and providing access to only the authorized person can help in maintaining the privacy of the image/ meta-data. Explore JPEG XT Privacy and Security [407].

P.A.11. In the conclusion section of [19] the authors state "The encoding or decoding hardware can be in fact designed based on a pair of existing JPEG coding chips, as shown in Figure 2, resulting in a minimal hardware change in the existing hardware infrastructure without influencing its real-time performances." Implement the encoding and decoding hardware.

P.A.12. See [381]. Implement this new framework of reversible data hiding in encrypted JPEG bit streams and confirm the conclusions.

P.A.13. See [423]. Explore the future work suggested in the conclusion section.

P.A.14. See [423]. The authors state that the proposed design paradigm can be undertaken for audio and video. Explore this for all the video coding standards such as H.264/AVC, HEVC, AV1 by AOM, DIRAC, etc. Please note each one of these is quite complex and requires group effort and also hardware facilities.

P.A.15. See [424]. As part of JPEG XT future extensions, JPEG privacy and security, Shah has developed an algorithm to encrypt and decrypt the images using C++ and OpenCV libraries. This algorithm performs well for protecting an individual's identity by blurring the image. By resolving the compatibility issues between the OpenCV libraries and JPEG XT code this algorithm can be incorporated in the JPEG XT code to provide the security feature automatically. Explore this.

P.A.16. See P.A.15. Shah suggests that the amount of blurring can be changed according to the user preference by changing the kernel size. This gives the user full control over how the image should be seen by others. Explore this in detail and draw some conclusions.

P.A.17. See [287] In conclusion the authors state "In future work, we will work on speeding up the proposed algorithm to make it more practical." Explore this.

P.A.18. See [1], [403] and [404]. Richter has proposed some optimization techniques to improve the legacy JPEG in terms of rate distortion performance close to JPEG XR (See Figure 5 in [1]). Using the software described in [403] and [404] confirm the enhanced JPEG rate distortion performance shown in [1].

P.A.19. JPEG XT standard defines a normative decoding procedure to reconstruct an HDR image from two JPEG regular code streams named the base layer (visible to legacy decoders) and an extension layer and how to merge them together to form one single image. See also [407]. The standard does not however, define the encoding procedure and leaves large freedoms to the encoder for defining the necessary decoder configuration. Mantiuk, Richter and Artusi [312], explored the whole space of possible configurations to achieve the best possible Rate-Distortion performance. Review this paper in detail and verify the performance results described in the figures [312].

P.A.20. See [401], [402] and [407].]. The JPEG XT encoder (review the encoder/decoder block diagrams) is both forward and backward compatible with legacy JPEG encoder. See references on JPEG. Verify the forward compatible characteristic of JPEG XT. This implies that the JPEG XT decoder should be able to decode and display the images based on the JPEG coded bit streams.

P.A.21. Verify the backward compatible characteristic of JPEG XT. This implies that the legacy JPEG decoder should be able to extract part of the JPEG XT coded bit stream, decode and display the JPEG images.

P.A.22. See [528]. Watanabe, Kobayashi and Kiya have proposed a two-layer coding method using the histogram packing technique with the backward compatibility to the legacy JPEG. The authors claim that this method has a better lossless compression performance than that of the JPEG XT. Using the test images verify this claim.

P.A.23. See [528]. The authors claim improved performance of lossless compression using histogram packing technique. Go thru the various references listed and verify this claim.

B Discrete Cosine Transform for Video Coding Standards

Video sequences contain significant statistical and subjective redundancy within (intra-frame) and between the frames (inter-frame). As in image compression case, the goal of the video coding is to reduce the amount of information (bit-rate) needed to represent the video for storage and transmission by removing both statistical and perceptual redundancy or repeated information.

Most of the video codecs conform to one of the international standards for video coding and they have to implement a similar set of basic coding and decoding functions such as inter-frame prediction to remove temporal redundancy, transform coding to remove spatial redundancy, motion vectors for motion estimation and motion compensation, and quantization and entropy coding to represent the coefficients after transformation with a reduced precision. The model is often described as a hybrid differential pulse code modulation (DPCM)/DCT codec with differences of detail between the standards and implementations [398].

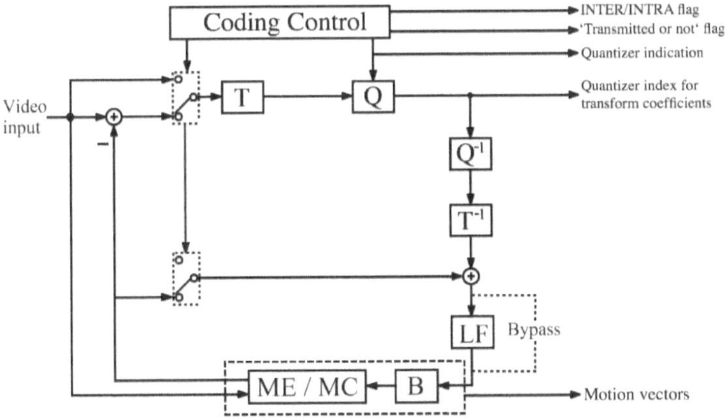

Figure B.1: Video encoder block diagram: T = transform, Q = quantizer, LF = loop filter, T^{-1} = inverse transform, Q^{-1} = inverse quantizer, B = buffer memory, ME = motion estimation, MC = motion compensation.

Figure B.1 shows a block diagram of a typical hybrid video enconder. Motion compensated prediction reduces temporal redundancies. The transform (typically DCT-based algorithms) is used to reduce the spatial redundancies. The quantized transform coefficients, motion vectors and side information are entropy coded using variable length codes (VLC). Notice that T = transform, Q = quantizer,

LF = loop filter, T^{-1} = inverse transform, Q^{-1} = inverse quantizer, B = buffer memory, ME = motion estimation, MC = motion compensation. Prediction errors are between original frames and motion-compensated predicted frames.

DCT IN COMMON VIDEO CODERS

H.261

The main applications of this standard are video telephony and teleconferencing over integrated services digital network (ISDN) lines on which data rates are multiples of 64 kb/s to operate at video bit rates of up to 2 Mb/s.

H.263

This standard was designed as a low-bitrate compressed format for videoconferencing and applications on the Internet. The codec is used for video on demand and streaming and it can be used for a wide range of simultaneous bit rates (20 Kb/s to 1 Mb/s), both high and low. Among other features this standard is based on macroblocks of size 16×16 and an 8×8 floating point DCT.

MPEG-1

The main application of this standard is video on digital storage media (CD-ROM) for a target bit rate of 1.5 Mb/S.

MPEG-2 / H.262

The main application of this standard is in digital television for a target bit rate of $1 - 20$ Mb/S. Among other features this standard is based on macroblocks of size 16×16, DC predictor for intra prediction and an 8×8 floating point DCT.

H.264 / MPEG-4 AVC

Is an improved video compression standard for a target bit rate of $10'$s kb/s to Mb/s. The standard is for a very broad application range that covers all forms of digital compressed video from low bit-rate Internet streaming applications to HDTV broadcast and digital cinema applications with nearly lossless coding. Among other features this standard is based on macroblocks of size 16×16, up to 9 predictors for intra prediction and IntDCT of 8×8 and 4×4. H.264/AVC is generally 40–50 % more bandwidth efficient than MPEG-2 and in some case higher than that for similar video quality.

H.265 / HEVC

For applications such as mobile video distribution and $4K$ Ultra-HD TV broadcast. Among other features this standard is based on coding units of size 8×8 to 64×64,

35 intra predictors and square IntDCT from 4×4 to 64×64 plus a DST luma intra 4×4. HEVC is in general more bandwidth efficient than H.264/AVC for similar video quality.

VERSATILE VIDEO CODING

The versatile video coding standard is under development at this moment [534], [359]. Its primary objective is to provide a significant improvement in compression performance over the existing HEVC [208], [387], [462]. Most of the improvements will come from refinements of the building blocks of HEVC and JEM to process larger block sizes. Some improvements in the transform building block are adaptive multiple core DCT/DST (maximum transform size up to 128×128 with high-frequency zeroing), transform matrices quantized more accurately, mode-dependent no-separable secondary transform-set selection for intra prediction modes, switchable primary transforms, mostly DCT/DST variants and secondary transforms targeting specific cases of prediction residual characteristics [534], [359]. The three sets mode-dependent transform for intra prediction are: Set 0 includes DST-VII and DCT-VIII, set 1 includes DST-VII and DST-I and set 2 includes DST-VII and DCT-VIII.

BJONTEGAARD DELTA PSNR

The Bjøntegaard [42] metric is used to compare the coding efficiency between two different compression algorithms. To fit a rate-distortion (R-D) curve with a third-order logarithmic polynomial based on experimental observations, Bjøntegaard approximates the distortion ($\hat{D}$) with four R-D values in PSNR as [173]

$$\hat{D}(r) = ar^3 + br^2 + cr + d \tag{B.1}$$

where a, b, c and d are fitting parameters, $r = log(R)$ and R is the bit rate. Therefore, the BD-PSNR between two fitted R-D curves $\hat{D}(r_1)$ and $\hat{D}(r_2)$ is

$$\Delta D = E[D_2 - D_1] \approx \frac{1}{r_H - r_L} \int_{r_L}^{r_H} \left[\hat{D}_2(r) - \hat{D}_1(r)\right] dr \tag{B.2}$$

The BD-PSNR is computed using N bit-rate values $(R_1, R_2, \cdots R_N)$ with $(D_1, D_2, \cdots D_N)$ distortions. Let $r_1 = [r_{1,1}, \cdots, r_{1,N_1}]$, $r_2 = [r_{2,1}, \cdots, r_{2,N_2}]$ then the integration bounds are,

$$\begin{aligned} r_L &= max\{min(r_1), min(r_2)\} \\ r_H &= min\{max(r_1), max(r_2)\} \end{aligned} \tag{B.3}$$

and the logarithm of the rate can be expressed as a function of the distortion using the third-order polynomial fitting [42]

$$\hat{r}(D) = aD^3 + bD^2 + cD + d \tag{B.4}$$

The BD-rate between two fitted R-D curves $\hat{r}_1(r)$ and $\hat{r}_2(r)$ can be approximated as [42], [173],

$$\Delta R = E\left[\frac{R_2 - R_1}{R_1}\right] = E\left[\frac{R_2}{R_1}\right] - 1 = E[10^{r_2 - r_1}] - 1$$
$$\Delta R \approx 10^{E[r_2 - r_1]} \approx 10^{\frac{1}{D_H - D_L}\int_{D_L}^{D_H}[\hat{r}_2(D) - \hat{r}_1(D)]dD} - 1 \tag{B.5}$$

Let $D_1 = [D_{1,1}, \cdots, D_{1,N_1}]$, $D_2 = [D_{2,1}, \cdots, D_{2,N_2}]$. Therefore, the integration bounds are [173],

$$D_L = max\{min(D_1), min(D_2)\}$$
$$D_H = min\{max(D_1), max(D_2)\} \tag{B.6}$$

Figures B.2a and B.2b show the plot of normal PSNR and the Bjøntegaard delta PSNR. In the first plot the difference between the curves is dominated by the high bitrates. In the second plot relation between the change in PSNR and the change in bitrate is well represented by 0.5 dB = 10% or 0.05 dB = 1% [42].

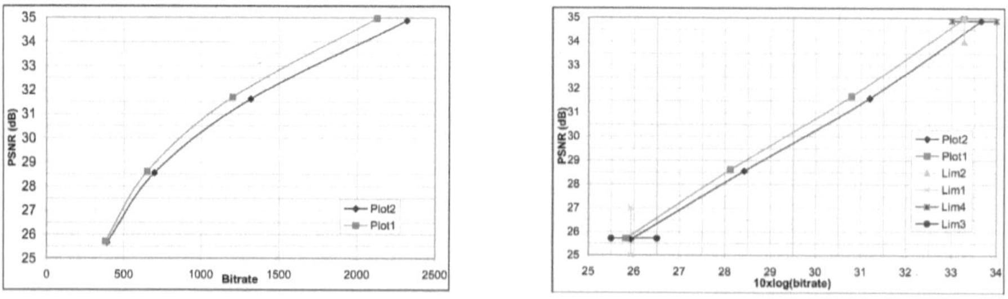

(a) Plot of PSNR. (b) BD-PSNR.

Figure B.2: Plots of (a) Rate-Distortion and (b) Bjøntegaard delta PSNR [42].

Figure B.3 explains clearly the BD-PSNR and BD-rate,

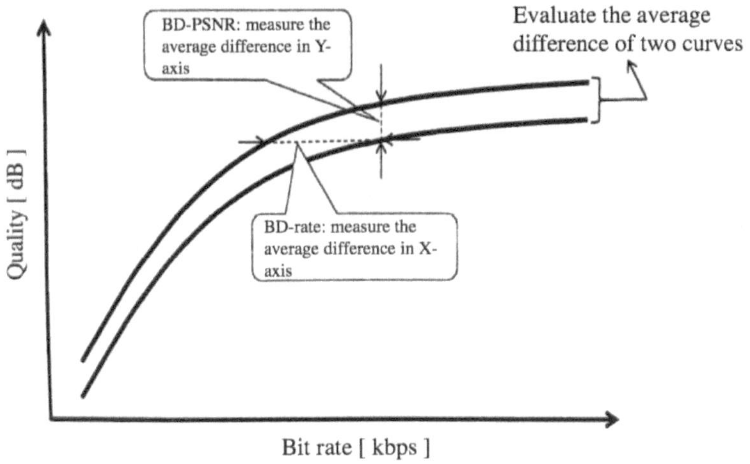

Figure B.3: General procedure to obtain the BD-PSNR and BD-rate.

SUMMARY

In the family of video coding standards, HEVC has the promise and potential to replace/supplement all the existing standards (MPEG and H.26x series including H.264/AVC). While the complexity of the HEVC encoder is several times that of the H.264/AVC, the decoder complexity is within the range of the latter. Due to the great success, researchers are exploring about reducing the HEVC encoder/decoder complexity while also exploring a new video coding standard called versatile video coding on top of the HEVC test model. All the standards have their success in the core transform which is a DCT. In this appendix, the DCT in common video coders are explained and some projects to show their performance.

PROJECTS

P.B.1. In [361], authors proposed a new video coding scheme optimized for high resolution video sources. This scheme follows the traditional block-based MC+DCT hybrid coding approach similar to H.264/AVC, HEVC and other ITU-T/VCEG and ISO/IEC MPEG standards. By introducing various technical optimizations in each functional block, they achieve roughly 26 % bit-rate savings on average compared to H.264/AVC high profile at the same PSNR. They also suggest improved measures for complexity comparison. Go through this paper in detail and implement the codec. Consider BD-PSNR and BD-rate and SSIM [SSIM1] as the comparison metrics. For evaluating the implementation complexity, consider both encoders and decoders. Can this video coding scheme be further improved? Explore all options.

P.B.2. See P.B.1. By bypassing some functional blocks such as transform/quantization/in loop deblocking filter, in [589] authors have implemented a HEVC lossless coding scheme. Can a similar lossless coding scheme be implemented in the codec proposed by Ono and Asai [361]. If so compare its performance with other lossless coding schemes.

P.B.3. Implement similar performance comparison between AVS China and HEVC (various profiles) as in [513].

P.B.4. In [358] authors have compared the HEVC coding efficiency with other video coding standards such as H.264/MPEG-4 AVC, H.264/MPEG-4 visual (part 2), H.262/MPEG-2 part 2 video and H.263. This comparison is based on several standard test sequences at various bit rates. The comparison metrics include PSNR and mean opinion score (MOS) based on double stimulus impairment scale (DSIS) both for interactive and entertainment applications. The comparison, however, does not include implementation complexity. Implement this metric for all the video coding standards and draw the conclusions.

P.B.5. See P.B.4. Extend the coding efficiency comparison between HEVC and DIRAC. Software and data for reproducing selected results can be found at ftp://ftp.hhi.de/ieee-tcsvt/2012.

P.B.6. See P.B.4. Extend the coding efficiency comparison between HEVC and VC-1.

P.B.7. See P.B.4. Extend the coding efficiency comparison between HEVC and AVS China.

P.B.8. See P.B.4. Extend the coding efficiency comparison between HEVC and VP9.

P.B.9. See [326]. In the conclusion section the authors state "we would like to extend this sketch attack framework to handle different video standards such as HEVC, AVS and Google VP9 as our future work." Extend this research to VP9.

C Definitions

RECURSIVE BLOCK CODING (RBC)

In recursive block coding (RBC), developed by Jain and Farrelle [219], the original data is divided into overlapping blocks, with one pel overlapping for practical images. In this scheme the original data is decomposed into two mutually uncorrected processes: a) boundary response, i.e., prediction of pels within the block based on overlapped pels and b) the residual process or prediction error, which is transform coded. Farrelle and Jain [132] have shown that the RBC reduces or even suppresses the block structure and simply blurs the edges, resulting in a sharper image compared with the DCT coding.

DISCRETE LEGENDRE TRANSFORM (DLT)

Miyahara and Kotani [327] have described the general requirements for a discrete transform that can suppress the block structure and have shown, based on simulation of test images, that the discrete Legendre transform (DLT) is superior to the DCT, purely from a block structure viewpoint. The DLT, however, has neither a fast algorithm nor a recursive structure, [496].

BLOCK OVERLAPPING

By overlapping adjacent blocks (see Figure 5.1), the block effect can be reduced. As neighborhood pels are common to adjacent blocks, both the processing and the bit rate increase. For pels common to adjacent blocks, after processing their averages represent the reconstructed values. In speech coding, overlapped trapezoidal windows have been adopted.

LAPPED ORTHOGONAL TRANSFORM (LOT)

The lapped orthogonal transform (LOT), [301], [302], [303], [304] combines the DCTs of adjacent blocks in such a way that the resulting transform has basis functions that overlap adjacent blocks. The basis functions decay smoothly towards zero at their boundaries, so that the LOT is virtually free from block boundary mismatch errors, as shown in [303] for image coding and in [302] for speech coding.

DCT/IDCT MISMATCH

The IDCT implementation at both the coder and decoder has to be exact if there is to be no difference between the reconstructed images at both ends in the absence of channel errors. However, because of the finite word-length effects (also different algorithms may be used for the IDCT at the coder and decoder), there can be mismatch between the two reconstructed images. The mismatch, therefore, is caused by different accuracies and/or different algorithms for the IDCT at the coder and the decoder. The forward DCT at the coder does not contribute to the mismatch. It has been shown that this mismatch progressively increases, leading to visible image degradations. To counteract the mismatch error accumulation at the decoder, after extensive simulations, specifications for implementing the IDCT have been formulated. These specifications not only can eliminate the incompatibility problem between the coder and decoder, but also can reduce the cost of the DCT/IDCT chip. Also, intra-frame DCT coding (cyclic refresh) can be periodically implemented so that the visible artifacts can be avoided. Too low a refresh rate would let the mismatch error dominate. On the other hand, too high a refresh rate requires too many bits for intra-frame coding. A refresh period between 45 and 135 frames at 15 frames/sec is a compromise between these two effects [390].

ADAPTIVE BLOCK DISTORTION EQUALIZATION

Whenever there is a large distortion gradient between adjacent blocks that shows up as a visible block structure, an adaptive technique that equalizes the distortion between the blocks is necessary to reduce blocky effect. The price for the improved image is the additional complexity, involving computation of distortion gradient, buffer monitoring, and distortion equalization.

ACTIVITY CLASSIFICATION IN ADAPTIVE TRANSFORM CODING

In adaptive transform coding of images, the activity of a block can be described in several ways. The general concept is to allocate more bits to a high activity class and vice versa. Assuming that an $N \times N$ image is divided into blocks of size $L \times L$, the 2D-DCT coefficients of each block are $X_{m,n}^{C(2)} \quad m,n = 0,1,\cdots,L-1$. Based on the test image or a class of images, the blocks can be divided into several classes for adaptive coding. Although the classification concepts described below are based on the 2D DCT, these schemes can, in general, be extended to the 3D DCT. Some of these classes are as follows:

AC ENERGY

Whereas the DC coefficient $X_{0,0}^{C(2)}$ represents the average brightness of a block, the AC energy (defined below) is an indication of activity or detail of a block.

$$\text{AC energy of a block} = \sum_{m=0}^{L-1}\sum_{n=0}^{L-1}\left[X_{m,n}^{C(2)}\right]^2 - \left[X_{0,0}^{C(2)}\right]^2. \tag{C.1}$$

The AC energies of all the blocks in an image or class of images can be computed and the blocks can be divided into several equally populated classes. Based on 2D DCT of these blocks and their AC energy distribution of K equally populated classes can been developed. The variances of the transform coefficients in each class can be computed, and, using this variance distribution, the bit allocation matrix for each class can be developed. Not only are the higher activity classes allocated more bits, but also the DC and low-frequency coefficients in each class, in general, are allocated more bits compared with the high-frequency coefficients, with some of the later even being discarded (zero bits). The number of bits allocated to each class can be based on the total AC energy of each class. Let,

$$E_k = \sum_{i=1}^{M_k} E_{k,i}$$

be the total energy of class k. $E_{k,i}$ is the AC energy of block i in class k, and M_k is the number of blocks in this class. The total AC energy of an image or set of images is then

$$E_{total} = \sum_{k=1}^{K} E_k,$$

If B_{total} is the total number of bits allocated to an image, then the number of bits allocated to any class can be based on the total energy of that class. Ferreyra, Picco, and Sobrero [133] have investigated this bit allocation based on the following three models;

$$\log(E_k + 1), \quad (E_k)^{1/3} \quad \text{and} \quad \frac{\log(E_k) + (E_k)^{1/2}}{2},$$

and have concluded that the middle one gave the best results from the image compression-quality viewpoint.

MAGNITUDE SUM OF THE AC COEFFICIENTS

This parameter can be expressed as

$$\text{MagAC} = \sum_{m=0}^{L-1}\sum_{n=0}^{L-1} \left|X_{m,n}^{C(2)}\right| - \left|X_{0,0}^{C(2)}\right|. \tag{C.2}$$

Compared with the AC energy, the magnitude sum of the AC coefficients of a block is computationally much simpler. Large values of MagAC indicate blocks with strong edges, whereas the magnitude of the DC coefficient can indicate dark blocks with little AC energy. MagAC and $X_{0,0}^{C(2)}$ formed the basis for classifying the 2D DCT of 8×8 blocks of the four-color space-yellow, magenta, cyan, and black—in printed-image coding [390]. This classification determined the number of bits allocated, the threshold (transform coefficients above this threshold were coded), and also the number of coefficients set to zero in a block. The activity index defined by (C.2) was originally proposed by Gimlett [159]. He has also suggested weighted activity measures such as

$$\text{weighted AC energy of a block} = \sum_{m=0}^{L-1}\sum_{n=0}^{L-1} w_{1,m,n}[X_{m,n}^{C(2)}]^2 - [X_{0,0}^{C(2)}]^2 \tag{C.3}$$

$$\text{weighted mag. sum of the AC coeff.} = \sum_{m=0}^{L-1}\sum_{n=0}^{L-1} w_{2,m,n}\left|X_{m,n}^{C(2)}\right| - X_{0,0}^{C(2)}, \tag{C.4}$$

where $w_{1,m,n}$ and $w_{2,m,n}$ are the weights for application in image coding.

SPECTRAL ENTROPY OF AC ENERGY

The spectral entropy is to indicate the degree of energy compaction (e.g., a few of the transform coefficients having the most energy of the block). The spectral entropy is defined as

$$E = -\sum_{m=0}^{L-1}\sum_{n=0}^{L-1} \frac{\left|X_{m,n}^{C(2)}\right|}{\text{MagAC}} \log_2\left[\frac{\left|X_{m,n}^{C(2)}\right|}{\text{MagAC}}\right], \tag{C.5}$$

where MagAC is defined in (C.2). Low entropy implies that the block energy is compacted into a few transform coefficients. Low entropy implies that the block energy is compacted into a few transform coefficients. Also, based on the ranges of MagAC and E a set of K classes can be derived and determine a maximum threshold that yields no visible degradation of the blocks belonging to that class. The magnitude of the AC coefficients below the threshold can be discarded [390].

MACE, DIRECTION, FINENESS

To reduce the differences between blocks belonging to the same class, Wu and Burge [538] proposed block classification based on middle frequency AC energy (MACE), directional features (DIR) and fineness of a sub-image (FIN) defined as

$$\text{MACE}_{p,q} = \sum_{m=0}^{p-1}\sum_{n=0}^{p-1}\left[X_{m,n}^{C(2)}\right]^2 - \sum_{m=0}^{q-1}\sum_{n=0}^{q-1}\left[X_{m,n}^{C(2)}\right]^2 \tag{C.6}$$

MACE represents the AC energy within a spatial frequency window with MACE $(L-1,0)$ as the AC energy of the block.

$$\text{DIR} = \frac{\sum_{m=0}^{12}\sum_{n=0}^{12} \tan^{-1}\left(\frac{m}{n}\right)\left|X_{m,n}^{C(2)}\right|}{\sum_{m=0}^{12}\sum_{n=0}^{12}\left|X_{m,n}^{C(2)}\right|}. \tag{C.7}$$

Note that (C.7) is defined for $m = n = 16$ and can be modified for other block sizes. DIR exhibits strong directional features.

$$\text{FIN} = \frac{\int_0^{\pi/2}\int_{\rho_1}^{\rho_2} X_{\rho,\theta}^{C(2)} d\rho d\theta \Big/ \int_0^{\pi/2}\int_{\rho_1}^{\rho_2} d\rho d\theta}{\int_0^{\pi/2}\int_{\rho_3}^{\rho_4} X_{\rho,\theta}^{C(2)} d\rho d\theta \Big/ \int_0^{\pi/2}\int_{\rho_3}^{\rho_4} d\rho d\theta}. \tag{C.8}$$

In (C.8), (ρ, θ) are the polar coordinates of the block in the transform domain. The computation of this feature is computational extensively applied to pattern recognition and feature classification in a compressed domain (reduced bit-rate transmission and storage) is desirable.

ADAPTIVE SEGMENTATION OF 2D-DCT BLOCKS INTO REGIONS

Another scheme proposed for activity classification is based on segmentation of 2D DCT of image blocks into regions such that the statistics for each region are almost stationary [390]. The regions are segmented such that the magnitudes of the transform coefficients in each region are within a threshold set for that region. Both the number of regions and the thresholds can be varied.

ACTIVITY CLASSIFICATION ACCORDING TO THE DOMINANT STRUCTURE

In this scheme the total AC energy in each region is used to classify the block (horizontal, vertical, and diagonal structures). This classification followed by further subclasses in each of these three regions, along with object matching, has effectively responded to translation, panning, and zooming of video, resulting in improved subjective quality of the processed images [390].

ACTIVITY CLASSIFICATION BASED ON MAXIMUM MAGNITUDE OF AC COEFFICIENTS

Block activity is determined by

$$\max \left| X_{m,n,l}^{C(2)} \right| \quad (m,n,l) \neq (0,0,0) \tag{C.9}$$

in a block. Based on this criterion the blocks are divided into four classes. Both the weighting function and the step size of the uniform quantizer for the AC coefficients are determined by the block classification. For the 2D DCT blocks, a criteria similar to (C.9) can be used. By multilevel thresholding the number of classes can be chosen as, for example, 4 or 8 [390].

COSINE MODULATED FILTER BANK

Figure C.1 shows the uniformly spaced, evenly stacked, critically sampled SSB analysis/synthesis. Figure C.2 shows the channel responses [379]. The center frequencies of the K channels are $\omega_k = \frac{2\pi k}{K}$, $k = 0, 1, \cdots, K-1$, n_0 is a time shift. The paths are multiplexed because $\sin(m\pi/2) = 0$, when m even and $\cos(m\pi/2) = 0$, when m odd. Only $K/2+1$ unique channels. Therefore, for critical sampling $M = K/2$. h_n and f_n are FIR filters and exist only in the interval $0 \leq n \leq P-1$. In other words, the analysis and synthesis window of length P and $K/2 \leq P \leq P$. When $P = K$ maximum overlaps occurs.

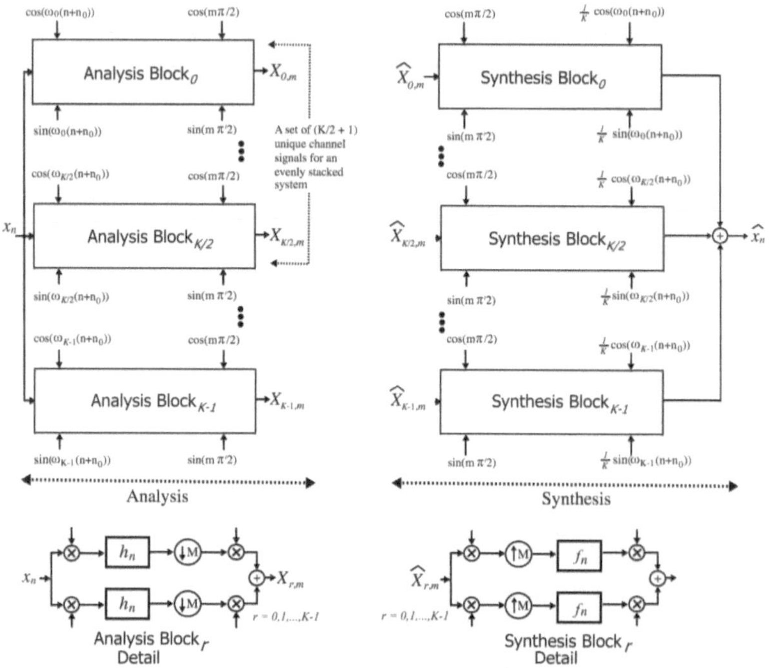

Figure C.1: A single-sideband analysis/synthesis system [379].

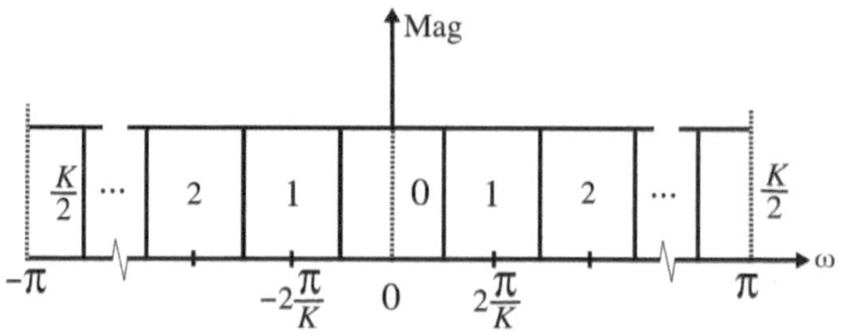

Figure C.2: Bank of real bandpass filters [379].

Half-width channels occur at $k = 0$ and $k = K/2$. If the full-width channels are decimated by M, then the halfwidth channels can be decimated by 2M. Critical sampling implies that the overall output sample rate equals the input sample rate, so

$$M = \frac{K}{2} \tag{C.10}$$

FAST DISCRETE COSINE TRANSFORM ALGORITHM

Figure C.3 shows the signal flowgraph of the fast discrete cosine transform (FDCT) proposed by Chen, Smith and Fralick [73] for $N = 32$. The structure can be extended to any value $N = 2^m \geq 2$. The signal flowgraph is the unnormalized forward DCT matrix $[DCT_N]$. The number of computational steps is less than $1/6$ of the conventional DCT algorithm employing a 2-sided FFT. The input samples are in natural order from top to bottom. For every N, the output transform coefficients are un bit-reversed order. As N increases the even coefficients of each successive transform are obtained directly from the coefficients of the prior transform by doubling the subscript of the prior coefficients. To normalize multipy the output coefficients by $2/N$. The inverse is obtained as $(N/2)[DCT_N]^T$.

Figure C.4 shows the 8-point DCT factorizations of Chen, Smith and Fralick [73] with series of butterflies and five plane rotations which can be implemented with 13 multiplications and 29 additions [282]. Figure C.5 shows the 8-point DCT factorizations of Loeffler, Ligtenberg and Moschytz [290]. The method is implemented using 11 multiplications and 29 additions. Figure C.6 shows the binDCT [282] implemented using 30 additions, 12 shifts, 8.77 dB coding gain which is quite close to the DCT.

SEPARABILITY OF THE DCT

Assuming that a dimensional signal $f_{m,n}$ is an $N \times N$ array and the corresponding transformation is given by,

$$X_{u,v} = \sum_{m=0}^{N-1} \sum_{n=0}^{N-1} f_{m,n} g_{m,n,u,v}, \tag{C.11}$$

where $g_{m,n,u,v}$ is called the forward transformation kernel. The inverse transform is given by,

$$f_{m,n} = \sum_{u=0}^{N-1} \sum_{v=0}^{N-1} X_{u,v} h_{m,n,u,v}, \tag{C.12}$$

where $h_{m,n,u,v}$ is called the inverse transformation kernel. This class of transformation will be separable if we can write $g(m,n,u,v) = g_{1,m,n} g_{2,m,n}$. Moreover, if we can write $g_{m,n,u,v} = g_{1,m,u} g_{1,n,v}$ we assume that g_1 and g_2 are functionally the same. Also, in

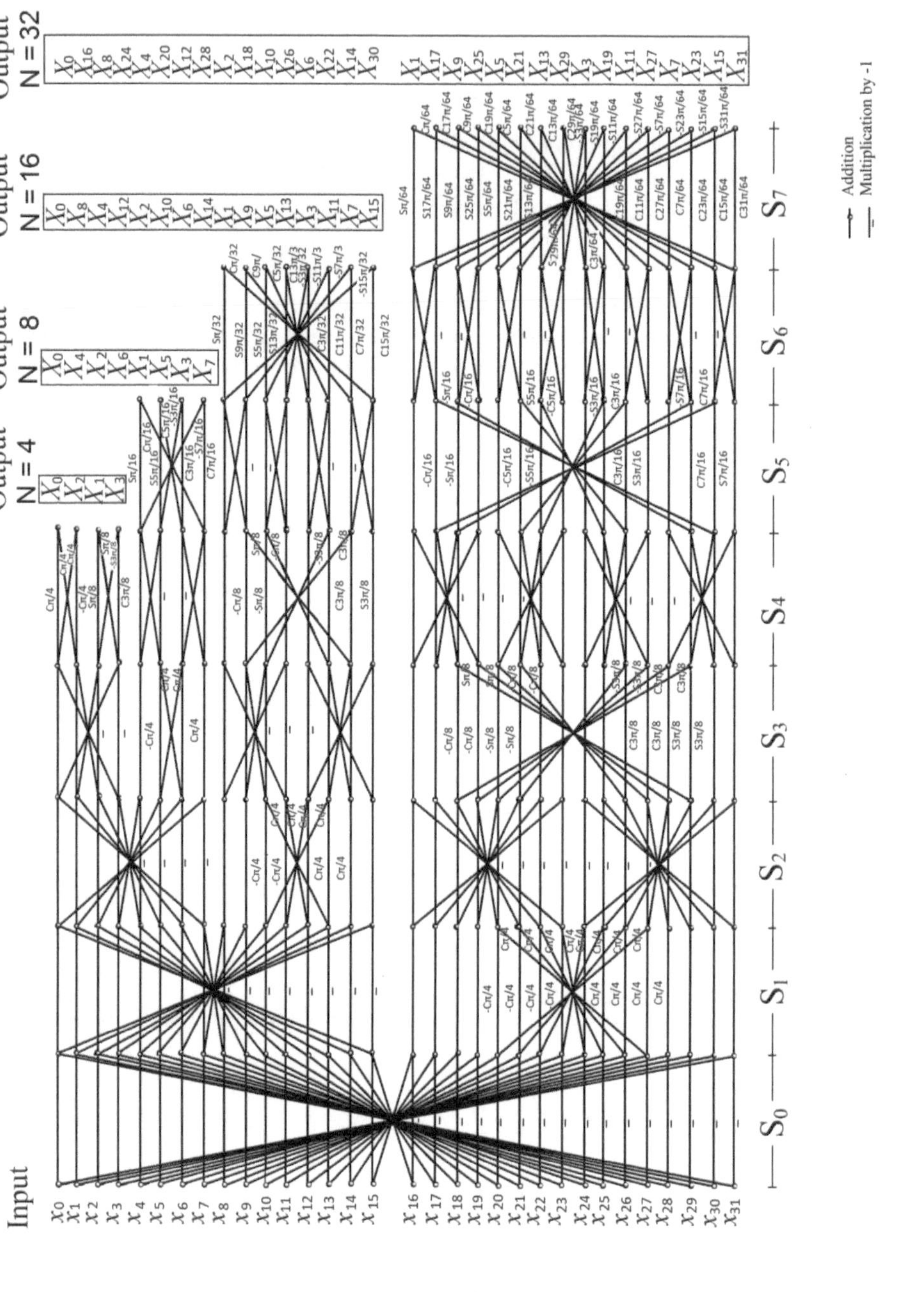

Figure C.3: Flowgraph of the FDCT for $N = 4$, $N = 8$, $N = 16$ and $N = 32$, $C_i = \cos i$, $S_i = \sin i$, [73].

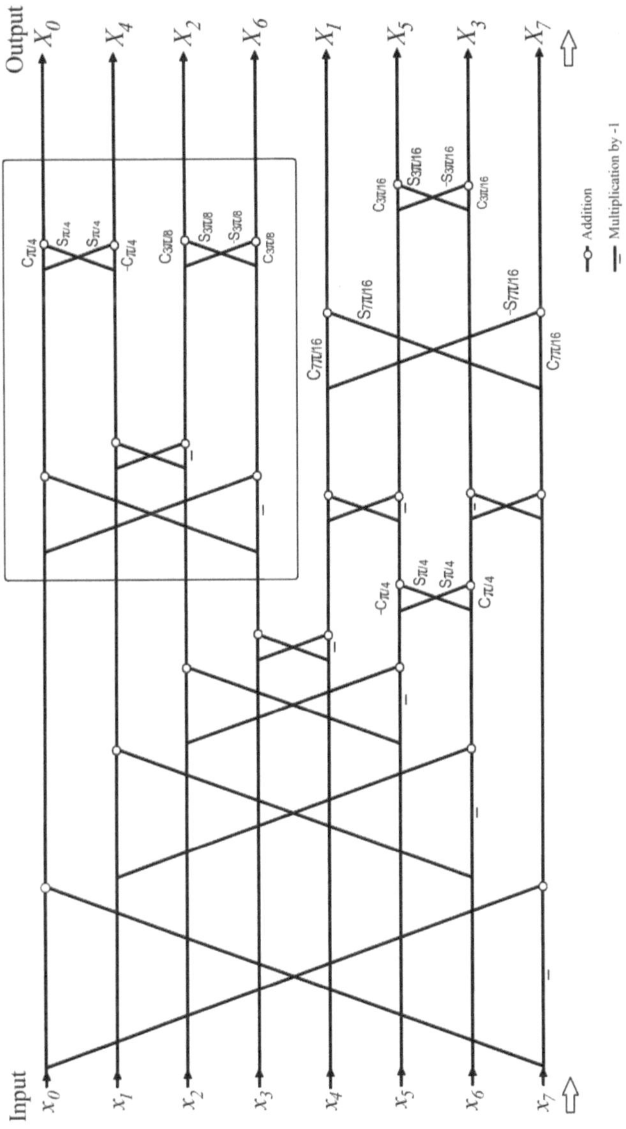

Figure C.4: Flowgraph of the 8-point DCT for Chen's factorization, $C_i = \cos i$, $S_i = \sin i$, [282].

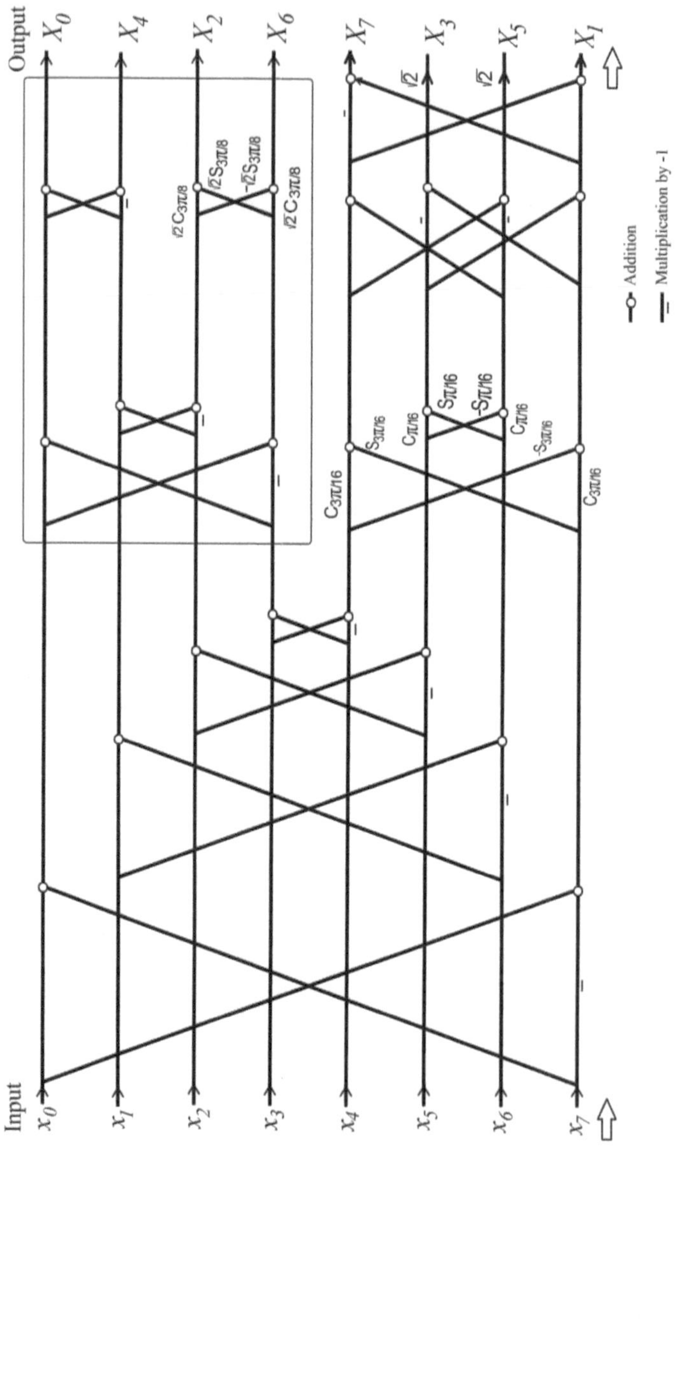

Figure C.5: Flowgraph of the 8-point DCT for Loeffler's factorization, $C_i = \cos i$, $S_i = \sin i$, [282].

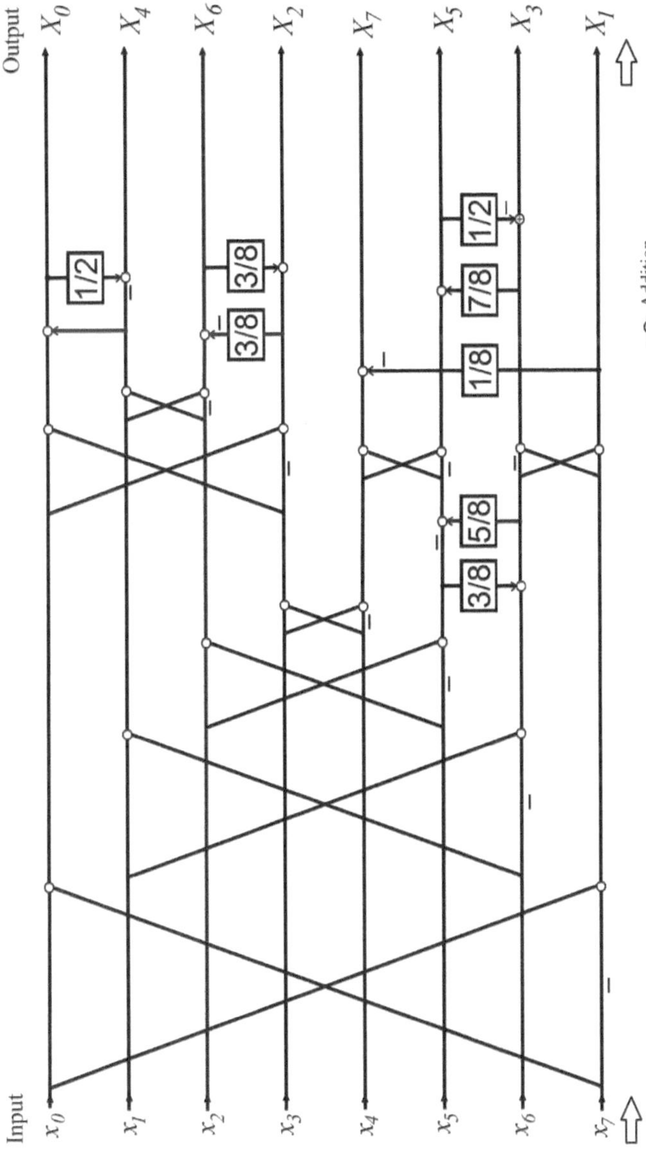

Figure C.6: Flowgraph of the 8-point DCT for binDCT factorization [282].

this case, the function will be called as symmetric. The DCT belongs to the class of separable and symmetric functions. Therefore, the DCT kernels are given by,

$$g_{m,n,u,v} = h_{m,n,u,v} = \alpha_u \alpha_v \cos\left[\frac{(2m+1)u\pi}{2N}\right] \cdot \cos\left[\frac{(2n+1)v\pi}{2N}\right] \qquad \text{(C.13)}$$

Therefore, this transformation is separable as well as symmetric and the forward and inverse transformation kernels are identical. $\alpha_u = 1/\sqrt{N}$ for $u = 0$ and $\alpha_u = \sqrt{2/N}$ for $u = 1, 2\cdots N-1$. $\alpha_v = 1/\sqrt{N}$ for $v = 0$ and $\alpha_v = \sqrt{2/N}$ for $v = 1, 2\cdots N-1$.

REFERENCES ORGANIZATION

Regular papers

4 J.I. Agbinya. Interpolation using the discrete cosine transform. *Electronics Letters,* 28(20):1927–1928, Sept 1992.

5 A. Ahmed, M.U. Shahid, and A. Rehman. N−point DCT VLSI architecture for emerging HEVC standard. *VLSI Design*, 2012:1–13, Jan 2012.

7 N. Ahmed, T. Natarajan, and K.R. Rao. Discrete cosine transform. *IEEE Transactions on Computers*, C-23(1):90–93, Jan 1974.

9 T. Ahonen, A. Hadid, and M. Pietikainen. Face description with local binary patterns: Application to face recognition. *IEEE Transactions on Pattern Analysis and Machine Intelligence*, 28(12):2037–2041, Dec 2006.

10 A.M. Alattar. Reversible watermark using the difference expansion of a generalized integer transform. *IEEE Transactions on Image Processing*, 13(8):1147–1156, Aug 2004.

12 Z. Alkachouh and M.G. Bellanger. Fast DCT-based spatial domain interpolation of blocks in images. *IEEE Transactions on Image Processing*, 9(4):729–732, Apr 2000.

15 A. Amani et al. Passive detection of image forgery using DCT and local binary pattern. *Signal, Image and Video Processing*, 11(1):81–88, Jan 2017.

16 T. Amornraksa and S. Tachaphetpiboon. Fingerprint recognition using DCT features. *Electronics Letters*, 42(9):522–523, April 2006.

20 A. Artusi et al. JPEG XT: A Compression Standard for HDR and WCG Images [Standards in a Nutshell]. *IEEE Signal Processing Magazine*, 33(2):118–124, March 2016.

21 M. Asikuzzaman et al. Robust DT CWT-based DIBR 3D video watermarking using chrominance embedding. *IEEE Transactions on Multimedia*, 18(9):1733–1748, Sept 2016.

23 R. Atta and M. Ghanbari. Low-memory requirement and efficient face recognition system based on DCT pyramid. *IEEE Transactions on Consumer Electronics*, 56(3):1542–1548, Aug 2010.

25 H. Baali et al. A transform-based feature extraction approach for motor imagery tasks classification. *IEEE Journal of Translational Engineering in Health and Medicine*, 3:1–8, Oct 2015.

27 S. Bacchelli and S. Papi. Image denoising using principal component analysis in the wavelet domain. *Journal of Computational and Applied Mathematics*, 189(1):606–621, 2006. Proc. of 11th Intl. Congress on Computational and Applied Mathematics.

28 S.H. Bae, J. Kim, and M. Kim. HEVC-based perceptually adaptive video coding using a DCT-based local distortion detection probability model. *IEEE Transactions on Image Processing*, 25(7):3343–3357, July 2016.

29 S.H. Bae and M. Kim. DCT-QM: A DCT-based quality degradation metric for image quality optimization problems. *IEEE Transactions on Image Processing*, 25(10):4916–4930, Oct 2016.

30 S.H. Bae and M. Kim. A DCT-based total JND profile for spatio-temporal and foveated masking effects. *IEEE Transactions on Circuits and Systems for Video Technology*, 27(6):1196–1207, June 2017.

33 R.H. Bamberger and M.J.T. Smith. A filter bank for the directional decomposition of images: theory and design. *IEEE Transactions on Signal Processing*, 40(4):882–893, Apr 1992.

34 D. Baron et al. Distributed compressive sensing. *ArXiv e-prints*, jan 2009.

36 F.M. Bayer and R.J. Cintra. DCT-like transform for image compression requires 14 additions only. *IET Electronics Letters*, 48(15):919–921, 2012.

38 G. Bhatnagar and Q.M.J. Wu. Biometrics inspired watermarking based on a fractional dual tree complex wavelet transform. *Future Generation Computer Systems*, 29(1):182 –195, 2013.

41 E.J.K. Bisherurwa and F.P. Coakley. New fast discrete sine transform with offset. *IEE Electronics Letters*, 17(21):803–805, October 1981.

45 S. Bouguezel, M.O. Ahmad, and M.N.S. Swamy. Binary discrete cosine and Hartley transforms. *IEEE Transactions on Circuits and Systems I: Regular Papers*, 60(4):989- 1002, April 2013.

46 N. Bozinovic and J. Konrad. Motion analysis in 3d dct domain and its application to video coding. *Signal Processing: Image Communication*, 20(6):510–528, 2005. Special Issue on Advanced Aspects of Motion Estimation.

48 V. Britanak. A unified approach to the fast computation of discrete sinusoidal transforms I: DCT and DST transforms. *Computers and Artificial Intelligence*, 17(6):583–607, Dec 1998.

49 V. Britanak. A note on the MDCT/MDST and pseudoinverse matrix. *Computing and Informatics*, 23(3):205–214, Feb 2004.

50 V. Britanak. An efficient computing of oddly stacked MDCT/MDST computation via evenly stacked MDCT/MDST and vice versa. *Signal Processing*, 85(7):1353–1374, July 2005.

51 V. Britanak and H.J.L. Arriens. Fast computational structures for an efficient implementation of the complete TDAC analysis/synthesis MDCT/MDST filter banks. *Signal Processing*, 89(7):1379–1394, July 2009.

54 M. Budagavi et al. Core transform design in the High Efficiency Video Coding (HEVC) standard. *IEEE Journal of Selected Topics in Signal Processing*, 7(6):1029–1041, Dec 2013.

56 J.B. Burl. Estimating the basis functions of the Karhunen-Loève transform. *IEEE Transactions on Acoustics, Speech, and Signal Processing*, 37(1):99–105, Jan 1989.

58 A.R. Calderbank et al. Wavelet transforms that map integers to integers. *Applied and computational harmonic analysis*, 5(3):332–369, 1998.

60 N.B. Chakrabarti and T.V.K.H. Rao. A recursive filter implementation of KL transform. *Signal Processing*, 44(3):269–284, 1995.

62 W.K. Cham and Y.T. Chan. An order-16 integer cosine transform. *IEEE Transactions on Signal Processing*, 39(5):1205–1208, May 1991.

63 W.K. Cham and R.J. Clarke. Application of the principle of dyadic symmetry to the generation of orthogonal transforms. *IEE Proc. F-Communications, Radar and Signal Processing*, 133(3):264–270, June 1986.

64 S.C. Chan and K.L. Ho. Direct methods for computing discrete sinusoidal transforms. *IEE Proc. F -Radar and Signal Processing*, 137(6):433–442, Dec 1990.

65 S. Chandrasekaran et al. An eigenspace update algorithm for image analysis. *Graphical Models and Image Processing*, 59(5):321–332, 1997.

66 C.L. Chang et al. Direction-adaptive partitioned block transform for color image coding. *IEEE Transactions on Image Processing*, 19(7):1740–1755, July 2010.

69 C.C.T. Chen, C.-Ta. Chen, and C.-M. Tsai. Hard-limited Karhunen-Loève transform for text independent speaker recognition. *Electronics Letters*, 33(24):2014–2016, Nov 1997.

70 H. Chen and B. Zeng. New transforms tightly bounded by DCT and KLT. *IEEE Signal Processing Letters*, 19(6):344–347, June 2012.

71 W.-K. Chen et al. Iris recognition based on bidimensional empirical mode decomposition and fractal dimension. *Information Sciences*, 221:439–451, 2013.

73 W.H. Chen, C.H. Smith, and S.C. Fralick. A fast computational algorithm for the discrete cosine transform. *IEEE Transactions on Communications*, 25(9):1004–1009, Sept 1977.

77 Y.-Y. Chen. Medical image compression using DCT-based subband decomposition and modified SPIHT data organization. *Intl. Journal of Medical Informatics*, 76(10):717–725, 2007.

78 Z. Chen, Q. Han, and W.-K. Cham. Low-complexity order-64 integer cosine transform design and its applications in HEVC. *IEEE Transactions on Circuits and Systems for Video Technology*, Early access 2018.

80 Y. Cheng et al. Robust face recognition based on illumination invariant in nonsub-sampled contourlet transform domain. *Neurocomputing*, 73(10):2217–2224, June 2010.

83 B. Chitprasert and K.R. Rao. Human visual weighted progressive image transmission. *IEEE Transactions on Communications*, 38(7):1040–1044, Jul 1990.

84 M.-R. Choi et al. Color image interpolation in the DCT domain using a wavelet-based differential value. *Multimedia Tools and Applications*, 77(16):21539–21556, Jan 2018.

86 V. Choudhary and P. Vodite. Extraction of region of interest in compressed domain. *Intl. Journal of Computer Science Issues*, 10(2):321–329, May 2013.

87 C. Christopoulos, A. Skodras, and T. Ebrahimi. The JPEG2000 still image coding system: an overview. *IEEE Transactions on Consumer Electronics*, 46(4):1103–1127, Nov 2000.

90 R.J. Cintra and F.M. Bayer. A DCT approximation for image compression. *IEEE Signal Processing Letters*, 18(10):579–582, Oct 2011.

91 R.J. Cintra, F.M. Bayer, and C.J. Tablada. Low-complexity 8-point DCT approximations based on integer functions. *Signal Processing*, 99(1):201–214, 2014.

95 J.W. Cooley and J.W. Tukey. An algorithm for the machine calculation of complex Fourier series. *Math of Comput.*, 19:297–301, 1965.

97 I.J. Cox et al. Secure spread spectrum watermarking for multimedia. *IEEE Transactions on Image Processing*, 6(12):1673–1687, Dec 1997.

99 A. Criminisi, P. Perez, and K. Toyama. Region filling and object removal by exemplar-based image inpainting. *IEEE Transactions on Image Processing*, 13(9):1200–1212, Sept 2004.

100 J. Cui et al. Hybrid all zero soft quantized block detection for HEVC. *IEEE Transactions on Image Processing*, 27(10):4987–5001, Oct 2018.

101 Z. Cvetkovic and M.V Popovic. New fast recursive algorithms for the computation of discrete cosine and sine transforms. *IEEE Transactions on Signal Processing*, 40(8):2083–2086, Aug 1992.

102 W. Dai and T.D. Tran. Regularity-constrained pre-and post-filtering for block DCT-based systems. *IEEE Transactions on Signal Processing*, 51(10):2568–2581, Oct 2003.

106 C. Deng et al. Performance analysis, parameter selection and extensions to H.264/AVC FRExt for high resolution video coding. *Journal of Visual Communication and Image Representation*, 22(8):749–759, 2011.

108 S. Dhamija and P. Jain. Comparative analysis for discrete sine transform as a suitable method for noise estimation. *Intl. Journal of Computer Science Issues*, 8(3):162–164, Sept 2011.

109 W. Ding et al. Fast mode dependent directional transform via buttery-style transform and integer lifting steps. *Journal of Visual Communication and Image Representation*, 22(8):721–726, 2011.

110 Y. Ding et al. Application of the Karhunen-Loeve transform temporal image filter to reduce noise in real-time cardiac cine MRI. *Physics in Medicine and Biology*, 54(12):3909, 2009.

112 Z. Dokur. A unified framework for image compression and segmentation by using an incremental neural network. *Expert Systems with Applications*, 34(1):611–619, 2008.

114 A.B. Doulah et al. DCT domain feature extraction scheme based on motor unit action potential of EMG signal for neuromuscular disease classification. *Healthcare Technology Letters*, 1(1):26–31, Jan 2014.

116 K. Duda. Accurate, guaranteed stable, sliding discrete Fourier transform. *IEEE Signal Processing Magazine*, 27(6):124–127, Nov 2010.

118 P. Duhamel. Implementation of "Split-radix" FFT algorithms for complex, real, and real-symmetric data. *IEEE Transactions on Acoustics, Speech, and Signal Processing*, 34(2):285–295, Apr 1986.

120 S.A. Dyer, N. Ahmed, and D.R. Hummels. Classification of vector cardiograms using Walsh and cosine orthogonal transforms. *IEEE Transactions on Electromagnetic Compatibility*, EMC-27(1):35–40, Feb 1985.

121 T. Ebrahimi et al. JPEG Pleno: Toward an efficient representation of visual reality. *IEEE MultiMedia*, 23(4):14–20, Oct 2016.

124 F. Ernawan and M.N. Kabir. A robust image watermarking technique with an optimal DCT-psychovisual threshold. *IEEE Access*, 6(1):20464–20480, April 2018.

126 M.M.M. Fahmy. Palmprint recognition based on Mel frequency cepstral coefficients feature extraction. *Ain Shams Engineering Journal*, 1(1):39–47, 2010.

128 M. Fallahpour et al. Tampering detection in compressed digital video using watermarking. *IEEE Transactions on Instrumentation and Measurement*, 63(5):1057–1072, May 2014.
129 C.P. Fan, C.W. Chang, and S.J. Hsu. Cost-effective hardware-sharing design of fast algorithm based multiple forward and inverse transforms for H.264/AVC, MPEG-1/2/4, AVS, and VC-1 video encoding and decoding applications. *IEEE Transactions on Circuits and Systems for Video Technology*, 24(4):714–720, April 2014.
130 C.P. Fan and G.A. Su. Efficient fast 1-D 8×8 inverse integer transform for VC-1 application. *IEEE Transactions on Circuits and Systems for Video Technology*, 19(4):584–590, April 2009.
131 G.M. Farinella et al. Representing scenes for real-time context classification on mobile devices. *Pattern Recognition*, 48(4):1086–1100, 2015.
132 P. Farrelle and A. K. Jain. Recursive block coding -a new approach to transform coding. *IEEE Transactions on Communications*, 34(2):161–179, Feb 1986.
134 A. Foi, V. Katkovnik, and K. Egiazarian. Pointwise shape-adaptive DCT for high quality denoising and deblocking of grayscale and color images. *IEEE Transactions on Image Processing*, 16(5):1395–1411, May 2007.
135 C.K. Fong, Q. Han, and W.K. Cham. Recursive integer cosine transform for HEVC and future video coding standards. *IEEE Transactions on Circuits and Systems for Video Technology*, 27(2):326–336, Feb 2017.
137 G. Fracastoro, S.M. Fosson, and E. Magli. Steerable discrete cosine transform. *IEEE Transactions on Image Processing*, 26(1):303–314, Jan 2017.
145 K. Fukunaga and W.L.G Koontz. Application of the Karhunen-Loeve expansion to feature selection and ordering. *IEEE Transactions on Computers*, C-19(4):311–318, April 1970.
149 Y. Gao et al. Encoder-driven inpainting strategy in multiview video compression. *IEEE Transactions on Image Processing*, 25(1):134–149, Jan 2016.
150 M.J. Garrido et al. A high performance FPGA-based architecture for the future video coding adaptive multiple core transform. *IEEE Transactions on Consumer Electronics*, 64(1):53–60, Feb 2018.
151 M. Gastpar, P.L. Dragotti, and M. Vetterli. The distributed Karhunen-Loeve transform. *IEEE Transactions on Information Theory*, 52(12):5177–5196, Dec 2006.
152 S. Gazor and W. Zhang. A soft voice activity detector based on a Laplacian-Gaussian model. *IEEE Transactions on Speech and Audio Processing*, 11(5):498–505, Sept 2003.
158 M. Gilge, T. Engelhardt, and R. Mehlan. Coding of arbitrarily shaped image segments based on a generalized orthogonal transform. *Signal Processing: Image Communication*, 1(2):153–180, 1989. 64 kbit/s Coding of Moving Video.
159 J. Gimlett. Use of "activity" classes in adaptive transform image coding. *IEEE Transactions on Communications*, 23(7):785–786, July 1975.
161 H.B. Golestani and M. Ghanbari. Minimisation of image watermarking side effects through subjective optimisation. *IET Image Processing*, 7(8):733–741, Nov 2013.
162 D. Gong, Y. He, and Z Cao. New cost-effective VLSI implementation of a 2-D discrete cosine transform and its inverse. *IEEE Transactions on Circuits and Systems for Video Technology*, 14(4):405–415, April 2004.
167 G. Hamarneh and T. Gustavsson. Combining snakes and active shape models for segmenting the human left ventricle in echocardiographic images. *In Computers in Cardiology*. Vol.27 (Cat. 00CH37163), pages 115-118, 2000.
168 M. Hamidi and J. Pearl. Comparison of the cosine and Fourier transforms of Markov-1 signals. *IEEE Transactions on Acoustics, Speech, and Signal Processing*, 24(5):428–429, Oct 1976.
169 J. Han et al. Real-time video content analysis tool for consumer media storage system. *IEEE Transactions on Consumer Electronics*, 52(3):870–878, Aug 2006.
170 J. Han et al. Jointly optimized spatial prediction and block transform for video and image coding. *IEEE Transactions on Image Processing*, 21(4):1874–1884, April 2012.
173 P. Hanhart and T. Ebrahimi. Calculation of average coding efficiency based on subjective quality scores. *Journal of Visual Communication and Image Representation*, 25(3):555–564, 2014.
174 R.M. Haralick. A storage efficient way to implement the discrete cosine transform. *IEEE Transactions on Computers*, C-25(7):764–765, July 1976.

175 F. Hartung and B. Girod. Watermarking of uncompressed and compressed video. *Signal Processing (Special Issue on Copyright Protection and Access Control for Multimedia Services)*, pages 283–301, 1998.

177 T.I. Haweel. A new square wave transform based on the DCT. *Signal Processing*, 81(11):2309–2319, 2001.

178 D. Hein and N. Ahmed. On a real-time Walsh-Hadamard/cosine transform image processor. *IEEE Transactions on Electromagnetic Compatibility*, EMC-20(3):453–457, Aug 1978.

181 V. Holub and J. Fridrich. Low-complexity features for JPEG steganalysis using undecimated DCT. *IEEE Transactions on Information Forensics and Security*, 10(2):219–228, Feb 2015.

183 H.S. Hou. A fast recursive algorithm for computing the discrete cosine transform. *IEEE Transactions on Acoustics, Speech, and Signal Processing*, 35(10):1455–1461, Oct 1987.

184 H.S. Hou. Correction to: The fast Hartley transform algorithm. *IEEE Transactions on Computers*, C-36(9):1135–1136, Sept 1987.

185 H.S. Hou. The fast Hartley transform algorithm. *IEEE Transactions on Computers*, C-36(2):147–156, Feb 1987.

187 H.C. Hsu et al. Architecture design of shape-adaptive discrete cosine transform and its inverse for MPEG-4 video coding. *IEEE Transactions on Circuits and Systems for Video Technology*, 18(3):375–386, March 2008.

188 H. Hu. Illumination invariant face recognition based on dual-tree complex wavelet transform. *IET Computer Vision*, 9(2):163–173, Feb 2015.

189 X.-Y. Hu and X.-H Yu. The Chebyshev polynomial fitting properties of discrete cosine transform. *Signal Processing: Image Communication*, 13(1):15–20, July 1998.

190 Y. Hua and W. Liu. Generalized Karhunen-Loeve transform. *IEEE Signal Processing Letters*, 5(6):141–142, June 1998.

191 F. Huang et al. Reversible data hiding in JPEG images. *IEEE Transactions on Circuits and Systems for Video Technology*, 26(9):1610–1621, Sept 2016.

192 H. Huang et al. Integer MDCT with enhanced approximation of the DCT-IV. *IEEE Transactions on Signal Processing*, 54(3):1156–1159, March 2006.

193 H. Huang et al. Blind integrity verification of medical images. *IEEE Transactions on Information Technology in Biomedicine*, 16(6):1122–1126, Nov 2012.

194 H.Y. Huang, C.H. Yang, and W.H. Hsu. A video watermarking technique based on pseudo-3-D DCT and quantization index modulation. *IEEE Transactions on Information Forensics and Security*, 5(4):625–637, Dec 2010.

197 B. Hunt and O Kubler. Karhunen-Lo'eve multispectral image restoration, part I: Theory. *IEEE Transactions on Acoustics, Speech, and Signal Processing*, 32(3):592–600, June 1984.

220 A.K. Jain. A fast Karhunen-Loeve transform for a class of random processes. *IEEE Transactions on Communications*, 24(9):1023–1029, Sept 1976.

221 A.K. Jain. A fast Karhunen-Loéve transform for digital restoration of images degraded by white and colored noise. *IEEE Transactions on Computers*, C-26(6):560–571, June 1977.

222 A.K. Jain. A sinusoidal family of unitary transforms. *IEEE Transactions on Pattern Analysis and Machine Intelligence*, PAMI-1(4):356–365, Oct 1979.

224 S. Javanbakhti, S. Zinger, and P.H.N. De With. Fast scene analysis for surveillance video databases. *IEEE Transactions on Consumer Electronics*, 63(3):325–333, August 2017.

228 X. Ji et al. Early determination of zero-quantized 8 x 8 DCT coefficients. *IEEE Transactions on Circuits and Systems for Video Technology*, 19(12):1755–1765, Dec 2009.

229 J. Jiang and G. Feng. The spatial relationship of DCT coefficients between a block and its sub-blocks. *IEEE Transactions on Signal Processing*, 50(5):1160–1169, May 2002.

230 X.-Y. Jing and D. Zhang. A face and palmprint recognition approach based on discriminant DCT feature extraction. *IEEE Transactions on Systems, Man, and Cybernetics, Part B (Cybernetics)*, 34(6):2405–2415, Dec 2004.

231 T. Jinshan, E. Peli, and S. Acton. Image enhancement using a contrast measure in the compressed domain. *IEEE Signal Processing Letters*, 10(10):289–292, Oct 2003.

232 A.W. Johnson and A.B. Bradley. Adaptive transform coding incorporating time domain aliasing cancellation. *Speech Comm.*, 6(4):299–308, Dec 1987.
233 C.R. Johnson et al. Image processing for artist identification. *IEEE Signal Processing Magazine*, 25(4):37–48, July 2008.
235 K. Josef and P.C. Young. A new approach to feature selection based on the Karhunen-Loéve expansion. *Pattern Recognition*, 5(4):335–352, 1973.
238 M. Jridi, A. Alfalou, and P.K. Meher. A generalized algorithm and reconfigurable architecture for efficient and scalable orthogonal approximation of DCT. *IEEE Transactions on Circuits and Systems I: Regular Papers*, 62(2):449–457, Feb 2015.
239 M. Jridi and P.K. Meher. Scalable approximate DCT architectures for efficient HEVC-compliant video coding. *IEEE Transactions on Circuits and Systems for Video Technology*, 27(8):1815–1825, Aug 2017.
240 E. Kalali et al. A low energy HEVC inverse transform hardware. *IEEE Transactions on Consumer Electronics*, 60(4):754–761, Nov 2014.
242 E. Kalali, A.C. Mert, and I. Hamzaoglu. A computation and energy reduction technique for HEVC discrete cosine transform. *IEEE Transactions on Consumer Electronics*, 62(2):166–174, May 2016.
246 M. Karczewicz et al. A hybrid video coder based on extended macroblock sizes, improved interpolation, and flexible motion representation. *IEEE Transactions on Circuits and Systems for Video Technology*, 20(12):1698–1708, Dec 2010.
247 P. Kauff and K. Schuur. Shape-adaptive DCT with block-based DC separation and Delta-DC correction. *IEEE Transactions on Circuits and Systems for Video Technology*, 8(3):237–242, Jun 1998.
250 D.N. Kim and K.R. Rao. Sequence mirroring properties of orthogonal transforms having even and odd symmetric vectors. *ECTI Transactions on Computer and Information Technology*, 3(2):1–5, Nov 2007.
252 W.H. Kim, K. Yi, and C.M. Kyung. Energy-efficient illumination-invariant change detection in DCT coefficient domain for vehicular black box camera. *Electronics Letters*, 51(11):822–824, May 2015.
253 M. Kirby and L. Sirovich. Application of the Karhunen-Loeve procedure for the characterization of human faces. *IEEE Transactions on Pattern Analysis and Machine Intelligence*, 12(1):103-/108, Jan 1990.
254 H. Kitajima. Energy packing efficiency of the Hadamard transform. *IEEE Transactions on Communications*, 24(11):1256–1258, November 1976.
255 H. Kitajima. A symmetric cosine transform. *IEEE Transactions on Computers*, C-29(4):317–323, April 1980.
257 W. Kou and T. Fjallbrant. Fast computation of transform coefficients for a subadjacent block for a transform family. *IEEE Transactions on Signal Processing*, 39:1695–1699, July 1991.
259 E. Kurniawati, C.T. Lau, and B. Premkumar. Time-domain aliasing cancellation role in error concealment. *IET Electronics Letter*, 40(12):781–783, June 2004.
262 S. Lai, C.H. Luo, and S. Lei. Common architecture design of novel recursive MDCT and IMDCT algorithms for applications to AAC, AAC in DRAM, and MP3 codecs. *IEEE Transactions on Circuits and Systems II: Express Brief*, 56(10):793–797, Oct 2009.
260 M. Laadjel, S.A. Maadeed, and A. Bouridane. Combining Fisher locality preserving projections and passband DCT for efficient palmprint recognition. *Neurocomputing*, 152:179–189, 2015.
263 Y.K. Lai and Y.F. Lai. A reconfigurable IDCT architecture for universal video decoders. *IEEE Transactions on Consumer Electronics*, 56(3):1872–1879, Aug 2010.
264 J. Lainema et al. Intra coding of the HEVC standard. *IEEE Transactions on Circuits and Systems for Video Technology*, 22(12):1792–1801, Dec 2012.
267 C. Lan et al. Variable block-sized signal-dependent transform for video coding. *IEEE Transactions on Circuits and Systems for Video Technology*, 28(8):1920–1933, Aug 2018.
268 B.G Lee. A new algorithm to compute the discrete cosine transform. *IEEE Transactions on Acoustics, Speech, and Signal Processing*, 32(6):1243–1245, Dec 1984.
270 B.G Lee. Input and output index mappings for a prime-factor-decomposed computation of discrete cosine transform. *IEEE Transactions on Acoustics, Speech, and Signal Processing*, 37(2):237–244, Feb 1989.

273 K. Lee et al. A novel algorithm for zero block detection in High Efficiency Video Coding. *IEEE Journal of Selected Topics in Signal Processing*, 7(6):1124–1134, Dec 2013.

274 A. Levey and M. Lindenbaum. Sequential Karhunen-Loeve basis extraction and its application to images. *IEEE Transactions on Image Processing*, 9(8):1371–1374, Aug 2000.

277 P. Li and K. Lo. A content-adaptive joint image compression and encryption scheme. *IEEE Transactions on Multimedia*, 20(8):1960–1972, Aug 2018.

280 T. Li et al. On integer MDCT for perceptual audio coding. *IEEE Transactions on Audio, Speech, and Language Processing*, 15(8):2236–2248, Nov 2007.

281 Y.R. Li, L. Shen, and B.W. Suter. Adaptive inpainting algorithm based on DCT induced wavelet regularization. *IEEE Transactions on Image Processing*, 22(2):752–763, Feb 2013.

282 J. Liang and T.D. Tran. Fast multiplierless approximations of the DCT with the lifting scheme. *IEEE Transactions on Signal Processing*, 49(12):3032–3044, Dec 2001.

285 Y.K. Lin. High capacity reversible data hiding scheme based upon discrete cosine transformation. *Journal of Systems and Software*, 85(10):2395–2404, 2012.

286 M. Liu, S. Liu, and Q. Zhao. Fingerprint orientation field reconstruction by weighted discrete cosine transform. *Information Sciences*, 268:65–77, 2014.

287 X. Liu et al. Random walk graph Laplacian-based smoothness prior for soft decoding of JPEG images. *IEEE Transactions on Image Processing*, 26(2):509–524, Feb 2017.

295 Z. Ma et al. A video watermarking DRM method based on H.264 compressed domain with low bit-rate increasement. *Chinese Journal of Electronics*, 25(4):641–647, Aug 2016.

299 T. Mahmood et al. A robust technique for copy-move forgery detection and localization in digital images via stationary wavelet and discrete cosine transform. *Journal of Visual Communication and Image Representation*, 53:202–214, 2018.

300 J. Makhoul. A fast cosine transform in one and two dimensions. *IEEE Transactions on Acoustics, Speech, and Signal Processing*, 28(1):27–34, Feb 1980.

304 H. S. Malvar and D. H. Staelin. The LOT: transform coding without blocking effects. *IEEE Transactions on Acoustics, Speech, and Signal Processing*, 37(4):553–559, April 1989.

305 H.S. Malvar. Fast computation of discrete cosine transform through fast Hartley transform. *Electronics Letters*, 22(7):352–353, March 1986.

306 H.S. Malvar. Fast computation of the discrete cosine transform and the discrete Hartley transform. *IEEE Transactions on Acoustics, Speech, and Signal Processing*, 35(10):1484–1485, Oct 1987.

307 H.S. Malvar. Corrections to, Fast computation of the discrete cosine transform and the discrete Hartley transform. *IEEE Transactions on Acoustics, Speech, and Signal Processing*, ASSP–36:610, April 1988.

308 H.S. Malvar et al. Low-complexity transform and quantization in H.264/AVC. *IEEE Transactions on Circuits and Systems for Video Technology*, 13(7):598–603, July 2003.

310 J.V. Manjon et al. New methods for MRI denoising based on sparseness and self-similarity. *Medical Image Analysis*, 16(1):18–27, Jan 2012.

311 A. Mansouri et al. A low complexity video watermarking in H.264 compressed domain. *IEEE Transactions on Information Forensics and Security*, 5(4):649–657, Dec 2010.

314 F. Marcelloni and M. Vecchio. An efficient lossless compression algorithm for tiny nodes of monitoring wireless sensor networks. *The Computer Journal*, 52(8):969–987, 2009.

315 G. Martinelli, L.P. Ricotti, and G. Marcone. Neural clustering for optimal KLT image compression. *IEEE Transactions on Signal Processing*, 41(4):1737–1739, April 1993.

316 E.E. Martínez et al. Directional transforms for video coding based on lifting on graphs. *IEEE Transactions on Circuits and Systems for Video Technology*, 28(4):933–946, April 2018.

319 M. Masera, M. Martina, and G. Masera. Adaptive approximated DCT architectures for HEVC. *IEEE Transactions on Circuits and Systems for Video Technology*, 27(12):2714–2725, Dec 2017.

321 K. Meenakshi, K.S. Prasad., and C.S. Rao. Development of low-complexity video watermarking with conjugate symmetric sequency - complex Hadamard transform. *IEEE Communications Letters*, 21(8):1779–1782, Aug 2017.

322 P.K. Meher et al. Efficient integer DCT architectures for HEVC. *IEEE Transactions on Circuits and Systems for Video Technology*, 24(1):168–178, Jan 2014.

323 R. Merris. Laplacian matrices of graphs: a survey. *Linear Algebra and its Applications*, 197-198:143–176, 1994.

324 A.C. Mert, E. Kalali, and I. Hamzaoglu. High performance 2D transform hardware for future video coding. *IEEE Transactions on Consumer Electronics*, 63(2):117–125, May 2017.

326 K. Minemura et al. A novel sketch attack for H.264/AVC format-compliant encrypted video. *IEEE Transactions on Circuits and Systems for Video Technology*, 27(11):2309–2321, Nov 2017.

327 M. Miyahara and K. Kotani. Block distortion in orthogonal transform coding-analysis, minimization and distortion measure. *IEEE Transactions on Communications*, 33(1):90–96, January 1985.

330 Y.H. Moon, J.H. Kim, and G.Y. Kim. An improved early detection algorithm for all-zero blocks in H.264 video encoding. *IEEE Transactions on Circuits and Systems for Video Technology*, 15(8):1053–1057, Aug 2005.

331 M. Moosazadeh and G. Ekbatanifard. An improved robust image watermarking method using DCT and YCoCg-R color space. *Optik - Intl. Journal for Light and Electron Optics*, 140:975–988, 2017.

334 R.J. Mstafa, K.M. Elleithy, and E. Abdelfattah. A robust and secure video steganography method in DWT-DCT domains based on multiple object tracking and ECC. *IEEE Access*, 5(1):5354–5365, April 2017.

335 R. Mukundan, S.H. Ong, and P.A. Lee. Image analysis by Tchebichef moments. *IEEE Transactions on Image Processing*, 10(9):1357–1364, Sept 2001.

338 V. Nagesha. Comments on "Fast computation of the discrete cosine transform and the discrete Hartley transform" by H.S. Malvar. *IEEE Transactions on Acoustics, Speech, and Signal Processing*, 37(3):439–440, March 1989.

339 V.P.S. Naidu. Hybrid DDCT-PCA based multi sensor image fusion. *Journal of Optics*, 43(1):48–61, Mar 2014.

340 K. Nakagaki and R. Mukundan. A fast 4×4 forward discrete Tchebichef transform algorithm. *IEEE Signal Processing Letters*, 14(10):684–687, Oct 2007.

341 M.J. Narasimha. Design of FIR filter banks for a 24-channel transmultiplexer. *IEEE Transactions on Communications*, 30(7):1506–1510, July 1982.

343 M.J. Narasimha and A. Peterson. On the computation of the discrete cosine transform. *IEEE Transactions on Communications*, 26(6):934–936, June 1978.

344 M.J. Narasimha and A. Peterson. Design of a 24-channel transmultiplexer. *IEEE Transactions on Acoustics, Speech, and Signal Processing*, 27(6):752–762, Dec 1979.

346 D.R. Nayak et al. Combining extreme learning machine with modified sine cosine algorithm for detection of pathological brain. *Computers and Electrical Engineering*, 68:366–380, 2018.

350 T. Nguyen and D. Marpe. Objective performance evaluation of the HEVC main still picture profile. *IEEE Transactions on Circuits and Systems for Video Technology*, 25(5):790–797, May 2015.

352 J. A. Nikara, J. H. Takala, and J. T. Astola. Discrete cosine and sine transforms-Regular algorithms and pipeline architectures. *Signal Processing*, 86(2):230–249, 2006.

358 J. Ohm et al. Comparison of the coding efficiency of video coding standards-including high efficiency video coding (HEVC). *IEEE Transactions on Circuits and Systems for Video Technology*, 22(12):1669–1684, Dec 2012.

360 P.A.M. Oliveira et al. Low-complexity image and video coding based on an approximate discrete Tchebichef transform. *IEEE Transactions on Circuits and Systems for Video Technology*, 27(5):1066–1076, May 2017.

365 N. Paramanandham and K. Rajendiran. Infrared and visible image fusion using discrete cosine transform and swarm intelligence for surveillance applications. *Infrared Physics and Technology*, 88:13–22, 2018.

366 J.-S. Park et al. 2-D large inverse transform (16×16, 32×32) for HEVC (high efficiency video coding). *Journal of Semiconductor Technology and Science*, 12(2):203-211, June 2012.

369 G. Pastuszak. Hardware architectures for the H.265/HEVC discrete cosine transform. *IET Image Processing*, 9(6):468–477, June 2015.

370 D.-B. Perng and S.-H. Chen. Directional textures auto-inspection using discrete cosine transform. *Intl. Journal of Production Research*, 49(23):7171–7187, 2011.

371 G. Plonka. A global method for invertible integer DCT and integer wavelet algorithms. *Applied and Computational Harmonic Analysis*, 16(2):90–110, 2004.

373 G. Plonka and M. Tasche. Fast and numerically stable algorithms for discrete cosine transforms. *Linear Algebra and its Applications*, 49(9):309–345, Sept 2005.

376 U.S. Potluri et al. Multiplier-free DCT approximations for RF multi-beam digital aperture-array space imaging and directional sensing. *Measurement Science and Technology*, 23(11):1–15, 2012.

377 M.T. Pourazad et al. HEVC: The new gold standard for video compression: How does HEVC compare with H.264/AVC? *IEEE Consumer Electronics Magazine*, 1(3):36–46, July 2012.

379 J. Princen and A. Bradley. Analysis/Synthesis filter bank design based on time domain aliasing cancellation. *IEEE Transactions on Acoustics, Speech, and Signal Processing*, 34(5):1153–1161, Oct 1986.

381 Z. Qian et al. New framework of reversible data hiding in encrypted JPEG bitstreams. *IEEE Transactions on Circuits and Systems for Video Technology*, (Early access), 2018.

383 R.L. Queiroz. Processing JPEG-compressed images and documents. *IEEE Transactions on Image Processing*, 7(12):1661–1672, Dec 1998.

384 G. Quer et al. Sensing, compression, and recovery for WSNs: Sparse signal modeling and monitoring framework. *IEEE Transactions on Wireless Communications*, 11(10):3447–3461, Oct 2012.

392 D. Ravi et al. Semantic segmentation of images exploiting DCT-based features and random forest. *Pattern Recognition*, 52:260–273, 2016.

393 W. Ray and R. Driver. Further decomposition of the Karhunen–Loève series representation of a stationary random process. *IEEE Transactions on Information Theory*, 16(6):663–668, Nov 1970.

394 I.S. Reed and L.S Lan. A fast approximate Karhunen–Loève transform (AKLT) for data compression. *Journal of Visual Communication and Image Representation*, 5(4):304–316, Dec 1994.

395 H. Reitboeck and T.P. Brody. A transformation with invariance under cyclic permutation for applications in pattern recognition. *Information and Control*, 15(2):130–154, 1969.

396 H. Ren, P. Yu, and P. Zhang. Illumination invariant feature extraction using relative gradient difference. *Optik - Intl. Journal for Light and Electron Optics*, 126(23): 4531–4533, 2015.

406 T. Richter, A. Artusi, and T. Ebrahimi. JPEG XT: A new family of JPEG backwardcompatible standards. *IEEE MultiMedia*, 23(3):80–88, July 2016.

410 L.R. Roldan et al. Watermarking-based color image authentication with detection and recovery capability. *IEEE Latin America Transactions*, 14(2):1050–1057, Feb 2016.

411 Sreedevi and E., Sherly. A novel approach for removal of pectoral muscles in digital mammogram. *Procedia Computer Science*, 46:1724–1731, 2015.

412 M.A. Saad, A.C. Bovik, and C. Charrier. A DCT statistics-based blind image quality index. *IEEE Signal Processing Letters*, 17(6):583–586, June 2010.

413 M.A. Saad, A.C. Bovik, and C. Charrier. Blind image quality assessment: A natural scene statistics approach in the DCT domain. *IEEE Transactions on Image Processing*, 21(8):3339–3352, Aug 2012.

414 J.A. Saghri, A.G. Tescher, and J.T. Reagan. Practical transform coding of multispectral imagery. *IEEE Signal Processing Magazine*, 12(1):32–43, Jan 1995.

415 L. Sangkil and L. Insung. A low-delay MDCT/IMDCT. *ETRI Journal*, 35(5):935–938, 2013.

417 B.N.S. Sarvesh, Y.N. Mamatha, and M.K. Subramanian. Fast and efficient computation of MDCT/IMDCT algorithms for MP3 applications. *Intl. Journal of Scientific and Engineering Research*, 4(6):1121–1125, 2013.

420 I.D. Schizas, G.B. Giannakis, and Z.Q. Luo. Distributed estimation using reduced dimensionality sensor observations. *IEEE Transactions on Signal Processing*, 55(8):4284–4299, Aug 2007.

421 I.W. Selesnick, R.G. Baraniuk, and N.C. Kingsbury. The dual-tree complex wavelet transform. *IEEE Signal Processing Magazine*, 22(6):123–151, Nov 2005.

423 A. Sengupta et al. Low-cost obfuscated JPEG CODEC IP core for secure CE hardware. *IEEE Transactions on Consumer Electronics, (Early access)*, 2018.

426 Q. Shang et al. Single-port SRAM-based transpose memory with diagonal data mapping for large size 2-D DCT/IDCT. *IEEE Transactions on Very Large Scale Integration (VLSI) Systems*, 22(11):2423–2427, Nov 2014.

427 H.R. Sheikh and A.C. Bovik. Image information and visual quality. *IEEE Transactions on Image Processing*, 15(2):430–444, Feb 2006.

428 H.R. Sheikh, A.C. Bovik, and G. Veciana. An information fidelity criterion for image quality assessment using natural scene statistics. *IEEE Transactions on Image Processing*, 14(12):2117–2128, Dec 2005.

429 B.H. Shekar and B. Pilar. Discrete cosine transformation and height functions based shape representation and classification. *Procedia Computer Science*, 58:714–722, 2015.

432 D.I. Shuman et al. The emerging field of signal processing on graphs: Extending high dimensional data analysis to networks and other irregular domains. *IEEE Signal Processing Magazine*, 30(3):83–98, May 2013.

433 T Sikora. Low complexity shape-adaptive DCT for coding of arbitrarily shaped image segments. *Signal Processing: Image Communication*, 7(4):381–395, 1995.

434 T. Sikora and B. Makai. Shape-adaptive DCT for generic coding of video. *IEEE Transactions on Circuits and Systems for Video Technology*, 5(1):59–62, Feb 1995.

435 H. Silverman. An introduction to programming the Winograd Fourier transform algorithm (WFTA). *IEEE Transactions on Acoustics, Speech, and Signal Processing*, 25(2):152–165, April 1977.

436 H. Silverman. Corrections and addendum to "An introduction to programming the Winograd Fourier transform algorithm (WFTA)." *IEEE Transactions on Acoustics, Speech, and Signal Processing*, 26(3):268–268, June 1978.

437 P. Singh and G. Bhatnagar. A new robust watermarking system in integer DCT domain. *Journal of Visual Communication and Image Representation*, 53:86–101, 2018.

438 P. Singh and R.S. Chadha. A survey of digital watermarking techniques, applications and attacks. *Intl. Journal of Engineering and Innovative Technology (IJEIT)*, 2(9):165–175, 2013.

441 A. Skodras, C. Christopoulos, and T. Ebrahimi. The JPEG 2000 still image compression standard. *IEEE Signal Processing Magazine*, 18(5):36–58, Sept 2001.

442 B.C. Smith and L.A. Rowe. Algorithms for manipulating compressed images. *IEEE Computer Graphics and Applications*, 13(5):34–42, Sept 1993.

447 H. Sorensen et al. On computing the discrete Hartley transform. *IEEE Transactions on Acoustics, Speech, and Signal Processing*, 33(5):1231–1238, Oct 1985.

449 S. Srinivasan et al. Windows media video 9: Overview and applications. *Signal Processing: Image Communication*, 19(9):851–875, 2004.

451 S. Srisuk. Robust face recognition based on texture analysis. *Intl. Journal of Advanced Robotic Systems*, 10(1):47, 2013.

454 G. Strang. The discrete cosine transform. *SIAM Review*, 41(1):135–147, 1999.

456 P.-C. Su et al. A practical design of digital video watermarking in H.264/AVC for content authentication. *Signal Processing: Image Communication*, 26(8):413–426, 2011.

458 B.N. Subudhi et al. Tumor or abnormality identification from magnetic resonance images using statistical region fusion based segmentation. *Magnetic Resonance Imaging*, 34(9):1292–1304, 2016.

459 N. Suehiro and M. Hatori. Fast algorithms for the DFT and other sinusoidal transforms. *IEEE Transactions on Acoustics, Speech, and Signal Processing*, 34(3):642–644, June 1986.

460 T. Sugimoto, Y. Nakayama, and T. Komori. 22.2 ch audio encoding/decoding hardware system based on MPEG-4 AAC. *IEEE Transactions on Broadcasting*, 63(2):426–432, June 2017.

461 M.A. Suhail and M.S Obaidat. Digital watermarking-based DCT and JPEG model. *IEEE Transactions on Instrumentation and Measurement*, 52(5):1640–1647, Oct 2003.

462 G.J. Sullivan et al. Overview of the high efficiency video coding (HEVC) standard. *IEEE Transactions on Circuits and Systems for Video Technology*, 22(12):1649–1668, Dec 2012.

463 C. Sun and E.H. Yang. An efficient DCT-based image compression system based on Laplacian transparent composite model. *IEEE Transactions on Image Processing*, 24(3):886–900, March 2015.

468 T. Suzuki and M. Ikehara. Integer DCT-based on direct-lifting of DCT-IDCT for lossless-to-lossy image coding. *IEEE Transactions on Image Processing*, 19(11):2958–2965, Nov 2010.

469 W. Sweldens. The lifting scheme: A construction of second generation wavelets. *SIAM Journal on Mathematical Analysis*, 29(2):511–546, Mar 1998.

470 D.L. Swets and J.J. Weng. Using discriminant eigenfeatures for image retrieval. *IEEE Transactions on Pattern Analysis and Machine Intelligence*, 18(8):831–836, Aug 1996.

473 S.-C. Tai, Y.-G. Wu, and C.-W. Lin. An adaptive 3-D discrete cosine transform coder for medical image compression. *IEEE Transactions on Information Technology in Biomedicine*, 4(3):259–263, Sept 2000.

474 C. Tang, X. Yang, and G. Zhai. Noise estimation of natural images via statistical analysis and noise injection. *IEEE Transactions on Circuits and Systems for Video Technology*, 25(8):1283–1294, Aug 2015.

478 G. Tech et al. Overview of the multiview and 3D extensions of high efficiency video coding. *IEEE Transactions on Circuits and Systems for Video Technology*, 26(1):35–49, Jan 2016.

479 H. Tian et al. Robust watermarking of mobile video resistant against barrel distortion. *China Communications*, 13(9):131–138, Sept 2016.

481 D. Tretter and C.A. Bouman. Optimal transforms for multispectral and multilayer image coding. *IEEE Transactions on Image Processing*, 4(3):296–308, Mar 1995.

482 B.D. Tseng and W.C. Miller. Comments on "An introduction to programming the Winograd Fourier transform algorithm (WFTA)." *IEEE Transactions on Acoustics, Speech, and Signal Processing*, 26(3):268–269, June 1978.

483 B.D. Tseng and W.C. Miller. On computing the discrete cosine transform. *IEEE Transactions on Computers*, C-27(10):966–968, Oct 1978.

486 U. Tuna, S. Peltonen, and U. Ruotsalainen. Gap-filling for the high-resolution PET sinograms with a dedicated DCT-domain filter. *IEEE Transactions on Medical Imaging*, 29(3):830–839, March 2010.

488 C. Turgay, O. Huseyin, and D. Hasan. Facial feature extraction using complex dualtree wavelet transform. *Computer Vision and Image Understanding*, 111(2):229–246, 2008.

493 V.D. Vaughn and T.S. Wilkinson. System considerations for multispectral image compression designs. *IEEE Signal Processing Magazine*, 12(1):19–31, Jan 1995.

496 S. Venkataraman et al. Discrete transforms via the Walsh–Hadamard transform. *Signal Processing*, 14(4):371–382, June 1988.

498 M. Vetterli and A. Ligtenberg. A discrete Fourier-cosine transform chip. *IEEE Journal on Selected Areas in Communications*, 4(1):49–61, Jan 1986.

499 M. Vetterli and H.J. Nussbaumer. Simple FFT and DCT algorithms with reduced number of operations. *Signal Processing*, 6(4):267–278, 1984.

500 Y. Voronenko and M. Püschel. Multiplierless multiple constant multiplication. *ACM Trans. Algorithms*, 3(2)1–39, May 2007.

501 G.K. Wallace. The JPEG still picture compression standard. *Communication of the ACM*, 34(4):30–44, April 1991.

502 R. Wallace et al. Cross-pollination of normalization techniques from speaker to face authentication using Gaussian mixture models. *IEEE Transactions on Information Forensics and Security*, 7(2):553–562, April 2012.

503 D. Wang et al. Perceptual differential energy watermarking for H.264/AVC. *Multimedia Tools and Applications*, 60(3):537–550, Oct 2012.

505 H.Wang et al. Early detection of all-zero 4 x 4 blocks in High Efficiency Video Coding. *Journal of Visual Communication and Image Representation*, 25(7):1784–1790, 2014.

506 H. Wang and S. Kwong. Hybrid model to detect zero quantized DCT coefficients in H.264. *IEEE Transactions on Multimedia*, 9(4):728–735, June 2007.

507 H. Wang and S. Kwong. Prediction of zero quantized DCT coefficients in H.264/AVC using Hadamard transformed information. *IEEE Transactions on Circuits and Systems for Video Technology*, 18(4):510–515, April 2008.

510 H. Wang, S. Kwong, and C.W. Kok. Efficient prediction algorithm of integer DCT coefficients for H.264/AVC optimization. *IEEE Transactions on Circuits and Systems for Video Technology*, 16(4):547–552, April 2006.

511 H. Wang, S. Kwong, and C.W. Kok. Efficient predictive model of zero quantized DCT coefficients for fast video encoding. *Image and Vision Computing*, 25(6):922–933, 2007.

512 W. Wang, J. Yang, and W. Gao. Modeling background and segmenting moving objects from compressed video. *IEEE Transactions on Circuits and Systems for Video Technology*, 18(5):670–681, May 2008.

513 X.-F. Wang and De-B. Zhao. Performance comparison of AVS and H.264/AVC video coding standards. *Journal of Computer Science and Technology*, 21(3):310–314, May 2006.
514 Y. Wang et al. DCT inspired feature transform for image retrieval and reconstruction. *IEEE Transactions on Image Processing*, 25(9):4 406–4420, Sept 2016.
515 Y.Wang et al. Parts for the Whole: The DCT norm for extreme visual recovery. *ArXiv e-prints*, abs/1604.05451, 2016.
516 Y. Wang et al. DCT regularized extreme visual recovery. *IEEE Transactions on Image Processing*, 26(7):3360–3371, July 2017.
517 Y. Wang and M. Vilermo. Modified discrete cosine transform - Its implications for audio coding and error concealment. *Journal of Audio Eng. Soc.*, 51(1):52–61, Feb 2003.
520 Z. Wang. A fast algorithm for the discrete sine transform implemented by the fast cosine transform. *IEEE Transactions on Acoustics, Speech, and Signal Processing*, 30(5):814–815, Oct 1982.
521 Z. Wang. Reconsideration of a fast computational algorithm for the discrete cosine transform. *IEEE Transactions on Communications*, 31(1):121–123, Jan 1983.
522 Z. Wang. Fast algorithms for the discrete W transform and for the discrete Fourier transform. *IEEE Transactions on Acoustics, Speech, and Signal Processing*, 32(4):803–816, Aug 1984.
523 Z. Wang. The discrete W transform. *Applied Mathematics and Computation*, 16:19–48, Jan 1985.
524 Z. Wang. Comments on a fast computational algorithm for the discrete sine transform. *IEEE Transactions on Communications*, 34(2):204–205, Feb 1986.
525 Z. Wang. Interpolation using the discrete cosine transform: Reconsideration. *Electronics Letters*, 29(2):198–200, Jan 1993.
526 Z. Wang et al. Image quality assessment: from error visibility to structural similarity. *IEEE Transactions on Image Processing*, 13(4):600–612, April 2004.
529 J. Wei and G. Li. Automated lung segmentation and image quality assessment for clinical 3-D/4-D-computed tomography. *IEEE Journal of Translational Engineering in Health and Medicine*, 2(1):1–10, Dec 2014.
538 J. K. Wu and R. E. Burge. Adaptive bit allocation for image compression. *Computer Graphics and Image Processing*, 19(4):392–400, 1982.
539 X.N. Wu and W. Sun. Robust copyright protection scheme for digital images using overlapping DCT and SVD. *Applied Soft Computing*, 13(2):1170–1182, 2013.
540 Y.-G. Wu. Medical image compression by sampling DCT coefficients. *IEEE Transactions on Information Technology in Biomedicine*, 6(1):86–94, March 2002.
541 Y.-G.Wu and S.-C. Tai. Medical image compression by discrete cosine transform spectral similarity strategy. *IEEE Transactions on Information Technology in Biomedicine*, 5(3):236–243, Sept 2001.
542 S. Xiang and J. He. Database authentication watermarking scheme in encrypted domain. *IET Information Security*, 12(1):42–51, Feb 2018.
543 Y. Xiang et al. Patchwork-based audio watermarking method robust to desynchronization attacks. *IEEE/ACM Transactions on Audio, Speech, and Language Processing*, 22(9):1413–1423, Sept 2014.
544 J.J. Xiao et al. Distributed compression-estimation using wireless sensor networks. *IEEE Signal Processing Magazine*, 23(4):27–41, July 2006.
545 Z. Xie et al. A general method for detecting all-zero blocks prior to DCT and quantization. *IEEE Transactions on Circuits and Systems for Video Technology*, 17(2):237–241, Feb 2007.
547 H. Xu, J. Xu, and F. Wu. Lifting-based directional DCT-like transform for image coding. *IEEE Transactions on Circuits and Systems for Video Technology*, 17(10):1325–1335, Oct 2007.
549 Z. Xuan, Y. Zhenghua, and Y. Songyu. Method for detecting all-zero DCT coefficients ahead of discrete cosine transformation and quantisation. *Electronics Letters*, 34(19):1839–1840, Sept 1998.
550 H.-Y. Yang et al. A geometric correction based robust color image watermarking scheme using quaternion exponent moments. *Optik–Intl. Journal for Light and Electron Optics*, 125(16):4456–4469, 2014.
551 J. Yang et al. Estimation of signal-dependent noise level function in transform domain via a sparse recovery model. *IEEE Transactions on Image Processing*, 24(5):1561–1572, May 2015.
552 J.-F. Yang, S.-C. Chang, and C.Y. Chen. Computation reduction for motion search in low rate video coders. *IEEE Transactions on Circuits and Systems for Video Technology*, 12(10):948–951, Oct 2002.

555 X. Yang et al. Adaptive Wiener filter based chrominance up-sampling enhancement method for video coding. *IEEE Transactions on Consumer Electronics*, 57(4):1851–1856, Nov 2011.
556 Y. Yemini and J. Pearl. Asymptotic properties of discrete unitary transforms. *IEEE Transactions on Pattern Analysis and Machine Intelligence*, PAMI–1(4):366–371, Oct 1979.
557 C. Yeo et al. Mode-dependent transforms for coding directional intra prediction residuals. *IEEE Transactions on Circuits and Systems for Video Technology*, 22(4):545–554, April 2012.
560 P. Yip and K.R. Rao. A fast computational algorithm for the discrete sine transform. *IEEE Transactions on Communications*, 28(2):304–307, Feb 1980.
561 P. Yip and K.R. Rao. Fast DIT algorithms for DSTs and DCTs. *Circuits, Systems and Signal Processing*, 3(4):387–408, 1984.
563 P. Yip and K.R. Rao. The decimation-in-frequency algorithms for a family of discrete sine and cosine transforms. *Circuits, Systems and Signal Processing*, 7(1):3–19, 1988.
565 G. Yu and G. Sapiro. DCT image denoising: A simple and effective image denoising algorithm. *Image Processing – On Line*, 1(1):292–296, 2011.
568 B. Zeng et al. Perceptual encryption of H.264 videos: Embedding sign–flips into the integer–based transforms. *IEEE Transactions on Information Forensics and Security*, 9(2):309–320, Feb 2014.
570 B. Zeng and J. Fu. Directional discrete cosine transforms – A new framework for image coding. *IEEE Transactions on Circuits and Systems for Video Technology*, 18(3):305–313, March 2008.
571 Y. Zeng et al. Integer DCTs and fast algorithms. *IEEE Transactions on Signal Processing*, 49(11):2774–2782, Nov 2001.
572 G. Zhai et al. Bayesian error concealment with DCT pyramid for images. *IEEE Transactions on Circuits and Systems for Video Technology*, 20(9):1224–1232, Sept 2010.
574 D. Zhang et al. Online palmprint identification. *IEEE Transactions on Pattern Analysis and Machine Intelligence*, 25(9):1041–1050, Sept 2003.
576 J. Zhang et al. Robust video watermarking of H.264/AVC. *IEEE Transactions on Circuits and Systems II: Express Briefs*, 54(2):205–209, Feb 2007.
578 T. Zhang et al. Signal dependent transform based on SVD for HEVC intracoding. *IEEE Transactions on Multimedia*, 19(11):2404–2414, Nov 2017.
579 W. Zhang and W.K. Cham. Hallucinating face in the DCT domain. *IEEE Transactions on Image Processing*, 20(10):2769–2779, Oct 2011.
582 Z. Zhang et al. Focus and blurriness measure using reorganized DCT coefficients for an autofocus application. *IEEE Transactions on Circuits and Systems for Video Technology*, 28(1):15–30, Jan 2018.
584 X. Zhao et al. Joint separable and non-separable transforms for next–generation video coding. *IEEE Transactions on Image Processing*, 27(5):2514–2525, May 2018.
585 Y.Q. Zhao et al. Tampered region detection of inpainting JPEG images. *Optik - Intl. Journal for Light and Electron Optics*, 124(16):2487–2492, Aug 2013.
586 Y.W. Zhao et al. The optimal digital filters of sine and cosine transforms for geophysical transient electromagnetic method. *Journal of Applied Geophysics*, 150:267–277, March 2018.
589 M. Zhou et al. HEVC lossless coding and improvements. *IEEE Transactions on Circuits and Systems for Video Technology*, 22(12):1839–1843, Dec 2012.
591 S. Zhu, S.K.A. Yeung, and B. Zeng. In search of better-than-DCT unitary transforms for encoding of residual signals. *IEEE Signal Processing Letters*, 17(11):961–964, Nov 2010.

Conference papers

2 H.M. Abbas and M.M. Fahmy. Neural model for Karhunen-Loève transform with application to adaptive image compression. *IEE Proceedings I -Communications, Speech and Vision*, 140(2):135–143, April 1993.

6 N. Ahmed and M.D. Flickner. Some considerations of the discrete cosine transform. *In Proc. of 16th Asilomar Conf. on Circuits. Systems and Computers*, Pacific Grove, CA, volume 1, pages 295–299, Nov 1982.

11 P.P. Alexander, W.P. Rosalind, and S. Scarloff. Photobook: Tools for content-based manipulation of image databases. *In Proc. of SPIE: Storage and Retrieval for Image and Video Databases II*, volume 2185, pages 34–47, April 1994.

13 E. Alshina, A. Alshin, and F.C. Fernandes. Rotational transform for image and video compression. *In Proc. of 18th Intl. Conf. on Image Processing (ICIP)*, pages 3689–3692, Sept 2011.

14 R.G.M. Alvarado et al. DCT-compressive sampling of multifrequency sparse audio signals. *In Proc. of IEEE Latin-America Conf. on Communications (LATINCOM)*, pages 1–5, Nov 2014.

17 A.E. Ansari, A. Mansouri, and A. Ahaitouf. An efficient VLSI architecture design for integer DCT in HEVC standard. *In Proc. of IEEE/ACS 13th Intl. Conf. of Computer Systems and Applications (AICCSA)*, pages 1–5, Nov 2016.

18 P. Arayacheeppreecha, S. Pumrin, and B. Supmonchai. Flexible input transform architecture for HEVC encoder on FPGA. *In Proc. of 12th Intl. Conf. on Electrical Engineering/Electronics, Computer, Telecommunications and Information Technology (ECTI-CON)*, pages 1–6, June 2015.

26 S.H. Babu et al. Face recognition using active illumination equalization and mirror image superposition as pre-processing techniques. *In Proc. of 5th Intl. Conf. on Signal and Image Processing*, pages 96–101, Jan 2014.

31 H. Bai, A. Wang, and M. Zhang. Compressive sensing for DCT image. *In Proc. of Int. Conf. on Computational Aspects of Social Networks*, pages 378–381, Sept 2010.

32 M.Z. Baig and A.V. Gokhale. Object tracking using mean shift algorithm with illumination invariance. *In Proc. of Intl. Conf. on Communication Systems and Network Technologies*, pages 550–553, April 2015.

35 M.M.A. Basiri and N.M. Sk. High performance integer DCT architectures for HEVC. *In Proc. of 30th Intl. Conf. on VLSI Design and 2017 16th Intl. Conf. on Embedded Systems (VLSID)*, pages 121–126, Jan 2017.

37 A. Bertrand. Applications and trends in wireless acoustic sensor networks: A signal processing perspective. *In Proc. of 18th Symp. on Communications and Vehicular Technology in the Benelux (SCVT)*, pages 1–6, Nov 2011.

40 J. Bigun. Unsupervised feature reduction in image segmentation by local Karhunen–Loeve transform. *In Proc. of 11th IAPR Intl. Conf. on Pattern Recognition*. Vol.II. Conf. B: Pattern Recognition Methodology and Systems, pages 79–83, Aug 1992.

43 S. Bouguezel, M.O. Ahmad, and M.N.S. Swamy. A fast 8×8 transform for image compression. *In Proc. of Intl. Conf. on Microelectronics -ICM*, pages 74–77, Dec 2009.

44 S. Bouguezel, M.O. Ahmad, and M.N.S. Swamy. A novel transform for image compression. *In Proc. of 53rd IEEE Intl. Midwest Symp. on Circuits and Systems*, pages 509–512, Aug 2010.

47 N. Brahimi and S. Bouguezel. An efficient fast integer DCT transform for images compression with 16 additions only. *In Proc. of Intl. Workshop on Systems, Signal Processing and their Applications, (WOSSPA)*, pages 71–74, May 2011.

55 M. Budagavi and V. Sze. Unified forward+inverse transform architecture for HEVC. *In Proc. of 19th IEEE Intl. Conf. on Image Processing*, pages 209–212, Sept 2012.

57 Q. Cai et al. Lossy and lossless intra coding performance evaluation: HEVC, H.264/AVC, JPEG 2000 and JPEG LS. *In Proc. of Asia Pacific Signal and Information Processing Association Annual Summit and Conference*, pages 1–9, Dec 2012.

59 C.K. Chaitanya et al. PET and MRI medical image fusion using STDCT and STSVD. *In 2017 Intl. Conf. on Computer Communication and Informatics (ICCCI)*, pages 1–4, Jan 2017.

61 W.K. Cham. Development of integer cosine transforms by the principle of dyadic symmetry. *In Proc. of Inst. Electr. Eng. I: Commun. Speech Vis.*, page 276–282, Aug 1989.

67 G. Chantas, S. Nikolopoulos, and I. Kompatsiaris. Sparse audio inpainting with variational Bayesian inference. *In Proc. of Intl. Conf. on Consumer Electronics (ICCE)*, pages 1–6, Jan 2018.

68 D.S. Chauhan et al. Double secret key based medical image watermarking for secure telemedicine in cloud environment. *In Proc. of 40th Intl. Conf. on Telecommunications and Signal Processing (TSP)*, pages 626–631, July 2017.

72 W.H. Chen and S.C. Fralick. Image enhancement using cosine transform filtering. *In Proc. of Symp. on Current Math. Problems in Image Science*, pages 186–192, Nov 1976.

74 W.M. Chen, C.J. Lai, and C.C. Chang. H.264 video watermarking with secret image sharing. *In Proc. of Intl. Symp. on Broadband Multimedia Systems and Broadcasting*, pages 1–7, May 2009.

75 Y. Chen and D. Mukherjee. Variable block-size overlapped block motion compensation in the next generation open-source video codec. *In Proc. of Intl. Conf. on Image Processing (ICIP)*, pages 938–942, Sept 2017.

76 Y.-J. Chen, S. Oraintara, and T. Nguyen. Video compression using integer DCT. *In Proc. of Intl. Conf. on Image Processing*, volume 2, pages 844–845, Sept 2000.

79 F. Cheng et al. Compressive sensing multi-focus image fusion. *In Proc. of Chinese Conf. on Pattern Recognition*, pages 107–116, Berlin, Heidelberg, Nov 2014. Springer.

81 P.T. Chiang and T.S. Chang. Fast zero block detection and early CU termination for HEVC Video Coding. *In Proc. of Intl. Symp. on Circuits and Systems (ISCAS)*, pages 1640–1643, May 2013.

82 B. Chitprasert and K.R. Rao. Discrete cosine transform filtering. *In Proc. of Intl. Conf. on Acoustics, Speech, and Signal Processing (ICASSP)*, volume 3, pages 1281–1284, Apr 1990.

85 S. Choi et al. Performance evaluation of JPEG XT standard for high dynamic range image coding. *In Proc. of the World Congress on Engineering (WCE)*, pages 1–4, July 2015.

86 F. Christos and L. King-Ip (David). FastMap: A fast algorithm for indexing, data mining and visualization of traditional and multimedia datasets. *In Proc. of SIGMOD Rec*, volume 24, pages 163–174, May 1995.

89 A. Ciancio et al. Energy-efficient data representation and routing for wireless sensor networks based on a distributed wavelet compression algorithm. *In Proc. of 5th Intl. Conf. on Information Processing in Sensor Networks, IPSN-06*, pages 309–316, New York, NY, USA, 2006. ACM.

96 M.S. Corrington. Implementation of fast cosine transforms using real arithmetic. *In Proc. of National Aerospace and Electronics Conf. (NAECON)*, pages 350–357, 1978.

103 A.D. Darji and R.P. Makwana. High-performance multiplierless DCT architecture for HEVC. *In Proc. of 19th Intl. Symp. on VLSI Design and Test*, pages 1–5, June 2015.

107 A. Descampe et al. JPEG XS, a new standard for visually lossless low-latency lightweight image compression. *In Proc of SPIE: Applications of Digital Image Processing XL*, volume 10396, pages 103960M1-103960M18, 2017.

111 M.N. Do and M. Vetterli. Contourlets: A new directional multiresolution image representation. *In Conf. Record of the 36th Asilomar Conf. on Signals, Systems and Computers*, volume 1, pages 497–501, Nov 2002.

113 L. Dong and S.C. Schwartz. DCT-based object tracking in compressed video. *In Proc. of Intl. Conf. on Acoustics Speech and Signal Processing Proceedings (ICASSP)*, volume 2, pages II–II, May 2006.

115 M.F. Duarte et al. Universal distributed sensing via random projections. *In Proc. of 5th Intl. Conf. on Information Processing in Sensor Networks*, IPSN-06, pages 177–185. ACM, 2006.

117 R. Dugad and N. Ahuja. A fast scheme for downsampling and upsampling in the DCT domain. *In Proc. of Intl. Conf. on Image Processing (ICIP)*, volume 2, pages 909–913, Oct 1999.

119 P. Duhamel and H. Hollmann. Implementation of "Split-radix" FFT algorithms for complex, real, and real symmetric data. *In Proc. of Intl. Conf. on Acoustics, Speech, and Signal Processing (ICASSP)*, volume 10, pages 784–787, Apr 1985.

125 F. Ernawan, M.N. Kabir, and Z. Mustaffa. A blind watermarking technique based on DCT psychovisual threshold for a robust copyright protection. *In Proc. of 12th Intl. Conf. for Internet Technology and Secured Transactions (ICITST)*, pages 92–97, Dec 2017.

138 G. Fracastoro and E. Magli. Steerable discrete cosine transform. *In Proc. of 17th Intl. Workshop on Multimedia Signal Processing (MMSP)*, pages 1–6, Oct 2015.

139 G. Fracastoro and E. Magli. Subspace-sparsifying steerable discrete cosine transform from graph Fourier transform. *In Proc. of Intl. Conf. on Image Processing (ICIP)*, pages 1534–1538, Sept 2016.

141 J. Fridrich, D. Soukal, and J. Lukas. Detection of copy-move forgery in digital images. *In Proc. of DFRWS Conf.*, pages 1–10, April 2003.

143 J. Fu and B. Zeng. Directional discrete cosine transforms: A theoretical analysis. *In Proc. of Intl. Conf. on Acoustics, Speech and Signal Processing (ICASSP)*, pages I–1105–I–1108, April 2007.

144 X. Fu and Y. Wan. Accurate image rotation using DCT transformation. *In Proc. of Advanced Information Technology, Electronic and Automation Control Conference (IAEAC)*, pages 537–541, Dec 2015.

146 S. Gaj, A. Sur, and P.K. Bora. A robust watermarking scheme against re-compression attack for H.265/HEVC. *In Proc. of 5th National Conf. on Computer Vision, Pattern Recognition, Image Processing and Graphics*, pages 1–4, Dec 2015.

148 L. Gao et al. A new reversible watermarking scheme based on Integer DCT for medical images. *In Proc. of Intl. Conf. on Wavelet Analysis and Pattern Recognition*, pages 33–37, July 2012.

153 R. Geiger et al. IntMDCT -A link between perceptual and lossless audio coding. *In Proc. of Intl. Conf. on Acoustics, Speech, and Signal Processing (ICASSP)*, volume 2, pages II–1813–II–1816, May 2002.

154 R. Geiger, Y. Yokotani, and G. Schuller. Improved integer transforms for lossless audio coding. *In Proc. of The 37th Asilomar Conf. on Signals, Systems Computers*, 2003, volume 2, pages 2119–2123, Nov 2003.

155 R. Geiger, Y. Yokotani, and G. Schuller. Audio data hiding with high data rates based on IntMDCT. *In Proc. of Intl. Conf. on Acoustics Speech and Signal Processing (ICASSP)*, volume 5, pages 205–208, May 2006.

156 G. Ghule and P. Mukherji. A novel approach for Marathi numeral recognition using Bark scale and discrete sine transform method. *In Proc. of Conf. on Advances in Signal Processing (CASP)*, pages 191–195, June 2016.

157 J.M. Gil-Cacho et al. Transform domain prediction error method for improved acoustic echo and feedback cancellation. *In Proc. of the 20th European Signal Processing Conf. (EUSIPCO)*, pages 2422–2426, Aug 2012.

163 S. Gupta et al. VLSI implementation of 3D integer DCT for video coding standards. *In Proc. of 2nd Intl. Conf. on Next Generation Computing Technologies (NGCT)*, pages 660–664, Oct 2016.

164 S. Guruprasad, M.Z. Kurian, and H.N. Suma. Fusion of CT and PET medical images using hybrid algorithm DWT-DCT-PCA. *In Proc. of 2nd Intl. Conf. on Information Science and Security (ICISS)*, pages 1–5, Dec 2015.

165 W. Hachicha, A. Beghdadi, and F.A. Cheikh. 1D directional DCT-based stereo residual compression. *In Proc. of 21st European Signal Processing Conf. (EUSIPCO)*, pages 1–5, Sept 2013.

166 M.N. Haggag, M. El-Sharkawy, and G. Fahmy. Modified efficient fast multiplication free integer transformation for the 2-D DCT H.265 standard. *In Proc. of Data Compression Conf. (DCC)*, pages 455–455, March 2011.

171 K.J. Han and A.H Tewk. Eigen-image based video segmentation and indexing. *In Proc. of Intl. Conf. on Image Processing (ICIP)*, volume 2, pages 538-541, Oct 1997.

172 M. Hangarge, K.C. Santosh, and R. Pardeshi. Directional discrete cosine transform for handwritten script identification. *In Proc. of 12th Intl. Conf. on Document Analysis and Recognition*, pages 344–348, Aug 2013.

176 M. Hassan, I. Osman, and M. Yahia. Discrete sine transform and alternative local linear regression for face recognition. *In Proc. of 5th IASTED Signal Processing, Pattern Recognition, and Applications (SPPRA)*, pages 99–103, Aug 2008.

180 A. Hilal, T. Hamzeh, and S. Chantaf. Copy-move forgery detection using principal component analysis and discrete cosine transform. *In Proc. of Sensors Networks Smart and Emerging Technologies (SENSET)*, pages 1–4, Sept 2017.

182 E. Hot and P. Sekulić. Compressed sensing MRI using masked DCT and DFT measurements. *In Proc. of 4th Mediterranean Conf. on Embedded Computing (MECO)*, pages 323–326, June 2015.

186 Y. Hozu et al. Segmentation method for phalanges in CR image by use of DCT. *In Proc. of Intl. Conf. on Control Automation and Systems (ICCAS)*, pages 2082–2085, Oct 2010.

195 X. Huang, I. Echizen, and A. Nishimura. A new approach of reversible acoustic steganography for tampering detection. *In Proc. of 6th Intl. Conf. on Intelligent Information Hiding and Multimedia Signal Processing*, pages 538–542, Oct 2010.

196 X. Huang et al. Scalable-to-lossless transform domain distributed video coding. *In Proc. of Intl. Workshop on Multimedia Signal Processing*, pages 327–332, Oct 2010.

198 T. Ichita et al. Directional discrete cosine transforms arising from discrete cosine and sine transforms for directional block-wise image representation. *In Proc. of Intl. Conf. on Acoustics, Speech and Signal Processing (ICASSP)*, pages 4536–4540, March 2017.

201 S. Ishwar, P.K. Meher, and M.N.S. Swamy. Discrete Tchebichef transform - A fast 4×4 algorithm and its application in image/video compression. *In Proc. of Intl. Symp. on Circuits and Systems*, pages 260–263, May 2008

218 S. Jadon, M. Kumar, and Y. Rathi. Face recognition using SOM neural network with DDCT facial feature extraction techniques. *In Proc. of Intl. Conf. on Communications and Signal Processing*, pages 1070–1074, April 2015.

219 A. K. Jain and Farrel P. M. Recursive block coding. *In Proc. of 16th Asilomar Conf. on Circuits, Systems and Computers*, pages 431–436, Nov 1982.

225 R. Jeni, T.S.R. Raj, and R.S. Roach. Design of high performance folded DIF FFT architecture using MMCM approach with Hcub algorithm. *In Proc. of Intl. Conf. on Circuits, Power and Computing Technologies (ICCPCT)*, pages 715–719, March 2013.

227 R. Jeske et al. Low cost and high throughput multiplierless design of a 16 point 1-D DCT of the new HEVC video coding standard. *In Proc. of VIII Southern Conf. on Programmable Logic*, pages 1–6, March 2012.

234 H.W. Jones, D.N. Hein, and S.C. Knauer. The Karhunen-Loeve, discrete cosine and related transforms via the Hadamard transform. *In Proc. of Intl. Telemetering Conf.*, pages 87–98, Nov 1978.

236 R. Joshi, Y.A. Reznik, and M. Karczewicz. Efficient large size transforms for high performance video coding. *In Proc. of SPIE: Optics and Photonics*, paper 7798-31, volume 7798, pages 77980W–1 – 77980W–7, Aug 2010.

241 E. Kalali and I. Hamzaoglu. FPGA implementations of HEVC inverse DCT using high-level synthesis. *In Proc. of Conf. on Design and Architectures for Signal and Image Processing (DASIP)*, pages 1–6, Sept 2015.

244 L. Kang and X.P. Cheng. Copy-move forgery detection in digital image. *In Proc. of 3rd Intl. Congress on Image and Signal Processing*, volume 5, pages 2419–2421, Oct 2010.

245 Q. Kang, K. Li, and J. Yang. A digital watermarking approach based on DCT domain combining QR code and chaotic theory. *In Proc. of 10th Intl. Conf. on Intelligent Computer Communication and Processing (ICCP)*, pages 331–337, Sept 2014.

248 C. Khirallah, L. Stankovic, and V. Stankovic. Low complexity multi-view video streaming over multipath fading channels. *In Proc. of 3rd Intl. Conf. on Information and Communication Technologies: From Theory to Applications*, pages 1–5, April 2008.

251 K. Kim et al. Reconstruction of dose distribution in in-beam PET for carbon therapy. *In Proc. of IEEE Nuclear Science Symp. and Medical Imaging Conf. Record (NSS/MIC)*, pages 2433–2436, Oct 2012.

256 U.V. Koc and K.J.R. Liu. Discrete-cosine/sine–transform based motion estimation. *In Proc. of 1st Intl. Conf. on Image Processing (ICIP)*, volume 3, pages 771–775, Nov 1994.

266 M. Lamarre and J.J. Clark. Background subtraction using competing models in the block-DCT domain. *In Proc. of object recognition supported by user interaction for service robots*, pages 299–302, 2002.

269 B.G. Lee. FCT–A fast cosine transform. *In Proc. of Intl. Conf. on Acoustics, Speech, and Signal Processing (ICASSP)*, volume 9, pages 477–480, Mar 1984.

271 J. Lee et al. A compressed-domain corner detection method for a DCT-based compressed image. *In Proc. of Intl. Conf. on Consumer Electronics (ICCE)*, pages 306–307, Jan 2017.

272 J. Lee, N. Vijaykrishnan, and M.J. Irwin. Efficient VLSI implementation of inverse discrete cosine transform. *In Proc. of Intl. Conf. on Acoustics, Speech, and Signal Processing (ICASSP)*, volume 5, pages V–177–180, May 2004.

275 L. Li and Z. Hou. Multiview video compression with 3D-DCT. *In Proc. of ITI 5th Intl. Conf. on Information and Communications Technology*, pages 59–61, Dec 2007.
276 M. Li, Z. Fan, and Z. Liu. An improved visual tracking algorithm based on particle filter. *In Proc. of Intl. Conf. Region 10 (TENCON)*, pages 1–4, Oct 2013.
278 Q. Li and H. Zhu. Block-based compressed sensing of image using directional Tchebichef transforms. *In Proc. of Intl. Conf. on Systems, Man, and Cybernetics (SMC)*, pages 2207–2212, Oct 2012.
279 Q. Li, H. Zhu, and Q. Liu. Image recognition by combined affine and blur Tchebichef moment invariants. *In Proc. of 4th Intl. Congress on Image and Signal Processing*, volume 3, pages 1517–1521, Oct 2011.
283 A. Ligtenberg and J.H. O'Neill. A single chip solution for an 8 by 8 two dimensional DCT. *In Proc. of Intl. Symp. on Circuits and Systems (ISCAS)*, pages 1128–1131, Mar 1987.
288 X.W. Liu et al. Infrared face recognition based on radiative energy and sub-block DCT in wavelet domain. *In Proc. of Intl. Conf. on Wavelet Analysis and Pattern Recognition*, pages 84–89, July 2010.
289 C. Loeffler, A. Ligtenberg, and G.S. Moschytz. Algorithm-architecture mapping for custom DSP chips. *In Proc. of Intl. Symp. on Circuits and Systems*, volume 2, pages 1953–1956, June 1988.
290 C. Loeffler, A. Ligtenberg, and G.S. Moschytz. Practical fast 1-D DCT algorithms with 11 multiplications. *In Proc. of Intl. Conf. on Acoustics, Speech, and Signal Processing (ICASSP)*, volume 2, pages 988–991, May 1989.
291 T.T. Lu, H.L. Hsu, and P.C. Chang. Blind video watermarking for H.264. *In Proc. of Canadian Conf. on Electrical and Computer Engineering*, pages 2353–2356, May 2006.
292 Y. Lu and M.N. Do. The finer directional wavelet transform. *In Proc. of Intl. Conf. on Acoustics, Speech, and Signal Processing (ICASSP)*, volume 4, pages iv/573–iv/576, March 2005.
293 W. Luo, G.L. Heileman, and C.E. Pizano. Fast and robust watermarking of JPEG files. *In Proc. of 5th Southwest Symp. on Image Analysis and Interpretation*, pages 158–162, April 2002.
294 H. Lv et al. A comparison of fractional-pel interpolation filters in HEVC and H.264/AVC. *In Proc. of Visual Communications and Image Processing*, pages 1–6, Nov 2012.
297 B.N. Madhukar and S. Jain. A duality theorem for the discrete sine transform (DST). *In Proc. of Intl. Conf. on Applied and Theoretical Computing and Communication Technology (iCATccT)*, pages 156–160, Oct 2015.
298 J.P. Maheshwari et al. Robust digital image watermarking using DCT-based pyramid transform via image compression. *In Proc. of Intl. Conf. on Communications and Signal Processing (ICCSP)*, pages 1059–1063, April 2015.
301 H. S. Malvar. The LOT: A link between block transform coding and multirate filter banks. *In Proc. of Intl. Symp. on Circuits and Systems (ISCAS)*, volume 1, pages 835–838, June 1988.
302 H. S. Malvar and R. Duarte. Transform/subband coding of speech with the lapped orthogonal transform. *In Proc. of Intl. Symp. on Circuits and Systems (ISCAS)*, volume 2, pages 1268–1271, May 1989.
303 H. S. Malvar and D. H. Staelin. Reduction of blocking effects in image coding with a lapped orthogonal transform. *In Proc. of Intl. Conf. on Acoustics, Speech, and Signal Processing (ICASSP)*, volume 2, pages 781–784, April 1988.
309 H.S. Malvar, G.J. Sullivan, and S. Srinivasan. Lifting-based reversible color transformations for image compression. *In Proc. of SPIE Applications of Digital Image Processing. Intl. Soc. for Optical Engr.*, volume 7073, pages 707307–1–707307–10, Aug 2008.
312 R. K. Mantiuk, T. Richter, and A. Artusi. Fine-tuning JPEG-XT compression performance using large-scale objective quality testing. *In Proc. of Intl. Conf. on Image Processing (ICIP)*, pages 2152–2156, Sept 2016.
317 L.W. Martinson. A 10 MHz image bandwidth compression model. *In Proc. of Conf.on Pattern Recognition and Image Processing*, pages 132–136, 1978.
318 S.A. Martucci. Image resizing in the discrete cosine transform domain. *In Proc. of Intl. Conf. on Image Processing (ICIP)*, volume 2, pages 244–247, Oct 1995.
325 S. Midtsokogen and J.M. Valin. The AV1 constrained directional enhancement filter. *In Proc. of Intl. Conf. on Acoustics, Speech and Signal Processing (ICASSP)*, pages 1193–1197, April 2018.
329 F. Mokhtarian and N. Jin. Human motion recognition based on statistical shape analysis. *In Proc. of Conf. on Advanced Video and Signal Based Surveillance (AVSS)*, pages 4–9, Sept 2005.

333 R.J. Mstafa and K.M. Elleithy. A DCT-based robust video steganographic method using BCH error correcting codes. *In Proc. of Long Island Systems, Applications and Technology Conference (LISAT)*, pages 1–6, April 2016.

336 R. Muralishankar, H.N. Shankar, and D. O'Shaughnessy. A performance analysis of features from complex cepstra of warped DST, DCT and DHT filters for phoneme recognition. *In Proc. of 15th Intl. Conference on Digital Signal Processing*, pages 591–594, July 2007.

337 S. Muramatsu and H. Hiki. Block-wise implementation of directional genlot. *In Proc. of 16th Intl. Conf. on Image Processing (ICIP)*, pages 3977–3980, Nov 2009.

342 M.J. Narasimha et al. The TM 800-MI: A 60-channel CCITT transmultiplexer. *In Proc. of Intl. Conf. on Communications*, pages 672–674, May 1984.

345 M.V. Narayanan et al. Application of the Karhunen-Loeve transform to 4D reconstruction of cardiac gated SPECT images. *In Proc. of Nuclear Science Symp. and Medical Imaging Conf. (Cat. No.98CH36255)*, volume 2, pages 1141–1145, 1998.

348 R.T. Ng and A. Sedighian. Evaluating multidimensional indexing structures for images transformed by principal component analysis. *In Proc. of SPIE*, volume 2670, pages 50–61, Jan 1996.

349 K.N. Ngan and R.J. Clarke. Lowpass filtering in the cosine transform domain. *In Proc. of Intl. Conf. Commun.*, pages 37.7.1–37.7.5, 1980.

351 T. Nguyen and D. Marpe. Future video coding technologies: A performance evaluation of AV1, JEM, VP9, AND HM. *In Proc. of Picture Coding Symposium (PCS)*, volume 2670, pages 50–61, July 2018.

353 V. Nikolajevic and G. Fettweis. Improved implementation of MDCT in MP3 audio coding. In Proc of 5th Intl. Symp. on Multi-Dimensional Mobile Communications Proceedings. *The Joint Conference of the 10th Asia-Pacific Conference on Communications*, volume 1, pages 309–312, Aug 2004.

355 P. Norhashimah, H. Fang, and J. Jiang. Video extraction in compressed domain. *In Proc. of Conf. on Advanced Video and Signal Based Surveillance*, pages 321–326, July 2003.

357 K. Ogawa and G. Ohtake. Watermarking for HEVC/H.265 stream. *In Proc. of Intl. Conf. on Consumer Electronics (ICCE)*, pages 102–103, Jan 2015.

359 J.-R. Ohm and G.J. Sullivan. Invited talk: Versatile video coding–Towards the next generation of video compression. *Picture Coding Symp.*, San Francisco, 2018.

363 J. Ou et al. A new image inpainting algorithm based on DCT similar patches features. *In Proc of. Intl. Conf. on Computational Intelligence and Security (CIS)*, pages 152–155, Dec 2016.

364 L.B. Ozer and W. Wolf. Real-time posture and activity recognition. *In Proc. of Workshop on Motion and Video Computing*, pages 133–138, Dec 2002.

367 J.S. Park and T. Ogunfunmi. A new approach for image quality assessment: FSM (Frequency Similarity Method). *In Proc. of 7th IEEE Conf. on Industrial Electronics and Applications (ICIEA)*, pages 996–1000, July 2012.

368 S.Y. Park and P.K. Meher. Flexible integer DCT architectures for HEVC. *In Proc. of Intl. Symp. on Circuits and Systems (ISCAS)*, pages 1376–1379, May 2013.

372 G. Plonka and Tasche M. Integer DCT-II by lifting steps. In Proc. of Advances in Multivariate Approximation, pages 235–252, 2002.

374 N. Ponomarenko et al. On between-coefficient contrast masking of DCT basis functions. *In Proc. of 3rd Intl. Workshop on Video Processing and Quality Metrics for Consumer Electronics (VPQM)*, pages 1–4, 2007.

378 S. Prattipati et al. A fast 8×8 integer Tchebichef transform and comparison with integer cosine transform for image compression. *In Proc. of 56th IEEE Intl. Midwest Symp. on Circuits and Systems (MWSCAS)*, pages 1294–1297, Aug 2013.

380 J. Princen, A. Johnson, and A. Bradley. Subband/Transform coding using filter bank designs based on time domain aliasing cancellation. *In Proc. of Intl. Conf. on Acoustics, Speech, and Signal Processing (ICASSP)*, volume 12, pages 2161–2164, April 1987.

382 Z. Qian, W. Wang, and T. Qiao. An edge detection method in DCT domain. *Procedia Engineering, Intl. Workshop on Information and Electronics Engineering*, 29:344–348, 2012.

385 M. Rabbat, J. Haupt, A. Singh, and et al. Decentralized compression and predistribution via randomized gossiping. *Proc. of 5th Intl. Conf. on Information Processing in Sensor Networks, (IPSN)*, pages 51–59, New York, NY, USA, 2006.

397 X. Ren and D. Fang. A sensor node compression algorithm based on H.264 integer DCT Transform in WSN. *In Proc. of IET Intl. Conf. on Wireless Sensor Network (IETWSN)*, pages 214–217, Nov 2010.
399 T. Richter. On the standardization of the JPEG XT image compression. *In Proc. of Picture Coding Symp. (PCS)*, pages 37–40, Dec 2013.
400 T Richter. On the integer coding profile of JPEG XT. *In Proc.of SPIE: Applications of Digital Image Processing XXXVII*, volume 9217, pages 9217–9217–19, 2014.
402 T. Richter. Lossless coding extensions for JPEG. *In Proc. of Data Compression Conf. (DCC)*, pages 143–152, April 2015.
403 T. Richter. JPEG on steroids: Common optimization techniques for JPEG image compression. *In Proc. of Intl. Conf. on Image Processing (ICIP)*, pages 61–65, Sept 2016.
404 T. Richter. Error bounds for HDR image coding with JPEG XT. *In Proc. of Data Compression Conference (DCC)*, pages 122–130, April 2017.
407 T. Richter et al. The JPEG XT suite of standards: Status and future plans. *In Proc. of SPIE: Applications of Digital Image Processing XXXVIII*, pages 1–18, Aug. 2015.
408 T. Richter et al. Entropy coding and entropy coding improvements of JPEG XS. *In Proc. of Data Compression Conference (DCC)*, pages 87–96, March 2018.
409 B.A. Ringnyu, A. Tangel, and E. Karabulut. Implementation of different architectures of forward 4×4 integer DCT for H:264/AVC encoder. *In Proc. of 10th Intl. Conf. on Electrical and Electronics Engineering (ELECO)*, pages 423–427, Nov 2017.
416 S. Saponara, L. Fanucci, and P. Terreni. Low-power VLSI architectures for 3D discrete cosine transform (DCT). *In Proc. of 46th Midwest Symp. on Circuits and Systems*, volume 3, pages 1567–1570, Dec 2003.
422 S.A. Selouani and D. O'Shaughnessy. Robustness of speech recognition using genetic algorithms and a mel-cepstral subspace approach. *In Proc. of Intl. Conf. on Acoustics, Speech, and Signal Processing (ICASSP)*, volume 1, pages 201–204, May 2004.
425 Z. Shahid, M. Chaumont, and W. Puech. Fast protection of H.264/AVC by selective encryption of CABAC for I and P frames. *In Proc. of 17th European Signal Processing Conf.*, pages 2201–2205, Aug 2009.
430 B. Shen and S.K. Ishwar. Inner-block Operations on Compressed Images. *In Proc. of the Third ACM Intl. Conf. on Multimedia*, pages 489–498, New York, NY, USA, 1995.
431 S. Shen et al. A unified 4/8/16/32-point integer IDCT architecture for multiple video coding standards. *In Proc. of Intl. Conf. on Multimedia and Expo*, pages 788–793, July 2012.
440 P. Sjövall et al. High-level synthesis implementation of HEVC 2-D DCT/DST on FPGA. *In Proc. of Intl. Conf. on Acoustics, Speech and Signal Processing (ICASSP)*, pages 1547–1551, March 2017.
444 J.A. Solomon, A.B. Watson, and Ahumada A. Visibility of DCT basis functions: Effects of contrast masking. *In Proc. of Data Compression Conference (DCC)*, pages 362–370, March 1994.
445 G. Song and X. Jiang. All-zero block detection algorithm based on quantitative parameters. *In Proc of 3rd Intl. Conf. on System Science, Engineering Design and Manufacturing Informatization*, pages 296–299, Oct 2012.
446 X. Song, S.Wang, and X. Niu. An integer DCT and affine transformation based image steganography method. *In Proc. of 8th Intl. Conf. on Intelligent Information Hiding and Multimedia Signal Processing*, pages 102–105, July 2012.
448 D. Sridhar and I. M. Krishna. Brain tumor classification using discrete cosine transform and probabilistic neural network. *In Proc. of Intl. Conf. on Signal Processing*, Image Processing Pattern Recognition, pages 92–96, Feb 2013.
452 D. Stanescu et al. Digital watermarking using Karhunen-Loeve transform. *In Proc. of 4th Intl. Symp. on Applied Computational Intelligence and Informatics*, pages 187–190, 2007.
464 C.W. Sun and P. Yip. Split-radix algorithms for DCT and DST. *In Proc. of 23rd Asilomar Conf. on Signals, Systems and Computers*, pages 508–512, Nov 1989.
465 Z. Sun et al. Ordinal palmprint represention for personal identification. *In Proc. of Computer Society Conf. on Computer Vision and Pattern Recognition (CVPR)*, volume 1, pages 279–284, June 2005.
466 Z. Sun et al. Removal of non-informative frames for wireless capsule endoscopy video segmentation. *In Proc. of Intl. Conf. on Automation and Logistics*, pages 294–299, Aug 2012.

467 N. Surve and A. Kulkarni. Iris recognition using discrete sine transform and neural network. *In Proc. of Intl. Conf. and Workshop on Emerging Trends in Technology, ICWET*, pages 750–755, New York, NY, USA, 2010.

472 Y. Tachizaki, M. Fujiyoshi, and H. Kiya. A DCT coefficient sign-based background model for moving objects detection from motion JPEG coded movies. *In Proc. of Intl. Symp. on Intelligent Signal Processing and Communication Systems (ISPACS)*, pages 37–40, Jan 2009.

475 H. Tang, M. Fan, and L. Yu. Encoder-embedded temporal-spatial Wiener filter for video encoding. *In Proc. of Picture Coding Symp.*, pages 361–364, May 2012.

476 H. Tang and H. Shi. Fast mode decision algorithm for H.264/AVC based on all-zero blocks predetermination. *In Proc. of Intl. Conf. on Measuring Technology and Mechatronics Automation*, volume 2, pages 780–783, April 2009.

477 X. Tang and W.K Stewart. Texture classification using principal component analysis techniques. *In Proc. of SPIE*, volume 2315, pages 22–35, Sept 1994.

484 U. Tuna, S. Peltonen, and U. Ruotsalainen. Data estimation for the ECAT HRRT sinograms by utilizing the DCT domain. *In Proc. of Nuclear Science Symp. Conference Record (NSS/MIC)*, pages 5076–5080, Oct 2008.

485 U. Tuna, S. Peltonen, and U. Ruotsalainen. Interpolation for the gap-filling of the HRRT PET sinograms by using the slices in the direction of the radial samples. *In Proc. of Nuclear Science Symp. Conf. Record (NSS/MIC)*, pages 3273–3279, Oct 2009.

487 P. Turcza, T. Zielinski, and M. Duplaga. Low complexity image coding algorithm for capsule endoscopy with Bayer color filter array. *In Proc. of Signal Processing Algorithms, Architectures, Arrangements, and Applications (SPA)*, pages 27–32, Sept 2007.

489 M.A. Turk and A.P. Pentl. Face recognition using eigenfaces. *In Proc. of Computer Society Conf. on Computer Vision and Pattern Recognition*, pages 586–591, Jun 1991.

491 P.T. Vanishree and A.M.V. Prakash. VLSI implementation of discrete cosine transform and intra prediction. *In Proc. of Intl. Conf. on Advances in Electronics Computers and Communications*, pages 1–6, Oct 2014.

492 A. Vashistha and A.M. Joshi. Fingerprint based biometric watermarking architecture using integer DCT. *In Proc. of Region 10 Conf. (TENCON)*, pages 2818–2821, Nov 2016.

494 V. Velisavljevic et al. Discrete multidirectional wavelet bases. *In Proc. of Intl. Conf. on Image Processing (Cat. No.03CH37429)*, volume 1, pages I–1025–8, Sept 2003.

495 S. Venkataraman et al. Discrete transforms via the Walsh-Hadamard transform. *In Proc. of 26th Midwest Symp. Circuits and Systems*, pages 74–78, Aug 1983.

497 A. Vetro and D. Tian. Analysis of 3D and multiview extensions of the emerging HEVC standard. *In Proc. of SPIE: Applications of Digital Image Processing XXXV*, volume 8499, pages 8499–1–8499–7, 2012.

504 H. Wang, H. Du, and J. Wu. Predicting zero coefficients for High Efficiency Video Coding. *In Proc. of Intl. Conf. on Multimedia and Expo (ICME)*, pages 1–6, July 2014.

508 H. Wang, S. Kwong, and C.W. Kok. Analytical model of zero quantized DCT coefficients for video encoder optimization. *In Proc. of Intl. Conf. on Multimedia and Expo*, pages 801–804, July 2006.

509 H. Wang, S. Kwong, and C.W. Kok. Effectively detecting all-zero DCT blocks for H.264 optimization. *In Proc. of Intl. Conf. on Image Processing (ICIP)*, pages 1329–1332, Oct 2006.

518 Y. Wang, L. Yaroslavsky, and M. Vilermo. On the relationship between MDCT, SDFT and DFT. *In Proc. of 5th Intl. Conf. on Signal Processing (ICSP). 16th World Computer Congress (WCC)*, volume 1, pages 44–47, Aug 2000.

519 Y.E. Wang et al. Some peculiar properties of the MDCT. *In Proc. of Intl. Conf. on Signal Processing (ICSP)*, pages 1–4, Aug 2000.

527 Z. Wang, E.P. Simoncelli, and A.C. Bovik. Multiscale structural similarity for image quality assessment. *In Proc. of 37th Asilomar Conf. on Signals, Systems Computers*, volume 2, pages 1398–1402, Nov 2003.

528 O.Watanabe, H. Kobayashi, and H. Kiya. Two-layer lossless HDR coding considering histogram sparseness with backward compatibility to JPEG. *In Proc. of Picture Coding Symp.* (PCS), pages 1–5, June 2018.

530 J. Wen et al. A compressive sensing image compression algorithm using quantized DCT and noiselet information. *In Proc. of Intl. Conf. on Acoustics, Speech and Signal Processing (ICASSP)*, pages 1294–1297, March 2010.
534 M.Wien. Versatile video coding Video compression beyond HEVC: Coding Tools for SDR and 360° Video. *University Lecture*, 2018.
535 A. Willème et al. Overview of the JPEG XS objective evaluation procedures. *In Proc. of SPIE: Applications of Digital Image Processing XL*, volume 10396, pages 103960N–1 – 103960N–18, 2017.
537 D.Wu et al. An adaptive thresholding technique for the detection of all-zeros blocks in H.264. *In Proc. of Intl. Conf. on Image Processing (ICIP)*, volume 5, pages V–329–V–332, Sept 2007.
546 Z. Xin and Z.Wei. Fast inter prediction block mode decision approach for H.264/AVC based on all-zero blocks detection. *In Proc. of 8th Conf. on Industrial Electronics and Applications (ICIEA)*, pages 896–899, June 2013.
553 L. Yang and Y. Wan. Accurate image noise level estimation through DCT transformation and approximation by Chi-Square distribution. *In Proc. of 7th Intl. Conf. on Intelligent Human-Machine Systems and Cybernetics*, volume 1, pages 387–390, Aug 2015.
554 P. Yang and M.J. Narasimha. Prime factor decomposition of the discrete cosine transform and its hardware realization. *In Proc. of Intl. Conf. on Acoustics, Speech, and Signal Processing (ICASSP)*, volume 10, pages 772–775, April 1985.
562 P. Yip and K.R. Rao. DIF Algorithms for DCT and DST. *In Proc. of Intl. Conf. on Acoustics, Speech, and Signal Processing (ICASSP)*, volume 10, pages 776–779, April 1985.
564 Y. Yokotani and S. Oraintara. Lossless audio compression using integer modified discrete cosine transform. *In Proc. of Intl. Symp. on Intelligent Signal Processing and Communication Systems*, pages 120–126, July 2003.
566 L. Yu et al. Overview of AVS-video: Tools, performance and complexity. *In Proc. of SPIE: Visual Communications and Image Processing*, volume 5960, pages 679–690, July 2006.
567 F. Zargari and S. Ghorbani. Fast calculation of 8 *times*8 integer DCT in the software implementation of H.264/AVC. *In Proc. of 7th Intl. Conf. on Application of Information and Communication Technologies*, pages 1–4, Oct 2013.
569 B. Zeng and J. Fu. Directional discrete cosine transforms for image coding. *In Proc. of Intl. Conf. on Multimedia and Expo*, pages 721–724, July 2006.
573 C.-X. Zhang et al. The technique of pre-scaled integer transform. *In Proc. of Intl. Symp. on Circuits and Systems*, volume 1, pages 316–319, May 2005.
575 H. Zhang and Z. Ma. Fast intra prediction for high efficiency video coding. *In Proc. of 13th Pacific-Rim Conf. on Advances in Multimedia Information Processing*, pages 568–577, Berlin, Heidelberg, 2012.
577 J. Zhang and A.T.S. Ho. An efficient digital image-in-image watermarking algorithm using the integer discrete cosine transform (IntDCT). *In Proc. of Fourth Intl. Conf. on Information, Communications and Signal Processing, 2003 and the Fourth Pacific RimConf. on Multimedia. Proc. of the 2003 Joint*, volume 2, pages 1163–1167, Dec 2003.
580 Y. Zhang. Digital watermarking technology: A review. *In Proc. of ETP Intl. Conf. on Future Computer and Communication*, pages 250–252, June 2009.
581 Y. Zhang and R. Hu. Scalable audio coding based on integer transform. *In Proc. of 1st. Intl. Conf. on Communications and Networking in China*, pages 1–5, Oct 2006.
583 W. Zhao, T. Onoye, and T. Song. High-performance multiplierless transform architecture for HEVC. *In Proc. of Intl. Symp. on Circuits and Systems (ISCAS)*, pages 1668–1671, May 2013.
587 D. Zhong and I. Defee. Pattern recognition by grouping areas in DCT compressed images. *In Proc. of 6th Nordic Signal Processing Symposium, (NORSIG)*, pages 312–315, June 2004.
588 G. Zhong, L. Lu, and N. Jiang. Fast mode decision based on all-zero block in H.264/AVC. *In Proc. of 2nd Intl. Conf. on Digital Manufacturing Automation*, pages 535–538, Aug 2011.
590 C. Zhu et al. Multi-level low-complexity coefficient discarding scheme for video encoder. *In Proc. of Intl. Symp. on Circuits and Systems (ISCAS)*, pages 5–8, June 2014.
592 N.A. Ziyal. Image representation, recovery and analysis using principal component analysis. *In Proc. of Intl. Conf. on Space Planes and Aircraft Technologies (ICSPAT)*, pages 321–324, Sept 1997.

Books

8 N. Ahmed and K.R. Rao. *Orthogonal transforms for digital signal processing*. New York, NY: Springer, 1975.

39 G. Bi and Y. Zeng. *Transforms and fast algorithms for signal analysis and representations*. Birkhauser, 2004.

52 V. Britanak, P.C. Yip, and K.R. Rao. *Discrete cosine and sine transforms: General properties, fast algorithms and integer approximations*. Elsevier, 2007.

92 R.J. Clarke. *Transform coding of images*. Academic Press, London, 1985.

98 J. Cox, L. Miller, and A. Bloom. *Digital Watermarking*. Morgan Kaufman Publishers, 2002.

104 W.B. Davenport and W.L. Root. *An introduction to the theory of random signals and noise*. Wiley-IEEE Press, 1987.

105 P. Davis. *Circulant matrices*. Wiley-IEEE Press, 1979.

122 D.F. Elliott and K.R. Rao. *Fast transforms algorithms, analyses, applications*. Academic Press, 1983.

133 H. D. Ferreyra, R. Picco, and L. A. Sobero. Analysis and project considerations on 2D DCT adaptive coding. In V. Cappelini and A. G. Constantinides, editors, *Digital signal processing-87*. Elsevier Science, 1987.

160 B. Gold, N. Morgan, and D. Ellis. *Perceptual audio coding*. Wiley Telecom, 2011.

223 A.K. Jain. *Fundamentals of digital image processing*. Prentice Hall, 1986.

226 A.J. Jerri. *The discrete sine (DST) and cosine (DCT) transforms for boundary value problems*. Springer, Boston, MA, 1996.

284 A. Ligtenberg, R.H. Wright, and J.H. O'Neill. *A VLSI orthogonal transform chip for realtime image compression*. Visual Commun. and Image Process. II, SPIE, Cambridge, MA,, Oct 1987.

296 B. Madhukar, A. Fuldseth, and G. Bjontegaard. *HEVC transform and quantization*. In V. Sze, M. Bugadavi, and G.J. Sullivan, editors, High efficiency video coding (HEVC): Algorithms and architectures, pages 141–169. Springer, 2014.

320 J.C. Mason and D.C. Handscomb. *Chebyshev polynomials*. CRC Press, 2003.

332 H. Moussa et al. *A robust-fragile dual watermarking system in the DCT domain*. In K. Rajiv, H. Robert, and J. Lakhmi, editors, *Proc. of Knowledge-Based Intelligent Information and Engineering Systems*, pages 548–553, Berlin, Heidelberg, 2005. Springer.

361 T. M. Ono and K. Asai. *Ubiquitous technology, "hybrid video coding MPEG-4 and H.264"* (In Japanese, title translated). Tokyo, Ohmsha Press, 2005.

387 K.R. Rao, J.J. Hwang, and D.N. Kim. *High efficiency video coding and other emerging standards*. River Publishers, 2017.

388 K.R. Rao, D.N. Kim, and J.J. Hwang. *Fast Fourier transform-Algorithms and applications*. Springer, 2010.

389 K.R. Rao, D.N. Kim, and J.J. Hwang. *Video coding standards: AVS China, H.264/MPEG-4 part 10, HEVC, VP6, DIRAC and VC-1*. Springer, 2014.

390 K.R. Rao and P. Yip. *Discrete cosine transform: Algorithms, advantages, applications*. Academic Press, 1990.

398 I.E. Richardson. *The H.264 advanced video compression standard*. Wiley, 2010.

443 I.N. Sneddon. *Special functions of mathematical physics and chemistry*. New York, NY: Longman Inc, 1980.

453 G. Strang. Every unit matrix is a LULU. *Linear Algebra and its Applications*, Elsevier, 1997.

455 G. Strang and T. Nguyen. *Wavelets and filter banks (Rev. ed)*. Wellesley-Cambridge Press, 1997.

471 W. Sze, M. Bugadavi, and G. J. Sullivan. *High efficiency video coding (HEVC): Algorithms and architectures*. Springer, 2014.

533 M. Wien. *High efficiency video coding: Coding tools and specification*. Springer, 2014.

558 P. Yip. Sine and cosine transforms. In P.D. Poularikas, editor, *Transforms and applications handbook*, pages 3–1 – 3–34. CRC Press, 2010.

Overview/Survey papers

19 A. Artusi et al. Overview and evaluation of the JPEG XT HDR image compression standard. *Journal of Real-Time Image Processing*, Dec 2015.

22 M. Asikuzzaman and M.R. Pickering. An overview of digital video watermarking. *IEEE Transactions on Circuits and Systems for Video Technology*, (Early access).

53 L.G. Brown. A survey of image registration techniques. *ACM Comput. Surv.*, 24(4):325–376, Dec 1992.

93 JPEG committee. Overview of JPEG Pleno. https://jpeg.org/jpegpleno/index.html, 2018. [Online: accessed July 2018].

94 JPEG committee. Overview of JPEG XR. https://jpeg.org/jpegxr/, 2018. [Online: accessed July 2018].

147 M. Galar et al. A survey of fingerprint classification Part I: Taxonomies on feature extraction methods and learning models. *Knowledge-Based Systems*, 81:76–97, 2015.

243 N. Kanagavalli and L. Latha. A survey of copy-move image forgery detection techniques. *In Proc. of Intl. Conf. on Inventive Systems and Control (ICISC)*, pages 1–6, Jan 2017.

265 V. Lakshmanan and R. Gomathi. A survey on image completion techniques in remote sensing images. *In Proc. of 4th Intl. Conf. on Signal Processing, Communication and Networking (ICSCN)*, pages 1–6, March 2017.

328 J. Mohan, V. Krishnaveni, and Y. Guo. A survey on the magnetic resonance image denoising methods. *Biomedical Signal Processing and Control*, 9:56–69, 2014.

450 T. Srisooksai et al. Practical data compression in wireless sensor networks: A survey. *Journal of Network and Computer Applications*, 35(1):37–59, 2012.

490 K.-H. Tzou. Progressive image transmission: A review and comparison of techniques. *Optical Engineering*, 26(7):581–589, 1987.

536 S.Wong et al. Radiologic image compression-A review. *Proc. of the IEEE*, 83(2):194–219, Feb 1995.

548 J. Xu, R. Joshi, and R.A. Cohen. Overview of the emerging HEVC screen content coding extension. *IEEE Transactions on Circuits and Systems for Video Technology*, 26(1):50–62, Jan 2016.

Theses

179 Z.G. Hernández. *Transformada cosenoidal direccionable y simulación en VHDL (in Spanish)*. B. S. thesis, Depto. de Ing. Eléctrica y Comp., Univ. Autónoma de Ciudad Juárez, 2018.

258 D. Kristie. *Assessing the benefits of DCT compressive sensing for computational electromagnetics*. M. S. thesis, Department of Electrical Engineering and Computer Science, Massachusetts Institute of Technology, 2011.

347 L. Netsch. *A robust telephone-based speaker verification system*. Ph. D. dissertation, Electrical Engineering Dept., Univ. of Texas at Arlington, 1993.

354 P. Nirajan. *Discrete sine and cosine transforms on parallel processors*. M. S. thesis, Faculty of Computing and Electrical Engineering, Tempere University of Technology, 2015.

424 M. Shah. *Future of JPEG XT: Scrambling to enhance privacy and security*. M. S. thesis, Electrical Engineering Dept., The Univ. of Texas at Arlington, 2016.

457 S. Subbarayappa. *Implementation and analysis of directional discrete cosine transform in H.264 for baseline profile*. M. S. thesis, Electrical Engineering Dept., The Univ. of Texas at Arlington, 2012.

Technical reports

42 G. Bjontegaard. Calculation of average PSNR differences between RD-Curves. Technical report, ITU-T/ISO/IEC, 2001. Report VCEG M–33 ITU–T SG16/Q6, Austin, Texas, USA.

391 S.K. Rao. Performance comparison of HEVC intra, JPEG, JPEG 2000, JPEG XR, JPEG LS and VP9 intra. Technical report, The University of Texas at Arlington, 2016.

418 A. Saxena and F.C. Fernandez. Jointly optimal intra prediction and adaptive primary transform. Technical report, ITU-T/ISO/IEC, 2010.

419 A. Saxena and F.C. Fernandez. CTVC-E125: CE7: Mode-dependent DCT/DST without full matrix multiplication for intra prediction. Technical report, ITU-T/ISO/IEC, 2011.

559 P. Yip and B.P. Agrawal. Theory and applications of Toeplitz matrices. Technical report, CRL Report, Jan 1979.

Standards

202 ISO/IEC 10918-1. *Information technology - Digital compression and coding of continuous-tone still images: Requirements and guidelines*, 1994. JPEG was reviewed and confirmed in 2017.

199 IEEE Std 1180-1990. *IEEE Standard specifications for the implementations of* 8×8 *inverse discrete cosine transform*, 1991. Withdrawn.

205 ISO/IEC 15444-1. *Information technology – JPEG 2000 image coding system: Core coding system*, 1994. JPEG2000 was reviewed and confirmed in 2016.

206 ISO/IEC 18477-1. *Information technology – Scalable compression and coding of continuous–tone still images - Part 1*, 2015. JPEG XT.

212 ISO/IEC DIS 21122-1. Information technology – Low–latency lightweight image coding system – Part 1: Core coding system, 2017.

209 ISO/IEC 23008-2:2015. *Information technology – High efficiency coding and media delivery in heterogeneous environments - Part 2: High efficiency video coding*, 2015.

213 ISO/IEC TR 29170-1. Information technology – Advanced image coding and evaluation – Part 1: Guidelines for image coding system evaluation, 2017.

210 ISO/IEC 29170-2. *Information technology – Advanced image coding and evaluation – Part 2: Evaluation procedure for nearly lossless coding*, 2015.

211 ISO/IEC 29199-2. *Information technology – JPEG XR image coding system – Part 2: Image coding specification*, 2012.

203 ISO/IEC 10918-5 (JPEG Part 5). *Information technology – Digital compression and coding of continuous–tone still images: JPEG File Interchange Format (JFIF)*, 1994. JPEG was reviewed and confirmed in 2017.

214 ITU-R BT.2020-2. *Parameter values for ultra-high definition television systems for production and international programme exchange*, 2015.

204 ITU-T and ISO/IEC. ITU-T Rec. H.264 and ISO/IEC 14496-10:2009: *Advanced video coding*, 2010.

208 ITU-T and ISO/IEC. ITU-T Rec. H.265 and ISO/IEC 23008-2: *High efficiency video coding*, 2013.

207 ISO/IEC Std 23002-1:2006 (MPEG-C). *Information technology - MPEG video technologies-Part 1: Accuracy requirements for implementation of integer - output* 8×8 *inverse discrete cosine transform*, 2006.

356 AVS Workgroup of China. *Information technology - Advanced coding of audio and video, Part 2:Video, GB/T 20090.2–2006*, 2006.

Websites

1 J. Aas. Mozilla JPEG encoder project. https://github.com/mozilla/mozjpeg/, 2015. [Online: accessed July 2018].

3 M. Adams. The JasPer Project. http://www.dimin.net/software/utils.html, 2018. [Online: accessed July 2018].

24 Avsgroup. AVS work documents. http://www.avs.org.cn/english/Document.asp, 2018. [Online: accessed July 2018].

93 JPEG committee. Overview of JPEG Pleno. https://jpeg.org/jpegpleno/index.html, 2018. [Online: accessed July 2018].
94 JPEG committee. Overview of JPEG XR. https://jpeg.org/jpegxr/, 2018. [Online: accessed July 2018].
123 Erik Van Bilsen. AIC-The Source code. http://bilsen.com/aic/source.shtml, 2011. [Online: accessed Aug 2018].
127 M. Fairchild. The HDR photographic survey. http://rit-mcsl.org/fairchild/HDR.html, 2015. [Online: accessed July 2017].
136 Joint Development Foundation. Alliance for Open Media. https://aomedia.org, 2018. [Online: accessed July 2018].
140 Heinrich-Hertz-Institut Fraunhofer-Institut fur Nachrichtentechnik. JVET software. https://jvet.hhi.fraunhofer.de/, 2018. [Online: accessed July 2018].
142 M. Frigo and S.G. Johnson. FFTW. http://www.fftw.org/, 2005. [Online: accessed July 2018].
200 IJG. Independent JPEG Group. http://www.ijg.org/, 1991. [Online: accessed Aug 2018].
215 ITU. ITU News. https://news.itu.int/versatile-video-coding-project-starts-strongly/, 2018. [Online: accessed July 2018].
216 ITU/ISO/IEC. Software for JPEG XR. https://jpeg.org/jpegxr/software.html, 2012. [Online: accessed Aug 2018].
237 JPEG2000 WG 1. JPEG 2000 software. https://jpeg.org/jpeg2000/software.html, 2018. [Online: accessed July 2018].
249 T. Kientzle. Algorithm alley. http://drdobbs.com/parallel/184410889/, 1999. [Online: accessed July 2018].
261 MP Lab. Implementation and Analysis of Directional Discrete Cosine Transform in Baseline Profile in H.264. http:www.uta.edu/faculty/krrao/dip/Courses/EE5359/index_tem.html, 2018. [Online: accessed July 2018].
313 R.K. Mantiuk. Pfstools : High dynamic range images and video. http://pfstools.sourceforge.net/download.html, 2015. [Online: accessed July 2017].
362 T. Ooura. General purpose FFT (fast Fourier/cosine/sine transform) package. http://www.kurims.kyoto-u.ac.jp/~ooura/fft.html, 2017. [Online: accessed Dec.2017].
375 Ponomarenko, N. PSNR-HVS-M download page. https://github.com/Rolinh/VQMT/blob/master/inc/PSNRHVS.hpp, 2009. [Online: accessed July 2018].
401 T. Richter. JPEG XT reference codec 1.31. http://www.jpeg.org/jpegxt/software.html, 2015. [Online: accessed July 2017].
405 T. Richter. libjpeg: A complete implementation of 10918-1 (JPEG). https://github.com/thorfdbg/libjpeg, 2017. [Online: accessed July 2018].
439 Digital Formats Web site. HD Photo, version 1.0 (Windows Media Photo). https://www.loc.gov/preservation/digital/formats/fdd/fdd000243.shtml, 2008. [Online: accessed July 2018].
480 Trellis Management Co. Ltd,. Software for JPEG XT. https://jpeg.org/jpegxt/software.html, 2015. [Online: accessed July 2018].
532 T. Wiegand. JPEG image coding standard. http://iphome.hhi.de/wiegand/assets/pdfs/DIC_JPEG_07.pdf, 2018. [Online: accessed July 2018].

Late additions

LA1 S. Ahmad and T. Somayeh. Low-power DCT-based compressor for wireless capsule endoscopy. *Signal Processing: Image Communication*, 59(1):83–95
LA2 S. Al-Azawi et al. New fast and area-efficient pipeline 3-D DCT architectures. *Digital Signal Processing*, Early access 2018.
LA3 K. J. Chandan, H. K. Maheshkumar. Electrocardiogram data compression using DCT-based discrete orthogonal Stockwell transform. *Biomedical Signal Processing and Control*, 46(1):174–181, Sept 2018.
LA4 S. Chandan and B. Anu. A DCT-based local and non-local fuzzy C-means algorithm for segmentation of brain magnetic resonance images. *Applied Soft Computing*, 68(1):447–457, July 2018.
LA5 S. Chatterjee and K. Sarawadekar. An optimized architecture of HEVC core transform using real-valued DCT coefficients. *IEEE Transactions on Circuits and Systems II: Express Briefs*, Early access.

LA6 M. Chen, Y. Zhang, and C. Lu. Efficient architecture of variable size HEVC 2D-DCT for FPGA platforms. *AEU - Intl. Journal of Electronics and Communications*, 73(1):1–8, March 2017.

LA7 D. F. Coelho et al. Computation of 2D 8 × 8 DCT-based on the Loeffler factorization using algebraic integer encoding. *IEEE Transactions on Computers*, Early access 2018.

LA8 D. Dong et al. Compressive sensing image recovery using dictionary learning and shape-adaptive DCT thresholding. *Magnetic Resonance Imaging*, Early access 2018.

LA9 G. Margelis et al. Efficient DCT-based secret key generation for the Internet of Things. *Ad Hoc Networks*, Early access 2018.

LA10 R.T. Haweel et al. Fast approximate DCT with GPU implementation for image compression. *Journal of Visual Communication and Image Representation*, 40:357–365, Oct 2016.

LA11 L. He et al. Video quality assessment by compact representation of energy in 3D-DCT domain. *Neurocomputing*, 269(1):108–116, Dec 2017.

LA12 K.W. Hung, K. Wang, and J. Jiang. Image up-sampling using deep cascaded neural networks in dual domains for images down-sampled in DCT domain. *Journal of Visual Communication and Image Representation*, Early access 2018.

LA13 B. Lee and M. Kim. A CU-level rate and distortion estimation scheme for RDO of hardware-friendly HEVC encoders using low-complexity integer DCTs. *IEEE Transactions on Image Processing*, 25(8):3787–3800, Aug 2016.

LA14 Y. Li et al. Low power area-efficient DCT implementation based on Markov random field-stochastic logic. *Int. Symp. on Circuits and Systems (ISCAS)*, pages 1–4, May 2018.

LA15 H. Mandal et al. FPGA based low power hardware implementation for quality access control of digital image using dither modulation, *Devices for Integrated Circuit (DevIC)*, pages 642 – 646, March 2017.

LA16 M. Masera, M. Martina, and G. Masera. Odd type DCT/DST for video coding: Relationships and low-complexity implementations. *Intl. Workshop on Signal Processing Systems (SiPS)*, pages 1–6, Oct 2017.

LA17 B. Mohamed et al. High-level synthesis hardware implementation and verification of HEVC DCT on SoC-FPGA. *13th Intl. Computer Engineering Conf. (ICENCO)*, pages 361–365, Dec 2017.

LA18 S. A. Parah et al. Robust and blind watermarking technique in DCT domain using inter-block coefficient differencing. *Digital Signal Processing*, 53(1): 11–24, June 2016.

LA19 A. A. Reem and A. E. Lamiaa. Text-image watermarking based on integer wavelet transform (IWT) and discrete cosine transform (DCT). *Applied Computing and Informatics*, Early access 2018.

LA20 V. Rekha and S. C. Sachin. Speech compression using multi-resolution hybrid wavelet using DCT and Walsh transforms. *Intl. Conf. on Computational Intelligence and Data Science (ICCIDS)*, pages 1404–1411, April 2018.

LA21 T. Rohit et al. An efficient medical image watermarking scheme based on FDCuT-DCT. *Engineering Science and Technology, an Intl. Journal*, 20(4):1366–1379, Aug 2017.

LA22 N. S. T. Sai. Truncated DCT and decomposed DWT SVD features for image retrieval. *Intl. Conf. on Communication, Computing and Virtualization*, pages 579 – 588, Feb. 2016.

LA23 Z. R. Saleemi and G. Raja. Requirement based transform coefficient coding architecture for DCT/DST for HEVC. *Intl. Conf. on Wireless Systems and Networks (ISWSN)*, pages 1–5, Nov 2017

LA24 O. SimYing et al. Fast recovery of unknown coefficients in DCT-transformed images. *Signal Processing: Image Communication*, 58(1):1–13, Oct 2017.

LA25 K. K. Singh and D. Pandey. Implementation of DCT and IDCT based image compression and decompression on FPGA. *Int. Conf. on Inventive Systems and Control (ICISC)*, pages 1–4, Jan 2017.

LA26 H. Sun et al. A low-cost approximate 32-point transform architecture. *Intl. Symp. on Circuits and Systems (ISCAS)*, pages 1–4, Sept. 2017.

LA27 C. J. Tablada et al. DCT approximations based on Chenś factorization. *Signal Processing: Image Communication*, 58(1):14–23, Oct 2017.

LA28 T. H. Thai, F. Retraint, and R. Cogranne. Camera model identification based on DCT coefficient statistics. *Digital Signal Processing*, 40(1):88–100, May 2015.

LA29 V. P. Vishwakarma and V. Sisaudia. Gray-scale image watermarking based on DE-KELM in DCT domain. *Intl. Conf. on Computational Intelligence and Data Science (ICCIDS)*, pages 1012 – 1020, April 2018.

Glossary

AC-3: An early name for Dolby Digital.

Algorithm: Set of finite and well-defined steps for solving a problem.

Basis functions: Are linearly independent functions of a function space that span all function space. In other words, the function in the function space can be represented as a linear combination of basis functions.

Basis vectors: Are linearly independent vectors in a vector space that span all the vector space. In other words, every vector in the vector space can be represented as a linear combination of basis vectors.

Bit allocation: After the signal is divided into classes, maybe by a transformation process, each class is assigned (allocated) a number of bits from the total bits available (budget) such that the recovered signal has the least possible distortion.

Chroma: Is the color portion of an image.

CIF: Common Interface Format or Common Image Format. The Common Interface Format was developed to support video conferencing. It has an active resolution of 352×288.

CODEC: Is a device or a program that is able to compress/decompress a signal. At the encoder side the codec compresses the data to reduce the bandwith of the original signal for faster transmission or storage. At the decoder side the compression operation is reversed to recover the signal or an approximation to it.

Coding tree unit: In HEVC standard is also called largest coding unit and it is the basic processing unit that is fed into the video enconder. The coding tree unit corresponds in structure to macroblock units of previous video coding standards.

Common image format: See CIF.

Common interface format: See CIF.

Compression ratio: Is a number used to measure how much information is squeezed out of an image when it has been compressed.

Correlation: Is the degree of similarity between two random variables or processes.

Covariance: Is the joint variability of two random variables.

Data compression: Is the process of reducing the transmission bandwidth and/or disk storage.

Decorrelation: Is a process to reduce the correlation within a signal or a set of signals.

DBV: Digital video broadcast, a method of transmitting digital audio and video (SDTV or HDTV resolution), based on MPEG-2.

Differential pulse code modulation (DPCM): Is a technique that quantizes the difference between the sampled value and its predicted value and then differences are encoded.

Dimension: The dimension is the number of basis vectors in a vector space S.

Distortion: Is the distance of the original signal to its approximation.

Dolby digital: An audio compression technique developed by Dolby. It is a multi-channel surround sound format used in DVD and HDTV.

Eigenvalues: Are the natural numbers of a system.

Eigenvectors: Are vectors that keep the same direction as the input vectors to the system. Eigenvectors are multiplied by their corresponding eigenvalues.

Entropy: Average amount of information required to describe the random variable.

Entropy coding: Is a type of compression where each symbol of a dictionary is assigned to a unique prefix code. The most common symbols are assigned to shorter codes. The codes are stored or sent to the decoder as their corresponding symbols appear at the input of the entropy encoder.

Expected value: Anticipated value expected for a given event at some point in the future.

Fast algorithm: Is one that for the same input, creates the same output as the original algorithm and takes fewer calculations.

Full HD: Is a display technology providing a frame resolution of 1920×1080 pixels.

Gray scale: Is the luma component of color video signals or black-and-white video signals.

H.261/H.263: Video compression standards developed to implement video conferencing over ISDN, LANs, regular phone lines, etc. H.261 supports video resolutions of 352×288 and 176×144 at up to 29.97 frames per second. H.263 supports video resolutions of 1408×1152, 704×576, 352×288, 176×144, and 128×96 at up to 29.97 frames per second.

H.264/AVC: Is a macroblock-based video coding standard, also known as MPEG-4 Part 10 that supports resolutions up to 8192×4320 @ 120 fps or 4096×2304 @ 300 fps.

High definition image: Image with a higher resolution than standard definition; typically defined for resolutions of up to 1280×720 and 1920×1080 pixels.

High efficiency video coding test model: Is the reference software for HEVC.

HEVC: Is a coding tree unit-based high efficiency video coding standard that supports resolutions up to 8192×4320 @ 120 fps or 3840×2160 @ 300 fps. HEVC is also known as H.265 and MPEG-H Part-2.

Image: Is a two-dimensional array of samples representing brightness levels.

Interpolation: Is a mathematical way of generating additional information.

Intra coding: Technique to encode a block of pixels without temporal reference.

Joint exploration model: Is the experimental software of the JVET containing the coding tools studied for achieving coding performance beyond HEVC.

JPEG: JPEG stands for Joint Photographic Experts Group. However, we refer to the temp JPEG as the image compression standard they developed. JPEG was developed to compress still images.

Lossless compression: Is when the decompressed or recovered data is exactly the same as the original data.

Lossy: Is when the decompressed or recovered data is approximated to the original data but it is not exactly the same.

Luma: Is the brightness in an image.

Macroblock: In image and video compression formats is a processing unit based on linear block transforms, such as the discrete cosine transform (DCT). A macroblock typically consists of 16×16 adjacent samples. It is further subdivided into transform blocks and may be further subdivided into prediction blocks.

Markov process: Is a random process whose future probabilities are determined by its most recent values. In a Markov process the future probabilities are independent of the past probabilities, given the current probabilities.

MPEG-1: Is a macroblock-based video coding standard. MPEG-1 was the first MPEG compression standard for real-time audio and video using video resolutions of 352×240 or 352×288 or higher and used for the video CD format. The maximum bit rate is about 1.5 Mbps.

MPEG-2: Is a macroblock-based video coding standard. MPEG-2 supports higher video resolutions than MPEG-1 to allow HDTV applications, and both progressive and interlaced video. MPEG 2 is used for the DVD–Video and SVCD formats, and also forms the basis for digital SDTV and HDTV.

MPEG-4: Uses an object-based approach. Scenes are modeled as compositions of natural and/or synthetic objects with a position in a two- or three-dimensional space. MPEG-4 allows user interaction. Similarly, audio objects are placed in a sound space.

Multispectral image: Collection of images of the same scene each captured with a different sensor sensitive to a different wavelength, to acquire the images at a specific frequencies of the electromagnetic spectrum. Each image is referred to as a band.

Prediction unit: In HEVC standard, the prediction unit is the basic unit used for carrying the information related to the prediction processes.

Profiles: In image and video coding is a set of capabilities targeting specific classes of applications.

Pulse code modulation (PCM): Method to represent sampled analog signals in digital form.

Quantization: In digital signal processing, quantization is a process to map many values to one resulting in a reduced number of values.

Random process: Collection of random variables.

Random variable: Is a rule or a function that assigns values to each of an outcome of a random experiment.

Rate: Is the number of bits per sample or per unit of time that are conveyed or processed.

Recursion: Is a technique used when a step of an algorithm calls the algorithm itself.

Resolution: Is the basic measurement of how much information is visible for an image. It is usually described as $h \times v$ where h is the horizontal resolution (across the display) and the v is the vertical resolution (down the display). The higher the numbers, the better, since that means there is more detail to see. If only one number is specified, it is the horizontal resolution.

RGC: Abbreviation for red, green, blue.

Separable kernel: A kernel separable if it can be expressed as a product of two transforms along the row and the column, i.e., $g(n_1, n_2, m_1, m_2) = g_1(n_1, m_1) g_2(n_2, m_2)$.

Symmetric kernel : A separable kernel is symmetric, if the kernels have the identical function along the row and along the column, i.e., $g(n_1, n_2, m_1, m_2) = g_1(n_1, m_1) g_1(n_2, m_2)$.

Systolic array: Consists of a large monolithic network, tightly coupled, of primitive identical computing cells (nodes) which can be hardwired or software configured for a specific application. The nodes are called data processing units (DPUs) or simply processors. Each processor is connected to a small number of nearest neighbors in a mesh like topology. Each processor performs a sequence of operations and the result passes it to its neighbor. Generally the operations will be the same in each cell.

Standard: A technical document that establishes guidelines universally understood and adopted.

Transform: In mathematics, a transform is a rule or a function to map from one space onto another or onto itself.

Transform unit: In HEVC video coding standard, a transform unit is the basic unit used for the transform and quantization processes.

Transform coding: Is a type of compression in which the input data are first transformed into another domain to be decorrelated before further processing.

Variance: Is variability of a random variable around its mean.

Versatile video coding: The primary objective of versatile video coding is to provide a significant improvement in compression performance over the existing HEVC standard, aiding in deployment of higher-quality video services and emerging applications such as 360° omnidirectional immersive multimedia and high-dynamic-range (HDR) video.

4K and 8K Ultra high definition: Are typical resolutions of 3840×2160 (8.3 megapixels) and 7680×4320 (33.2 megapixels) respectively used in the consumer media and display industries. $4K$ and $8K$ refers to a horizontal screen display resolution in the order of 4,000 pixels and 8,000 pixels respectively. $8K$ is also known as Super Hi-Vision. $4K$ UHD is exactly four times of the resolution of Full HD and $8K$ UHD is sixteen times the resolution of Full HD, and four times of $4K$ UHD.

References

1. J. Aas. Mozilla JPEG encoder project. https://github.com/mozilla/mozjpeg/, 2015. [Online: accessed July 2018].
2. H.M. Abbas and M.M. Fahmy. Neural model for Karhunen-Loève transform with application to adaptive image compression. *IEE Proceedings I - Communications, Speech and Vision*, 140(2):135–143, April 1993.
3. M. Adams. The JasPer Project. http://www.dimin.net/software/utils.html, 2018. [Online: accessed July 2018].
4. J.I. Agbinya. Interpolation using the discrete cosine transform. *Electronics Letters*, 28(20):1927–1928, Sept 1992.
5. A. Ahmed, M.U. Shahid, and A. Rehman. N-point DCT VLSI architecture for emerging HEVC standard. *VLSI Design*, 2012:1–13, Jan 2012.
6. N. Ahmed and M.D. Flickner. Some considerations of the discrete cosine transform. In *Proc. of 16th Asilomar Conf. on Circuits. Systems and Computers, Pacific Grove, CA*, volume 1, pages 295–299, Nov 1982.
7. N. Ahmed, T. Natarajan, and K.R. Rao. Discrete cosine transform. *IEEE Transactions on Computers*, C-23(1):90–93, Jan 1974.
8. N. Ahmed and K.R. Rao. *Orthogonal transforms for digital signal processing*. New York, NY: Springer, 1975.
9. T. Ahonen, A. Hadid, and M. Pietikainen. Face description with local binary patterns: Application to face recognition. *IEEE Transactions on Pattern Analysis and Machine Intelligence*, 28(12):2037–2041, Dec 2006.
10. A.M. Alattar. Reversible watermark using the difference expansion of a generalized integer transform. *IEEE Transactions on Image Processing*, 13(8):1147–1156, Aug 2004.
11. P.P. Alexander, W.P. Rosalind, and S. Scarloff. Photobook: Tools for content-based manipulation of image databases. In *Proc. of SPIE: Storage and Retrieval for Image and Video Databases II*, volume 2185, pages 34 – 47, April 1994.
12. Z. Alkachouh and M.G. Bellanger. Fast DCT-based spatial domain interpolation of blocks in images. *IEEE Transactions on Image Processing*, 9(4):729–732, April 2000.
13. E. Alshina, A. Alshin, and F.C. Fernandes. Rotational transform for image and video compression. In *Proc. of 18th Intl. Conf. on Image Processing (ICIP)*, pages 3689–3692, Sept 2011.
14. R.G.M. Alvarado et al. DCT-compressive sampling of multifrequency sparse audio signals. In *Proc. of IEEE Latin-America Conf. on Communications (LATINCOM)*, pages 1–5, Nov 2014.
15. A. Amani et al. Passive detection of image forgery using DCT and local binary pattern. *Signal, Image and Video Processing*, 11(1):81–88, Jan 2017.
16. T. Amornraksa and S. Tachaphetpiboon. Fingerprint recognition using DCT features. *Electronics Letters*, 42(9):522–523, April 2006.

17. A.E. Ansari, A. Mansouri, and A. Ahaitouf. An efficient VLSI architecture design for integer DCT in HEVC standard. In *Proc. of IEEE/ACS 13th Intl. Conf. of Computer Systems and Applications (AICCSA)*, pages 1–5, Nov 2016.
18. P. Arayacheeppreecha, S. Pumrin, and B. Supmonchai. Flexible input transform architecture for HEVC encoder on FPGA. In *Proc. of 12th Intl. Conf. on Electrical Engineering/Electronics, Computer, Telecommunications and Information Technology (ECTI-CON)*, pages 1–6, June 2015.
19. A. Artusi et al. Overview and evaluation of the JPEG XT HDR image compression standard. *Journal of Real-Time Image Processing*, Dec 2015.
20. A. Artusi et al. JPEG XT: A Compression Standard for HDR and WCG Images [Standards in a Nutshell]. *IEEE Signal Processing Magazine*, 33(2):118–124, March 2016.
21. M. Asikuzzaman et al. Robust DT CWT-based DIBR 3D video watermarking using chrominance embedding. *IEEE Transactions on Multimedia*, 18(9):1733–1748, Sept 2016.
22. M. Asikuzzaman and M.R. Pickering. An overview of digital video watermarking. *IEEE Transactions on Circuits and Systems for Video Technology*, (Early access), 2017.
23. R. Atta and M. Ghanbari. Low-memory requirement and efficient face recognition system based on DCT pyramid. *IEEE Transactions on Consumer Electronics*, 56(3):1542–1548, Aug 2010.
24. Avsgroup. AVS work documents. http://www.avs.org.cn/english/Document.asp, 2018. [Online: accessed July 2018].
25. H. Baali et al. A transform-based feature extraction approach for motor imagery tasks classification. *IEEE Journal of Translational Engineering in Health and Medicine*, 3:1–8, Oct 2015.
26. S.H. Babu et al. Face recognition using active illumination equalization and mirror image superposition as pre-processing techniques. In *Proc. of 5th Intl. Conf. on Signal and Image Processing*, pages 96–101, Jan 2014.
27. S. Bacchelli and S. Papi. Image denoising using principal component analysis in the wavelet domain. *Journal of Computational and Applied Mathematics*, 189(1):606 – 621, 2006. Proc. of 11th Intl. Congress on Computational and Applied Mathematics.
28. S.H. Bae, J. Kim, and M. Kim. HEVC-based perceptually adaptive video coding using a DCT-based local distortion detection probability model. *IEEE Transactions on Image Processing*, 25(7):3343–3357, July 2016.
29. S.H. Bae and M. Kim. DCT-QM: A DCT-based quality degradation metric for image quality optimization problems. *IEEE Transactions on Image Processing*, 25(10):4916–4930, Oct 2016.
30. S.H. Bae and M. Kim. A DCT-based total JND profile for spatio-temporal and foveated masking effects. *IEEE Transactions on Circuits and Systems for Video Technology*, 27(6):1196–1207, June 2017.
31. H. Bai, A. Wang, and M. Zhang. Compressive sensing for DCT image. In *Proc. of Int. Conf. on Computational Aspects of Social Networks*, pages 378–381, Sept 2010.
32. M.Z. Baig and A.V. Gokhale. Object tracking using mean shift algorithm with illumination invariance. In *Proc. of Intl. Conf. on Communication Systems and Network Technologies*, pages 550–553, April 2015.
33. R.H. Bamberger and M.J.T. Smith. A filter bank for the directional decomposition of images: Theory and design. *IEEE Transactions on Signal Processing*, 40(4):882–893, April 1992.
34. D. Baron et al. Distributed compressive sensing. *ArXiv e-prints*, Jan 2009.

35. M.M.A. Basiri and N.M. Sk. High performance integer DCT architectures for HEVC. In *Proc. of 30th Intl. Conf. on VLSI Design and 2017 16th Intl. Conf. on Embedded Systems (VLSID)*, pages 121–126, Jan 2017.
36. F.M. Bayer and R.J. Cintra. DCT-like transform for image compression requires 14 additions only. *IET Electronics Letters*, 48(15):919 – 921, 2012.
37. A. Bertrand. Applications and trends in wireless acoustic sensor networks: A signal processing perspective. In *Proc. of 18th Symp. on Communications and Vehicular Technology in the Benelux (SCVT)*, pages 1–6, Nov 2011.
38. G. Bhatnagar and Q.M.J. Wu. Biometrics inspired watermarking based on a fractional dual tree complex wavelet transform. *Future Generation Computer Systems*, 29(1):182 – 195, 2013.
39. G. Bi and Y. Zeng. *Transforms and fast algorithms for signal analysis and representations*. Birkhäuser, 2004.
40. J. Bigun. Unsupervised feature reduction in image segmentation by local Karhunen-Loeve transform. In *Proc. of 11th IAPR Intl. Conf. on Pattern Recognition. Vol.II. Conf. B: Pattern Recognition Methodology and Systems*, pages 79–83, Aug 1992.
41. E.J.K. Bisherurwa and F.P. Coakley. New fast discrete sine transform with offset. *IEE Electronics Letters*, 17(21):803–805, October 1981.
42. G. Bjontegaard. Calculation of average PSNR differences between RD-Curves. Technical report, ITU-T/ISO/IEC, 2001. Report VCEG M-33 ITU-T SG16/Q6, Austin, Texas, USA.
43. S. Bouguezel, M.O. Ahmad, and M.N.S. Swamy. A fast 8 x 8 transform for image compression. In *Proc. of Intl. Conf. on Microelectronics - ICM*, pages 74–77, Dec 2009.
44. S. Bouguezel, M.O. Ahmad, and M.N.S. Swamy. A novel transform for image compression. In *Proc. of 53rd IEEE Intl. Midwest Symp. on Circuits and Systems*, pages 509–512, Aug 2010.
45. S. Bouguezel, M.O. Ahmad, and M.N.S. Swamy. Binary discrete cosine and Hartley transforms. *IEEE Transactions on Circuits and Systems I: Regular Papers*, 60(4):989–1002, April 2013.
46. N. Bozinovic and J. Konrad. Motion analysis in 3d dct domain and its application to video coding. *Signal Processing: Image Communication*, 20(6):510 – 528, 2005. Special Issue on Advanced Aspects of Motion Estimation.
47. N. Brahimi and S. Bouguezel. An efficient fast integer DCT transform for images compression with 16 additions only. In *Proc. of Intl. Workshop on Systems, Signal Processing and their Applications, WOSSPA*, pages 71–74, May 2011.
48. V. Britanak. A unified approach to the fast computation of discrete sinusoidal transforms I: DCT and DST transforms. *Computers and Artificial Intelligence*, 17(6):583–607, Dec 1998.
49. V. Britanak. A note on the MDCT/MDST and pseudoinverse matrix. *Computing and Informatics*, 23(3):205–214, Feb 2004.
50. V. Britanak. An efficient computing of oddly stacked MDCT/MDST computation via evenly stacked MDCT/MDST and vice versa. *Signal Processing*, 85(7):1353 – 1374, July 2005.
51. V. Britanak and H.J.L. Arriens. Fast computational structures for an efficient implementation of the complete TDAC analysis/synthesis MDCT/MDST filter banks. *Signal Processing*, 89(7):1379 – 1394, July 2009.

52. V. Britanak, P.C. Yip, and K.R. Rao. *Discrete cosine and sine transforms: General properties, fast algorithms and integer approximations*. Elsevier, 2007.
53. L.G. Brown. A survey of image registration techniques. *ACM Comput. Surv.*, 24(4):325–376, Dec 1992.
54. M. Budagavi et al. Core transform design in the High Efficiency Video Coding (HEVC) standard. *IEEE Journal of Selected Topics in Signal Processing*, 7(6):1029–1041, Dec 2013.
55. M. Budagavi and V. Sze. Unified forward+inverse transform architecture for HEVC. In *Proc. of 19th Intl. Conf. on Image Processing (ICIP)*, pages 209–212, Sept 2012.
56. J.B. Burl. Estimating the basis functions of the Karhunen-Loève transform. *IEEE Transactions on Acoustics, Speech, and Signal Processing*, 37(1):99–105, Jan 1989.
57. Q. Cai et al. Lossy and lossless intra coding performance evaluation: HEVC, H.264/AVC, JPEG 2000 and JPEG LS. In *Proc. of Asia Pacific Signal and Information Processing Association Annual Summit and Conference*, pages 1–9, Dec 2012.
58. A.R. Calderbank et al. Wavelet transforms that map integers to integers. *Applied and Computational Harmonic Analysis*, 5(3):332–369, 1998.
59. C.K. Chaitanya et al. PET and MRI medical image fusion using STDCT and STSVD. In *2017 Intl. Conf. on Computer Communication and Informatics (ICCCI)*, pages 1–4, Jan 2017.
60. N.B. Chakrabarti and T.V.K.H. Rao. A recursive filter implementation of KL transform. *Signal Processing*, 44(3):269 – 284, 1995.
61. W.K. Cham. Development of integer cosine transforms by the principle of dyadic symmetry. In *Proc. of Inst. Electr. Eng. I: Commun. Speech Vis.*, page 276– 282, Aug 1989.
62. W.K. Cham and Y.T. Chan. An order-16 integer cosine transform. *IEEE Transactions on Signal Processing*, 39(5):1205–1208, May 1991.
63. W.K. Cham and R.J. Clarke. Application of the principle of dyadic symmetry to the generation of orthogonal transforms. *IEE Proc. F - Communications, Radar and Signal Processing*, 133(3):264–270, June 1986.
64. S.C. Chan and K.L. Ho. Direct methods for computing discrete sinusoidal transforms. *IEE Proc. F - Radar and Signal Processing*, 137(6):433–442, Dec 1990.
65. S. Chandrasekaran et al. An eigenspace update algorithm for image analysis. *Graphical Models and Image Processing*, 59(5):321 – 332, 1997.
66. C.L. Chang et al. Direction-adaptive partitioned block transform for color image coding. *IEEE Transactions on Image Processing*, 19(7):1740–1755, July 2010.
67. G. Chantas, S. Nikolopoulos, and I. Kompatsiaris. Sparse audio inpainting with variational Bayesian inference. In *Proc. of Intl. Conf. on Consumer Electronics (ICCE)*, pages 1–6, Jan 2018.
68. D.S. Chauhan et al. Double secret key based medical image watermarking for secure telemedicine in cloud environment. In *Proc. of 40th Intl. Conf. on Telecommunications and Signal Processing (TSP)*, pages 626–631, July 2017.
69. C.C.T. Chen, C.-Ta. Chen, and C.-M. Tsai. Hard-limited Karhunen-Loève transform for text independent speaker recognition. *Electronics Letters*, 33(24):2014–2016, Nov 1997.
70. H. Chen and B. Zeng. New transforms tightly bounded by DCT and KLT. *IEEE Signal Processing Letters*, 19(6):344–347, June 2012.
71. W.-K. Chen et al. Iris recognition based on bidimensional empirical mode decomposition and fractal dimension. *Information Sciences*, 221:439 – 451, 2013.

72. W.H. Chen and S.C. Fralick. Image enhancement using cosine transform filtering. In *Proc. of Symp. on Current Math. Problems in Image Science*, pages 186–192, Nov 1976.
73. W.H. Chen, C.H. Smith, and S.C. Fralick. A fast computational algorithm for the discrete cosine transform. *IEEE Transactions on Communications*, 25(9):1004–1009, Sept 1977.
74. W.M. Chen, C.J. Lai, and C.C. Chang. H.264 video watermarking with secret image sharing. In *Proc. of Intl. Symp. on Broadband Multimedia Systems and Broadcasting*, pages 1–7, May 2009.
75. Y. Chen and D. Mukherjee. Variable block-size overlapped block motion compensation in the next generation open-source video codec. In *Proc. of Intl. Conf. on Image Processing (ICIP)*, pages 938–942, Sept 2017.
76. Y.-J. Chen, S. Oraintara, and T. Nguyen. Video compression using integer DCT. In *Proc. of Intl. Conf. on Image Processing (ICIP)*, volume 2, pages 844–845, Sept 2000.
77. Y.-Y. Chen. Medical image compression using DCT-based subband decomposition and modified SPIHT data organization. *Intl. Journal of Medical Informatics*, 76(10):717 – 725, 2007.
78. Z. Chen, Q. Han, and W.-K. Cham. Low-complexity order-64 integer cosine transform design and its applications in HEVC. *IEEE Transactions on Circuits and Systems for Video Technology*, Early access 2018.
79. F. Cheng et al. Compressive sensing multi-focus image fusion. In *Proc. of Chinese Conf. on Pattern Recognition*, pages 107–116, Berlin, Heidelberg, Nov 2014. Springer.
80. Y. Cheng et al. Robust face recognition based on illumination invariant in nonsubsampled contourlet transform domain. *Neurocomputing*, 73(10):2217 – 2224, June 2010.
81. P.T. Chiang and T.S. Chang. Fast zero block detection and early CU termination for HEVC Video Coding. In *Proc. of Intl. Symp. on Circuits and Systems (ISCAS)*, pages 1640–1643, May 2013.
82. B. Chitprasert and K.R. Rao. Discrete cosine transform filtering. In *Proc. of Intl. Conf. on Acoustics, Speech, and Signal Processing*, volume 3, pages 1281–1284, April 1990.
83. B. Chitprasert and K.R. Rao. Human visual weighted progressive image transmission. *IEEE Transactions on Communications*, 38(7):1040–1044, Jul 1990.
84. M.-R. Choi et al. Color image interpolation in the DCT domain using a wavelet-based differential value. *Multimedia Tools and Applications*, 77(16):21539 — 21556, Jan 2018.
85. S. Choi et al. Performance evaluation of JPEG XT standard for high dynamic range image coding. In *Proc. of the World Congress on Engineering (WCE)*, pages 1–4, July 2015.
86. V. Choudhary and P. Vodite. Extraction of region of interest in compressed domain. *Intl. Journal of Computer Science Issues*, 10(2):321 – 329, May 2013.
87. C. Christopoulos, A. Skodras, and T. Ebrahimi. The JPEG2000 still image coding system: An overview. *IEEE Transactions on Consumer Electronics*, 46(4):1103–1127, Nov 2000.
88. F. Christos and L. King-Ip (David). FastMap: A fast algorithm for indexing, datamining and visualization of traditional and multimedia datasets. In *Proc. of SIGMOD Rec*, volume 24, pages 163 – 174, May 1995.
89. A. Ciancio et al. Energy-efficient data representation and routing for wireless sensor networks based on a distributed wavelet compression algorithm. In *Proc. of 5th Intl. Conf. on Information Processing in Sensor Networks*, IPSN-06, pages 309–316, New York, NY, USA, 2006. ACM.

90. R.J. Cintra and F.M. Bayer. A DCT approximation for image compression. *IEEE Signal Processing Letters*, 18(10):579 – 582, Oct 2011.
91. R.J. Cintra, F.M. Bayer, and C.J. Tablada. Low-complexity 8-point DCT approximations based on integer functions. *Signal Processing*, 99(1):201 – 214, 2014.
92. R.J. Clarke. *Transform coding of images*. Academic Press, London, 1985.
93. JPEG committee. Overview of JPEG Pleno. https://jpeg.org/jpegpleno/index.html, 2018. [Online: accessed July 2018].
94. JPEG committee. Overview of JPEG XR. https://jpeg.org/jpegxr/, 2018. [Online: accessed July 2018].
95. J.W. Cooley and J.W. Tukey. An algorithm for the machine calculation of complex Fourier series. *Math of Comput.*, 19:297–301, 1965.
96. M.S. Corrington. Implementation of fast cosine transforms using real arithmetic. In *Proc. of National Aerospace and Electronics Conf. (NAECON)*, pages 350 – 357, 1978.
97. I.J. Cox et al. Secure spread spectrum watermarking for multimedia. *IEEE Transactions on Image Processing*, 6(12):1673–1687, Dec 1997.
98. J. Cox, L. Miller, and A. Bloom. *Digital Watermarking*. Morgan Kaufman Publishers, 2002.
99. A. Criminisi, P. Perez, and K. Toyama. Region filling and object removal by exemplar-based image inpainting. *IEEE Transactions on Image Processing*, 13(9):1200–1212, Sept 2004.
100. J. Cui et al. Hybrid all zero soft quantized block detection for HEVC. *IEEE Transactions on Image Processing*, 27(10):4987–5001, Oct 2018.
101. Z. Cvetkovic and M.V Popovic. New fast recursive algorithms for the computation of discrete cosine and sine transforms. *IEEE Transactions on Signal Processing*, 40(8):2083–2086, Aug 1992.
102. W. Dai and T.D. Tran. Regularity-constrained pre- and post-filtering for block DCT-based systems. *IEEE Transactions on Signal Processing*, 51(10):2568–2581, Oct 2003.
103. A.D. Darji and R.P. Makwana. High-performance multiplierless DCT architecture for HEVC. In *Proc. of 19th Intl. Symp. on VLSI Design and Test*, pages 1–5, June 2015.
104. W.B. Davenport and W.L. Root. *An introduction to the theory of random signals and noise*. Wiley-IEEE Press, 1987.
105. P. Davis. *Circulant matrices*. Wiley-IEEE Press, 1979.
106. C. Deng et al. Performance analysis, parameter selection and extensions to H.264/AVC FRExt for high resolution video coding. *Journal of Visual Communication and Image Representation*, 22(8):749 – 759, 2011.
107. A. Descampe et al. JPEG XS, a new standard for visually lossless low-latency lightweight image compression. In *Proc. of SPIE: Applications of Digital Image Processing XL*, volume 10396, pages 103960M–1 – 103960M–18, 2017.
108. S. Dhamija and P. Jain. Comparative analysis for discrete sine transform as a suitable method for noise estimation. *Intl. Journal of Computer Science Issues*, 8(3):162 – 164, Sept 2011.
109. W. Ding et al. Fast mode dependent directional transform via butterfly-style transform and integer lifting steps. *Journal of Visual Communication and Image Representation*, 22(8):721 – 726, 2011.
110. Y. Ding et al. Application of the Karhunen-Loeve transform temporal image filter to reduce noise in real-time cardiac cine MRI. *Physics in Medicine and Biology*, 54(12):3909, 2009.

111. M.N. Do and M. Vetterli. Contourlets: A new directional multiresolution image representation. In *Conf. Record of the 36th Asilomar Conf. on Signals, Systems and Computers*, volume 1, pages 497–501, Nov 2002.
112. Z. Dokur. A unified framework for image compression and segmentation by using an incremental neural network. *Expert Systems with Applications*, 34(1):611 – 619, 2008.
113. L. Dong and S.C. Schwartz. DCT-based object tracking in compressed video. In *Proc. of Intl. Conf. on Acoustics Speech and Signal Processing (ICASSP)*, volume 2, pages II–II, May 2006.
114. A.B. Doulah et al. DCT domain feature extraction scheme based on motor unit action potential of EMG signal for neuromuscular disease classification. *Healthcare Technology Letters*, 1(1):26–31, Jan 2014.
115. M.F. Duarte et al. Universal distributed sensing via random projections. In *Proc. of 5th Intl. Conf. on Information Processing in Sensor Networks*, IPSN-06, pages 177–185. ACM, 2006.
116. K. Duda. Accurate, guaranteed stable, sliding discrete Fourier transform. *IEEE Signal Processing Magazine*, 27(6):124–127, Nov 2010.
117. R. Dugad and N. Ahuja. A fast scheme for downsampling and upsampling in the DCT domain. In *Proc. of Intl. Conf. on Image Processing (ICIP)*, volume 2, pages 909 – 913, Oct 1999.
118. P. Duhamel. Implementation of "Split-radix" FFT algorithms for complex, real, and real-symmetric data. *IEEE Transactions on Acoustics, Speech, and Signal Processing*, 34(2):285–295, April 1986.
119. P. Duhamel and H. Hollmann. Implementation of "Split-radix" FFT algorithms for complex, real, and real symmetric data. In *Proc. of Intl. Conf. on Acoustics, Speech, and Signal Processing (ICASSP)*, volume 10, pages 784–787, April 1985.
120. S.A. Dyer, N. Ahmed, and D.R. Hummels. Classification of vector cardiograms using Walsh and cosine orthogonal transforms. *IEEE Transactions on Electromagnetic Compatibility*, EMC-27(1):35–40, Feb 1985.
121. T. Ebrahimi et al. JPEG Pleno: Toward an efficient representation of visual reality. *IEEE MultiMedia*, 23(4):14–20, Oct 2016.
122. D.F. Elliott and K.R. Rao. *Fast transforms algorithms, analyses, applications*. Academic Press, 1983.
123. Erik Van Bilsen. AIC - The Source code. http://www.bilsen.com/aic/source.shtml, 2011. [Online: accessed Aug 2018].
124. F. Ernawan and M.N. Kabir. A robust image watermarking technique with an optimal DCT-psychovisual threshold. *IEEE Access*, 6(1):20464–20480, April 2018.
125. F. Ernawan, M.N. Kabir, and Z. Mustaffa. A blind watermarking technique based on DCT psychovisual threshold for a robust copyright protection. In *Proc. of 12th Intl. Conf. for Internet Technology and Secured Transactions (ICITST)*, pages 92–97, Dec 2017.
126. M.M.M. Fahmy. Palmprint recognition based on Mel frequency cepstral coefficients feature extraction. *Ain Shams Engineering Journal*, 1(1):39 – 47, 2010.
127. M. Fairchild. The HDR photographic survey. http://rit-mcsl.org/fairchild/HDR.html, 2015. [Online: accessed July 2017].
128. M. Fallahpour et al. Tampering detection in compressed digital video using watermarking. *IEEE Transactions on Instrumentation and Measurement*, 63(5):1057–1072, May 2014.

129. C.P. Fan, C.W. Chang, and S.J. Hsu. Cost-effective hardware-sharing design of fast algorithm based multiple forward and inverse transforms for H.264/AVC, MPEG-1/2/4, AVS, and VC-1 video encoding and decoding applications. *IEEE Transactions on Circuits and Systems for Video Technology*, 24(4):714–720, April 2014.
130. C.P. Fan and G.A. Su. Efficient fast 1-D 8×8 inverse integer transform for VC-1 application. *IEEE Transactions on Circuits and Systems for Video Technology*, 19(4):584–590, April 2009.
131. G.M. Farinella et al. Representing scenes for real-time context classification on mobile devices. *Pattern Recognition*, 48(4):1086 – 1100, 2015.
132. P. Farrelle and A. K. Jain. Recursive block coding–A new approach to transform coding. *IEEE Transactions on Communications*, 34(2):161–179, Feb 1986.
133. H. D. Ferreyra, R. Picco, and L. A. Sobero. Analysis and project considerations on 2D DCT adaptive coding. In V. Cappelini and A. G. Constantinides, editors, *Digital signal processing-87*. Elsevier Science, 1987.
134. A. Foi, V. Katkovnik, and K. Egiazarian. Pointwise shape-adaptive DCT for high-quality denoising and deblocking of grayscale and color images. *IEEE Transactions on Image Processing*, 16(5):1395–1411, May 2007.
135. C.K. Fong, Q. Han, and W.K. Cham. Recursive integer cosine transform for HEVC and future video coding standards. *IEEE Transactions on Circuits and Systems for Video Technology*, 27(2):326–336, Feb 2017.
136. Joint Development Foundation. Alliance for Open Media. https://aomedia.org, 2018. [Online: accessed July 2018].
137. G. Fracastoro, S.M. Fosson, and E. Magli. Steerable discrete cosine transform. *IEEE Transactions on Image Processing*, 26(1):303–314, Jan 2017.
138. G. Fracastoro and E. Magli. Steerable discrete cosine transform. In *Proc. of 17th Intl. Workshop on Multimedia Signal Processing (MMSP)*, pages 1–6, Oct 2015.
139. G. Fracastoro and E. Magli. Subspace-sparsifying steerable discrete cosine transform from graph Fourier transform. In *Proc. of Intl. Conf. on Image Processing (ICIP)*, pages 1534–1538, Sept 2016.
140. Heinrich-Hertz-Institut Fraunhofer-Institut für Nachrichtentechnik. JVET software. https://jvet.hhi.fraunhofer.de/, 2018. [Online: accessed July 2018].
141. J. Fridrich, D. Soukal, and J. Lukas. Detection of copy-move forgery in digital images. In *Proc. of DFRWS Conf.*, pages 1–10, April 2003.
142. M. Frigo and S.G. Johnson. FFTW. http://www.fftw.org/, 2005. [Online: accessed July 2018].
143. J. Fu and B. Zeng. Directional discrete cosine transforms: A theoretical analysis. In *Proc. of Intl. Conf. on Acoustics, Speech and Signal Processing (ICASSP)*, pages I-1105–I-1108, April 2007.
144. X. Fu and Y. Wan. Accurate image rotation using DCT transformation. In *Proc. of Advanced Information Technology, Electronic and Automation Control Conference (IAEAC)*, pages 537–541, Dec 2015.
145. K. Fukunaga and W.L.G Koontz. Application of the Karhunen-Loeve expansion to feature selection and ordering. *IEEE Transactions on Computers*, C-19(4):311–318, April 1970.
146. S. Gaj, A. Sur, and P.K. Bora. A robust watermarking scheme against re-compression attack for H.265/HEVC. In *Proc. of Fifth National Conf. on Computer Vision, Pattern Recognition, Image Processing and Graphics*, pages 1–4, Dec 2015.
147. M. Galar et al. A survey of fingerprint classification Part I: Taxonomies on feature extraction methods and learning models. *Knowledge-Based Systems*, 81:76 – 97, 2015.

148. L. Gao et al. A new reversible watermarking scheme based on Integer DCT for medical images. In *Proc. of Intl. Conf. on Wavelet Analysis and Pattern Recognition*, pages 33–37, July 2012.
149. Y. Gao et al. Encoder-driven inpainting strategy in multiview video compression. *IEEE Transactions on Image Processing*, 25(1):134–149, Jan 2016.
150. M.J. Garrido et al. A high performance FPGA-based architecture for the future video coding adaptive multiple core transform. *IEEE Transactions on Consumer Electronics*, 64(1):53–60, Feb 2018.
151. M. Gastpar, P.L. Dragotti, and M. Vetterli. The distributed Karhunen-Loeve transform. *IEEE Transactions on Information Theory*, 52(12):5177–5196, Dec 2006.
152. S. Gazor and W. Zhang. A soft voice activity detector based on a Laplacian-Gaussian model. *IEEE Transactions on Speech and Audio Processing*, 11(5):498–505, Sept 2003.
153. R. Geiger et al. IntMDCT - A link between perceptual and lossless audio coding. In *Proc. of Intl. Conf. on Acoustics, Speech, and Signal Processing (ICASSP)*, volume 2, pages II–1813–II–1816, May 2002.
154. R. Geiger, Y. Yokotani, and G. Schuller. Improved integer transforms for lossless audio coding. In *Proc. of The Thrity-Seventh Asilomar Conf. on Signals, Systems Computers, 2003*, volume 2, pages 2119–2123, Nov 2003.
155. R. Geiger, Y. Yokotani, and G. Schuller. Audio data hiding with high data rates based on IntMDCT. In *Proc. of Intl. Conf. on Acoustics Speech and Signal Processing (ICASSP)*, volume 5, pages 205–208, May 2006.
156. G. Ghule and P. Mukherji. A novel approach for Marathi numeral recognition using Bark scale and discrete sine transform method. In *Proc. of Conf. on Advances in Signal Processing (CASP)*, pages 191–195, June 2016.
157. J.M. Gil-Cacho et al. Transform domain prediction error method for improved acoustic echo and feedback cancellation. In *Proc. of the 20th European Signal Processing Conf. (EUSIPCO)*, pages 2422–2426, Aug 2012.
158. M. Gilge, T. Engelhardt, and R. Mehlan. Coding of arbitrarily shaped image segments based on a generalized orthogonal transform. *Signal Processing: Image Communication*, 1(2):153 – 180, 1989. 64 kbit/s Coding of Moving Video.
159. J. Gimlett. Use of "activity" classes in adaptive transform image coding. *IEEE Transactions on Communications*, 23(7):785 – 786, July 1975.
160. B. Gold, N. Morgan, and D. Ellis. *Perceptual audio coding*. Wiley Telecom, 2011.
161. H.B. Golestani and M. Ghanbari. Minimisation of image watermarking side effects through subjective optimisation. *IET Image Processing*, 7(8):733–741, Nov 2013.
162. D. Gong, Y. He, and Z Cao. New cost-effective VLSI implementation of a 2-D discrete cosine transform and its inverse. *IEEE Transactions on Circuits and Systems for Video Technology*, 14(4):405–415, April 2004.
163. S. Gupta et al. VLSI implementation of 3D integer DCT for video coding standards. In *Proc. of 2nd Intl. Conf. on Next Generation Computing Technologies (NGCT)*, pages 660–664, Oct 2016.
164. S. Guruprasad, M.Z. Kurian, and H.N. Suma. Fusion of CT and PET medical images using hybrid algorithm DWT-DCT-PCA. In *Proc. of 2nd Intl. Conf. on Information Science and Security (ICISS)*, pages 1–5, Dec 2015.
165. W. Hachicha, A. Beghdadi, and F.A. Cheikh. 1D directional DCT-based stereo residual compression. In *Proc. of 21st European Signal Processing Conf. (EUSIPCO)*, pages 1–5, Sept 2013.

166. M.N. Haggag, M. El-Sharkawy, and G. Fahmy. Modified efficient fast multiplication-free integer transformation for the 2-D DCT H.265 standard. In *Proc. of Data Compression Conf. (DCC)*, pages 455–455, March 2011.
167. G. Hamarneh and T. Gustavsson. Combining snakes and active shape models for segmenting the human left ventricle in echocardiographic images. In *Computers in Cardiology Vol.27 (Cat. 00CH37163)*, pages 115–118, 2000.
168. M. Hamidi and J. Pearl. Comparison of the cosine and Fourier transforms of Markov-1 signals. *IEEE Transactions on Acoustics, Speech, and Signal Processing*, 24(5):428–429, Oct 1976.
169. J. Han et al. Real-time video content analysis tool for consumer media storage system. *IEEE Transactions on Consumer Electronics*, 52(3):870–878, Aug 2006.
170. J. Han et al. Jointly optimized spatial prediction and block transform for video and image coding. *IEEE Transactions on Image Processing*, 21(4):1874–1884, April 2012.
171. K.J. Han and A.H Tewfik. Eigen-image based video segmentation and indexing. In *Proc. of Intl. Conf. on Image Processing (ICIP)*, volume 2, pages 538–541, Oct 1997.
172. M. Hangarge, K.C. Santosh, and R. Pardeshi. Directional discrete cosine transform for handwritten script identification. In *Proc. of 12th Intl. Conf. on Document Analysis and Recognition*, pages 344–348, Aug 2013.
173. P. Hanhart and T. Ebrahimi. Calculation of average coding efficiency based on subjective quality scores. *Journal of Visual Communication and Image Representation*, 25(3):555 – 564, 2014.
174. R.M. Haralick. A storage efficient way to implement the discrete cosine transform. *IEEE Transactions on Computers*, C-25(7):764–765, July 1976.
175. F. Hartung and B. Girod. Watermarking of uncompressed and compressed video. *Signal Processing (Special Issue on Copyright Protection and Access Control for Multimedia Services)*, pages 283–301, 1998.
176. M. Hassan, I. Osman, and M. Yahia. Discrete sine transform and alternative local linear regression for face recognition. In *Proc. of 5th IASTED Signal Processing, Pattern Recognition, and Applications (SPPRA)*, pages 99–103, Aug 2008.
177. T.I. Haweel. A new square wave transform based on the DCT. *Signal Processing*, 81(11):2309 – 2319, 2001.
178. D. Hein and N. Ahmed. On a real-time Walsh-Hadamard/cosine transform image processor. *IEEE Transactions on Electromagnetic Compatibility*, EMC-20(3):453–457, Aug 1978.
179. Z.G. Hernández. *Transformada cosenoidal direccionable y simulación en VHDL (in Spanish)*. B. S. thesis, Depto. de Ing. Eléctrica y Comp., Univ. Autónoma de Ciudad Juarez, 2018.
180. A. Hilal, T. Hamzeh, and S. Chantaf. Copy-move forgery detection using principal component analysis and discrete cosine transform. In *Proc. of Sensors Networks Smart and Emerging Technologies (SENSET)*, pages 1–4, Sept 2017.
181. V. Holub and J. Fridrich. Low-complexity features for JPEG steganalysis using undecimated DCT. *IEEE Transactions on Information Forensics and Security*, 10(2):219–228, Feb 2015.
182. E. Hot and P. Sekulić. Compressed sensing MRI using masked DCT and DFT measurements. In *Proc. of 4th Mediterranean Conf. on Embedded Computing (MECO)*, pages 323–326, June 2015.
183. H.S. Hou. A fast recursive algorithm for computing the discrete cosine transform. *IEEE Transactions on Acoustics, Speech, and Signal Processing*, 35(10):1455–1461, Oct 1987.

184. H.S. Hou. Correction to: The fast Hartley transform algorithm. *IEEE Transactions on Computers*, C-36(9):1135–1136, Sept 1987.
185. H.S. Hou. The fast Hartley transform algorithm. *IEEE Transactions on Computers*, C-36(2):147–156, Feb 1987.
186. Y. Hozu et al. Segmentation method for phalanges in CR image by use of DCT. In *Proc. of Intl. Conf. on Control Automation and Systems (ICCAS)*, pages 2082–2085, Oct 2010.
187. H.C. Hsu et al. Architecture design of shape-adaptive discrete cosine transform and its inverse for MPEG-4 video coding. *IEEE Transactions on Circuits and Systems for Video Technology*, 18(3):375–386, March 2008.
188. H. Hu. Illumination invariant face recognition based on dual-tree complex wavelet transform. *IET Computer Vision*, 9(2):163–173, Feb 2015.
189. X.-Y. Hu and X.-H Yu. The Chebyshev polynomial fitting properties of discrete cosine transform. *Signal Processing: Image Communication*, 13(1):15 – 20, July 1998.
190. Y. Hua and W. Liu. Generalized Karhunen-Loeve transform. *IEEE Signal Processing Letters*, 5(6):141–142, June 1998.
191. F. Huang et al. Reversible data hiding in JPEG images. *IEEE Transactions on Circuits and Systems for Video Technology*, 26(9):1610–1621, Sept 2016.
192. H. Huang et al. Integer MDCT with enhanced approximation of the DCT-IV. *IEEE Transactions on Signal Processing*, 54(3):1156–1159, March 2006.
193. H. Huang et al. Blind integrity verification of medical images. *IEEE Transactions on Information Technology in Biomedicine*, 16(6):1122–1126, Nov 2012.
194. H.Y. Huang, C.H. Yang, and W.H. Hsu. A video watermarking technique based on pseudo-3-D DCT and quantization index modulation. *IEEE Transactions on Information Forensics and Security*, 5(4):625–637, Dec 2010.
195. X. Huang, I. Echizen, and A. Nishimura. A new approach of reversible acoustic steganography for tampering detection. In *Proc. of 6th Intl. Conf. on Intelligent Information Hiding and Multimedia Signal Processing*, pages 538–542, Oct 2010.
196. X. Huang et al. Scalable-to-lossless transform domain distributed video coding. In *Proc. of Intl. Workshop on Multimedia Signal Processing*, pages 327–332, Oct 2010.
197. B. Hunt and O Kubler. Karhunen-Loève multispectral image restoration, part I: Theory. *IEEE Transactions on Acoustics, Speech, and Signal Processing*, 32(3):592–600, June 1984.
198. T. Ichita et al. Directional discrete cosine transforms arising from discrete cosine and sine transforms for directional block-wise image representation. In *Proc. of Intl. Conf. on Acoustics, Speech and Signal Processing (ICASSP)*, pages 4536–4540, March 2017.
199. IEEE-STD-1180-1990. *IEEE Standard specifications for the implementations of 8 x 8 inverse discrete cosine transform*, 1991. Withdrawn.
200. IJG. Independent JPEG Group. http://www.ijg.org/, 1991. [Online: accessed Aug 2018].
201. S. Ishwar, P.K. Meher, and M.N.S. Swamy. Discrete Tchebichef transform - A fast 4x4 algorithm and its application in image/video compression. In *Proc. of Intl. Symp. on Circuits and Systems*, pages 260–263, May 2008.
202. ISO/IEC-10918-1. *Information technology – Digital compression and coding of continuous-tone still images: Requirements and guidelines*, 1994. JPEG was reviewed and confirmed in 2017.

203. ISO/IEC-10918-5. *Information technology – Digital compression and coding of continuous-tone still images: JPEG File Interchange Format (JFIF)*, 1994. JPEG was reviewed and confirmed in 2017.
204. ISO/IEC-14496-10. *ITU-T Rec. H.264 and ISO/IEC 14496-10:2009: Advanced video coding*, 2010.
205. ISO/IEC-15444-1. *Information technology – JPEG 2000 image coding system: Core coding system*, 1994. JPEG2000 was reviewed and confirmed in 2016.
206. ISO/IEC-18477-1. *Information technology – Scalable compression and coding of continuous-tone still images – Part 1*, 2015. JPEG XT.
207. ISO/IEC-23002-1. *Information technology - MPEG video technologies - Part 1: Accuracy requirements for implementation of integer-output 8 x 8 inverse discrete cosine transform*, 2006.
208. ISO/IEC-23008-2. *ITU-T Rec. H.265 and ISO/IEC 23008-2: High efficiency video coding*, 2013.
209. ISO/IEC-23008-2. *Information technology – High efficiency coding and media delivery in heterogeneous environments – Part 2: High efficiency video coding*, 2015.
210. ISO/IEC-29170-2. *Information technology – Advanced image coding and evaluation – Part 2: Evaluation procedure for nearly lossless coding*, 2015.
211. ISO/IEC-29199-2. *Information technology – JPEG XR image coding system – Part 2: Image coding specification*, 2012.
212. ISO/IEC-DIS21122-1. *Information technology – Low-latency lightweight image coding system – Part 1: Core coding system*, 2017.
213. ISO/IEC-TR29170-1. *Information technology – Advanced image coding and evaluation – Part 1: Guidelines for image coding system evaluation*, 2017.
214. ITU-R-BT.2020-2. *Parameter values for ultra-high definition television systems for production and international programme exchange*, 2015.
215. ITU-T. ITU News. https://news.itu.int/versatile-video-coding-project-starts-strongly/, 2018. [Online: accessed July 2018].
216. ITU/ISO/IEC. Software for JPEG XR. https://jpeg.org/jpegxr/software.html, 2012. [Online: accessed Aug 2018].
217. ITU/ISO/IEC. Software for JPEG XT. https://jpeg.org/jpegxt/software.html, 2015. [Online: accessed July 2018].
218. S. Jadon, M. Kumar, and Y. Rathi. Face recognition using SOM neural network with DDCT facial feature extraction techniques. In *Proc. of Intl. Conf. on Communications and Signal Processing*, pages 1070–1074, April 2015.
219. A. K. Jain and Farrel P. M. Recursive block coding. In *Proc. of 16th Asilomar Conf. on Circuits, Systems and Computers*, pages 431–436, Nov 1982.
220. A.K. Jain. A fast Karhunen-Loeve transform for a class of random processes. *IEEE Transactions on Communications*, 24(9):1023–1029, Sept 1976.
221. A.K. Jain. A fast Karhunen-Loeve transform for digital restoration of images degraded by white and colored noise. *IEEE Transactions on Computers*, C-26(6):560–571, June 1977.
222. A.K. Jain. A sinusoidal family of unitary transforms. *IEEE Transactions on Pattern Analysis and Machine Intelligence*, PAMI-1(4):356–365, Oct 1979.
223. A.K. Jain. *Fundamentals of digital image processing*. Prentice Hall, 1986.
224. S. Javanbakhti, S. Zinger, and P.H.N. De With. Fast scene analysis for surveillance video databases. *IEEE Transactions on Consumer Electronics*, 63(3):325–333, August 2017.

225. R. Jeni, T.S.R. Raj, and R.S. Roach. Design of high performance folded DIF FFT architecture using MMCM approach with Hcub algorithm. In *Proc. of Intl. Conf. on Circuits, Power and Computing Technologies (ICCPCT)*, pages 715–719, March 2013.
226. A.J. Jerri. *The discrete sine (DST) and cosine (DCT) transforms for boundary value problems*. Springer, Boston, MA, 1996.
227. R. Jeske et al. Low cost and high throughput multiplierless design of a 16 point 1-D DCT of the new HEVC video coding standard. In *Proc. of VIII Southern Conf. on Programmable Logic*, pages 1–6, March 2012.
228. X. Ji et al. Early determination of zero-quantized 8 x 8 DCT coefficients. *IEEE Transactions on Circuits and Systems for Video Technology*, 19(12):1755–1765, Dec 2009.
229. J. Jiang and G. Feng. The spatial relationship of DCT coefficients between a block and its sub-blocks. *IEEE Transactions on Signal Processing*, 50(5):1160–1169, May 2002.
230. X.-Y. Jing and D. Zhang. A face and palmprint recognition approach based on discriminant DCT feature extraction. *IEEE Transactions on Systems, Man, and Cybernetics, Part B (Cybernetics)*, 34(6):2405 – 2415, Dec 2004.
231. T. Jinshan, E. Peli, and S. Acton. Image enhancement using a contrast measure in the compressed domain. *IEEE Signal Processing Letters*, 10(10):289–292, Oct 2003.
232. A.W. Johnson and A.B. Bradley. Adaptive transform coding incorporating time domain aliasing cancellation. *Speech Comm.*, 6(4):299–308, Dec 1987.
233. C.R. Johnson et al. Image processing for artist identification. *IEEE Signal Processing Magazine*, 25(4):37–48, July 2008.
234. H.W. Jones, D.N. Hein, and S.C. Knauer. The Karhunen-Loeve, discrete cosine and related transforms via the Hadamard transform. In *Proc. of Intl. Telemetering Conf.*, pages 87–98, Nov 1978.
235. K. Josef and P.C. Young. A new approach to feature selection based on the Karhunen-Loeve expansion. *Pattern Recognition*, 5(4):335 – 352, 1973.
236. R. Joshi, Y.A. Reznik, and M. Karczewicz. Efficient large size transforms for high performance video coding. In *Proc. of SPIE: Optics and Photonics, paper 7798-31*, volume 7798, pages 77980W–1 – 77980W–7, Aug 2010.
237. JPEG 2000 WG 1. JPEG 2000 software. https://jpeg.org/jpeg2000/software.html, 2018. [Online: accessed July 2018].
238. M. Jridi, A. Alfalou, and P.K. Meher. A generalized algorithm and reconfigurable architecture for efficient and scalable orthogonal approximation of DCT. *IEEE Transactions on Circuits and Systems I: Regular Papers*, 62(2):449–457, Feb 2015.
239. M. Jridi and P.K. Meher. Scalable approximate DCT architectures for efficient HEVC-compliant video coding. *IEEE Transactions on Circuits and Systems for Video Technology*, 27(8):1815–1825, Aug 2017.
240. E. Kalali et al. A low energy HEVC inverse transform hardware. *IEEE Transactions on Consumer Electronics*, 60(4):754–761, Nov 2014.
241. E. Kalali and I. Hamzaoglu. FPGA implementations of HEVC inverse DCT using high-level synthesis. In *Proc. of Conf. on Design and Architectures for Signal and Image Processing (DASIP)*, pages 1–6, Sept 2015.
242. E. Kalali, A.C. Mert, and I. Hamzaoglu. A computation and energy reduction technique for HEVC discrete cosine transform. *IEEE Transactions on Consumer Electronics*, 62(2):166–174, May 2016.
243. N. Kanagavalli and L. Latha. A survey of copy-move image forgery detection techniques. In *Proc. of Intl. Conf. on Inventive Systems and Control (ICISC)*, pages 1–6, Jan 2017.

244. L. Kang and X.P. Cheng. Copy-move forgery detection in digital image. In *Proc. of 3rd Intl. Congress on Image and Signal Processing*, volume 5, pages 2419–2421, Oct 2010.
245. Q. Kang, K. Li, and J. Yang. A digital watermarking approach based on DCT domain combining QR code and chaotic theory. In *Proc. of 10th Intl. Conf. on Intelligent Computer Communication and Processing (ICCP)*, pages 331–337, Sept 2014.
246. M. Karczewicz et al. A hybrid video coder based on extended macroblock sizes, improved interpolation, and flexible motion representation. *IEEE Transactions on Circuits and Systems for Video Technology*, 20(12):1698–1708, Dec 2010.
247. P. Kauff and K. Schuur. Shape-adaptive DCT with block-based DC separation and Delta-DC correction. *IEEE Transactions on Circuits and Systems for Video Technology*, 8(3):237–242, Jun 1998.
248. C. Khirallah, L. Stankovic, and V. Stankovic. Low complexity multi-view video streaming over multipath fading channels. In *Proc. of 3rd Intl. Conf. on Information and Communication Technologies: From Theory to Applications*, pages 1–5, April 2008.
249. T. Kientzle. Algorithm alley. http://drdobbs.com/parallel/184410889/, 1999. [Online: accessed July 2018].
250. D.N. Kim and K.R. Rao. Sequence mirroring properties of orthogonal transforms having even and odd symmetric vectors. *ECTI Transactions on Computer and Information Technology*, 3(2):1 – 5, Nov 2007.
251. K. Kim et al. Reconstruction of dose distribution in in-beam PET for carbon therapy. In *Proc. of Nuclear Science Symp. and Medical Imaging Conf. Record (NSS/MIC)*, pages 2433–2436, Oct 2012.
252. W.H. Kim, K. Yi, and C.M. Kyung. Energy-efficient illumination-invariant change detection in DCT coefficient domain for vehicular black box camera. *Electronics Letters*, 51(11):822–824, May 2015.
253. M. Kirby and L. Sirovich. Application of the Karhunen-Loeve procedure for the characterization of human faces. *IEEE Transactions on Pattern Analysis and Machine Intelligence*, 12(1):103–108, Jan 1990.
254. H. Kitajima. Energy packing efficiency of the Hadamard transform. *IEEE Transactions on Communications*, 24(11):1256–1258, November 1976.
255. H. Kitajima. A symmetric cosine transform. *IEEE Transactions on Computers*, C-29(4):317–323, April 1980.
256. U.V. Koc and K.J.R. Liu. Discrete-cosine/sine-transform based motion estimation. In *Proc. of 1st Intl. Conf. on Image Processing (ICIP)*, volume 3, pages 771–775, Nov 1994.
257. W. Kou and T. Fjallbrant. Fast computation of transform coefficients for a subadjacent block for a transform family. *IEEE Transactions on Signal Processing*, 39:1695–1699, July 1991.
258. D. Kristie. *Assessing the benefits of DCT compressive sensing for computational electromagnetics*. M. S. thesis, Department of Electrical Engineering and Computer Science, Massachusetts Institute of Technology, 2011.
259. E. Kurniawati, C.T. Lau, and B. Premkumar. Time-domain aliasing cancellation role in error concealment. *IET Electronics Letter*, 40(12):781–783, June 2004.
260. M. Laadjel, S.A. Maadeed, and A. Bouridane. Combining Fisher locality preserving projections and passband DCT for efficient palmprint recognition. *Neurocomputing*, 152:179 – 189, 2015.

261. MP Lab. Implementation and Analysis of Directional Discrete Cosine Transform in Baseline Profile in H.264. http://www.uta.edu/faculty/krrao/dip/Courses/EE5359/index_tem.html, 2018. [Online: accessed July 2018].
262. S. Lai, C.H. Luo, and S. Lei. Common architecture design of novel recursive MDCT and IMDCT algorithms for applications to AAC, AAC in DRAM, and MP3 codecs. *IEEE Transactions on Circuits and Systems II: Express Brief*, 56(10):793–797, Oct 2009.
263. Y.K. Lai and Y.F. Lai. A reconfigurable IDCT architecture for universal video decoders. *IEEE Transactions on Consumer Electronics*, 56(3):1872–1879, Aug 2010.
264. J. Lainema et al. Intra coding of the HEVC standard. *IEEE Transactions on Circuits and Systems for Video Technology*, 22(12):1792–1801, Dec 2012.
265. V. Lakshmanan and R. Gomathi. A survey on image completion techniques in remote sensing images. In *Proc. of 4th Intl. Conf. on Signal Processing, Communication and Networking (ICSCN)*, pages 1–6, March 2017.
266. M. Lamarre and J.J. Clark. Background subtraction using competing models in the block-DCT domain. In *Proc. of object recognition supported by user interaction for service robots*, pages 299–302, 2002.
267. C. Lan et al. Variable block-sized signal-dependent transform for video coding. *IEEE Transactions on Circuits and Systems for Video Technology*, 28(8):1920–1933, Aug 2018.
268. B.G Lee. A new algorithm to compute the discrete cosine transform. *IEEE Transactions on Acoustics, Speech, and Signal Processing*, 32(6):1243–1245, Dec 1984.
269. B.G. Lee. FCT – A fast cosine transform. In *Proc. of Intl. Conf. on Acoustics, Speech, and Signal Processing (ICASSP)*, volume 9, pages 477–480, Mar 1984.
270. B.G Lee. Input and output index mappings for a prime-factor-decomposed computation of discrete cosine transform. *IEEE Transactions on Acoustics, Speech, and Signal Processing*, 37(2):237–244, Feb 1989.
271. J. Lee et al. A compressed-domain corner detection method for a DCT-based compressed image. In *Proc. of Intl. Conf. on Consumer Electronics (ICCE)*, pages 306–307, Jan 2017.
272. J. Lee, N. Vijaykrishnan, and M.J. Irwin. Efficient VLSI implementation of inverse discrete cosine transform. In *Proc. of Intl. Conf. on Acoustics, Speech, and Signal Processing (ICASSP)*, volume 5, pages V–177–180, May 2004.
273. K. Lee et al. A novel algorithm for zero block detection in High Efficiency Video Coding. *IEEE Journal of Selected Topics in Signal Processing*, 7(6):1124–1134, Dec 2013.
274. A. Levey and M. Lindenbaum. Sequential Karhunen-Loeve basis extraction and its application to images. *IEEE Transactions on Image Processing*, 9(8):1371–1374, Aug 2000.
275. L. Li and Z. Hou. Multiview video compression with 3D-DCT. In *Proc. of ITI 5th Intl. Conf. on Information and Communications Technology*, pages 59–61, Dec 2007.
276. M. Li, Z. Fan, and Z. Liu. An improved visual tracking algorithm based on particle filter. In *Proc. of Intl. Conf. IEEE Region 10 (TENCON)*, pages 1–4, Oct 2013.
277. P. Li and K. Lo. A content-adaptive joint image compression and encryption scheme. *IEEE Transactions on Multimedia*, 20(8):1960–1972, Aug 2018.
278. Q. Li and H. Zhu. Block-based compressed sensing of image using directional Tchebichef transforms. In *Proc. of Intl. Conf. on Systems, Man, and Cybernetics (SMC)*, pages 2207–2212, Oct 2012.

279. Q. Li, H. Zhu, and Q. Liu. Image recognition by combined affine and blur Tchebichef moment invariants. In *Proc. of 4th Intl. Congress on Image and Signal Processing*, volume 3, pages 1517–1521, Oct 2011.
280. T. Li et al. On integer MDCT for perceptual audio coding. *IEEE Transactions on Audio, Speech, and Language Processing*, 15(8):2236–2248, Nov 2007.
281. Y.R. Li, L. Shen, and B.W. Suter. Adaptive inpainting algorithm based on DCT induced wavelet regularization. *IEEE Transactions on Image Processing*, 22(2):752–763, Feb 2013.
282. J. Liang and T.D. Tran. Fast multiplierless approximations of the DCT with the lifting scheme. *IEEE Transactions on Signal Processing*, 49(12):3032–3044, Dec 2001.
283. A. Ligtenberg and J.H. O'Neill. A single chip solution for an 8 by 8 two dimensional DCT. In *Proc. of Intl. Symp. on Circuits and Systems (ISCAS)*, pages 1128–1131, Mar 1987.
284. A. Ligtenberg, R.H. Wright, and J.H. O'Neill. *A VLSI orthogonal transform chip for realtime image compression*. Visual Commun. and Image Process. II, SPIE, Cambridge, MA,, Oct 1987.
285. Y.K. Lin. High capacity reversible data hiding scheme based upon discrete cosine transformation. *Journal of Systems and Software*, 85(10):2395 – 2404, 2012.
286. M. Liu, S. Liu, and Q. Zhao. Fingerprint orientation field reconstruction by weighted discrete cosine transform. *Information Sciences*, 268:65 – 77, 2014. New Sensing and Processing Technologies for Hand-based Biometrics Authentication.
287. X. Liu et al. Random walk graph Laplacian-based smoothness prior for soft decoding of JPEG images. *IEEE Transactions on Image Processing*, 26(2):509–524, Feb 2017.
288. X.W. Liu et al. Infrared face recognition based on radiative energy and sub-block DCT in wavelet domain. In *Proc. of Intl. Conf. on Wavelet Analysis and Pattern Recognition*, pages 84–89, July 2010.
289. C. Loeffler, A. Ligtenberg, and G.S. Moschytz. Algorithm-architecture mapping for custom DSP chips. In *Proc. of Intl. Symp. on Circuits and Systems*, volume 2, pages 1953–1956, June 1988.
290. C. Loeffler, A. Ligtenberg, and G.S. Moschytz. Practical fast 1-D DCT algorithms with 11 multiplications. In *Proc. of Intl. Conf. on Acoustics, Speech, and Signal Processing (ICASSP)*, volume 2, pages 988–991, May 1989.
291. T.T. Lu, H.L. Hsu, and P.C. Chang. Blind video watermarking for H.264. In *Proc. of Canadian Conf. on Electrical and Computer Engineering*, pages 2353–2356, May 2006.
292. Y. Lu and M.N. Do. The finer directional wavelet transform. In *Proc. of Intl. Conf. on Acoustics, Speech, and Signal Processing (ICASSP)*, volume 4, pages iv/573–iv/576, March 2005.
293. W. Luo, G.L. Heileman, and C.E. Pizano. Fast and robust watermarking of JPEG files. In *Proc. of 5th Southwest Symp. on Image Analysis and Interpretation*, pages 158–162, April 2002.
294. H. Lv et al. A comparison of fractional-pel interpolation filters in HEVC and H.264/AVC. In *Proc. of Visual Communications and Image Processing*, pages 1–6, Nov 2012.
295. Z. Ma et al. A video watermarking DRM method based on H.264 compressed domain with low bit-rate increasement. *Chinese Journal of Electronics*, 25(4):641–647, Aug 2016.

296. B. Madhukar, A. Fuldseth, and G. Bjontegaard. HEVC transform and quantization. In V. Sze, M. Bugadavi, and G.J. Sullivan, editors, *High efficiency video coding (HEVC): Algorithms and architectures*, pages 141–169. Springer, 2014.
297. B.N. Madhukar and S. Jain. A duality theorem for the discrete sine transform (DST). In *Proc. of Intl. Conf. on Applied and Theoretical Computing and Communication Technology (iCATccT)*, pages 156–160, Oct 2015.
298. J.P. Maheshwari et al. Robust digital image watermarking using DCT-based pyramid transform via image compression. In *Proc. of Intl. Conf. on Communications and Signal Processing (ICCSP)*, pages 1059–1063, April 2015.
299. T. Mahmood et al. A robust technique for copy-move forgery detection and localization in digital images via stationary wavelet and discrete cosine transform. *Journal of Visual Communication and Image Representation*, 53:202 – 214, 2018.
300. J. Makhoul. A fast cosine transform in one and two dimensions. *IEEE Transactions on Acoustics, Speech, and Signal Processing*, 28(1):27–34, Feb 1980.
301. H. S. Malvar. The LOT: A link between block transform coding and multirate filter banks. In *Proc. of Intl. Symp. on Circuits and Systems (ISCAS)*, volume 1, pages 835–838, June 1988.
302. H. S. Malvar and R. Duarte. Transform/subband coding of speech with the lapped orthogonal transform. In *Proc. of Intl. Symp. on Circuits and Systems (ISCAS)*, volume 2, pages 1268–1271, May 1989.
303. H. S. Malvar and D. H. Staelin. Reduction of blocking effects in image coding with a lapped orthogonal transform. In *Proc. of Intl. Conf. on Acoustics, Speech, and Signal Processing (ICASSP)*, volume 2, pages 781–784, April 1988.
304. H. S. Malvar and D. H. Staelin. The LOT: Transform coding without blocking effects. *IEEE Transactions on Acoustics, Speech, and Signal Processing*, 37(4):553–559, April 1989.
305. H.S. Malvar. Fast computation of discrete cosine transform through fast Hartley transform. *Electronics Letters*, 22(7):352–353, March 1986.
306. H.S. Malvar. Fast computation of the discrete cosine transform and the discrete Hartley transform. *IEEE Transactions on Acoustics, Speech, and Signal Processing*, 35(10):1484–1485, Oct 1987.
307. H.S. Malvar. Corrections to, Fast computation of the discrete cosine transform and the discrete Hartley transform. *IEEE Transactions on Acoustics, Speech, and Signal Processing*, ASSP-36:610, April 1988.
308. H.S. Malvar et al. Low-complexity transform and quantization in H.264/AVC. *IEEE Transactions on Circuits and Systems for Video Technology*, 13(7):598–603, July 2003.
309. H.S. Malvar, G.J. Sullivan, and S. Srinivasan. Lifting-based reversible color transformations for image compression. In *Proc. of SPIE Applications of Digital Image Processing*, volume 7073, pages 707307–1 – 707307–10. Intl. Soc. for Optical Engr., Aug 2008.
310. J.V. Manjon et al. New methods for MRI denoising based on sparseness and self-similarity. *Medical Image Analysis*, 16(1):18–27, Jan 2012.
311. A. Mansouri et al. A low complexity video watermarking in H.264 compressed domain. *IEEE Transactions on Information Forensics and Security*, 5(4):649–657, Dec 2010.
312. R. K. Mantiuk, T. Richter, and A. Artusi. Fine-tuning JPEG-XT compression performance using large-scale objective quality testing. In *Proc. of Intl. Conf. on Image Processing (ICIP)*, pages 2152–2156, Sept 2016.

313. R.K. Mantiuk. Pfstools: HDR images and video. http://pfstools.sourceforge.net/, 2015. [Online: accessed July 2017].
314. F. Marcelloni and M. Vecchio. An efficient lossless compression algorithm for tiny nodes of monitoring wireless sensor networks. *The Computer Journal*, 52(8):969–987, 2009.
315. G. Martinelli, L.P. Ricotti, and G. Marcone. Neural clustering for optimal KLT image compression. *IEEE Transactions on Signal Processing*, 41(4):1737–1739, April 1993.
316. E.E. Martínez et al. Directional transforms for video coding based on lifting on graphs. *IEEE Transactions on Circuits and Systems for Video Technology*, 28(4):933–946, April 2018.
317. L.W. Martinson. A 10 MHz image bandwidth compression model. In *Proc. of Conf. on Pattern Recognition and Image Processing*, pages 132–136, 1978.
318. S.A. Martucci. Image resizing in the discrete cosine transform domain. In *Proc. of Intl. Conf. on Image Processing (ICIP)*, volume 2, pages 244 – 247, Oct 1995.
319. M. Masera, M. Martina, and G. Masera. Adaptive approximated DCT architectures for HEVC. *IEEE Transactions on Circuits and Systems for Video Technology*, 27(12):2714–2725, Dec 2017.
320. J.C. Mason and D.C. Handscomb. *Chebyshev polynomials*. CRC Press, 2003.
321. K. Meenakshi, K.S. Prasad., and C.S. Rao. Development of low-complexity video watermarking with conjugate symmetric sequency - complex Hadamard transform. *IEEE Communications Letters*, 21(8):1779–1782, Aug 2017.
322. P.K. Meher et al. Efficient integer DCT architectures for HEVC. *IEEE Transactions on Circuits and Systems for Video Technology*, 24(1):168–178, Jan 2014.
323. R. Merris. Laplacian matrices of graphs: A survey. *Linear Algebra and its Applications*, 197-198:143 – 176, 1994.
324. A.C. Mert, E. Kalali, and I. Hamzaoglu. High performance 2D transform hardware for future video coding. *IEEE Transactions on Consumer Electronics*, 63(2):117–125, May 2017.
325. S. Midtsokogen and J.M. Valin. The AV1 constrained directional enhancement filter (CDEF). In *Proc. of Intl. Conf. on Acoustics, Speech and Signal Processing (ICASSP)*, pages 1193–1197, April 2018.
326. K. Minemura et al. A novel sketch attack for H.264/AVC format-compliant encrypted video. *IEEE Transactions on Circuits and Systems for Video Technology*, 27(11):2309–2321, Nov 2017.
327. M. Miyahara and K. Kotani. Block distortion in orthogonal transform coding—analysis, minimization and distortion measure. *IEEE Transactions on Communications*, 33(1):90–96, January 1985.
328. J. Mohan, V. Krishnaveni, and Y. Guo. A survey on the magnetic resonance image denoising methods. *Biomedical Signal Processing and Control*, 9:56 – 69, 2014.
329. F. Mokhtarian and N. Jin. Human motion recognition based on statistical shape analysis. In *Proc. of Conf. on Advanced Video and Signal Based Surveillance (AVSS)*, pages 4–9, Sept 2005.
330. Y.H. Moon, J.H. Kim, and G.Y. Kim. An improved early detection algorithm for all-zero blocks in H.264 video encoding. *IEEE Transactions on Circuits and Systems for Video Technology*, 15(8):1053–1057, Aug 2005.
331. M. Moosazadeh and G. Ekbatanifard. An improved robust image watermarking method using DCT and YCoCg-R color space. *Optik - Intl. Journal for Light and Electron Optics*, 140:975–988, 2017.

332. H. Moussa et al. A robust-fragile dual watermarking system in the DCT domain. In K. Rajiv, H. Robert, and J. Lakhmi, editors, *Proc. of Knowledge-Based Intelligent Information and Engineering Systems*, pages 548 – 553, Berlin, Heidelberg, 2005. Springer.
333. R.J. Mstafa and K.M. Elleithy. A DCT-based robust video steganographic method using BCH error correcting codes. In *Proc. of Long Island Systems, Applications and Technology Conference (LISAT)*, pages 1–6, April 2016.
334. R.J. Mstafa, K.M. Elleithy, and E. Abdelfattah. A robust and secure video steganography method in DWT-DCT domains based on multiple object tracking and ECC. *IEEE Access*, 5(1):5354–5365, April 2017.
335. R. Mukundan, S.H. Ong, and P.A. Lee. Image analysis by Tchebichef moments. *IEEE Transactions on Image Processing*, 10(9):1357–1364, Sept 2001.
336. R. Muralishankar, H.N. Shankar, and D. O'Shaughnessy. A performance analysis of features from complex cepstra of warped DST, DCT and DHT filters for phoneme recognition. In *Proc. of 15th Intl. Conference on Digital Signal Processing*, pages 591–594, July 2007.
337. S. Muramatsu and H. Hiki. Block-wise implementation of directional genlot. In *Proc. of 16th Intl. Conf. on Image Processing (ICIP)*, pages 3977–3980, Nov 2009.
338. V. Nagesha. Comments on "Fast computation of the discrete cosine transform and the discrete Hartley transform" by H.S. Malvar. *IEEE Transactions on Acoustics, Speech, and Signal Processing*, 37(3):439–440, March 1989.
339. V.P.S. Naidu. Hybrid DDCT-PCA based multi sensor image fusion. *Journal of Optics*, 43(1):48–61, Mar 2014.
340. K. Nakagaki and R. Mukundan. A fast 4×4 forward discrete Tchebichef transform algorithm. *IEEE Signal Processing Letters*, 14(10):684–687, Oct 2007.
341. M.J. Narasimha. Design of FIR filter banks for a 24-channel transmultiplexer. *IEEE Transactions on Communications*, 30(7):1506–1510, July 1982.
342. M.J. Narasimha et al. The TM 800-MI: A 60-channel CCITT transmultiplexer. In *Proc. of Intl. Conf. on Communications*, pages 672–674, May 1984.
343. M.J. Narasimha and A. Peterson. On the computation of the discrete cosine transform. *IEEE Transactions on Communications*, 26(6):934 – 936, June 1978.
344. M.J. Narasimha and A. Peterson. Design of a 24-channel transmultiplexer. *IEEE Transactions on Acoustics, Speech, and Signal Processing*, 27(6):752–762, Dec 1979.
345. M.V. Narayanan et al. Application of the Karhunen-Loeve transform to 4D reconstruction of cardiac gated SPECT images. In *Proc. of Nuclear Science Symp. and Medical Imaging Conf. (Cat. No.98CH36255)*, volume 2, pages 1141–1145, 1998.
346. D.R. Nayak et al. Combining extreme learning machine with modified sine cosine algorithm for detection of pathological brain. *Computers and Electrical Engineering*, 68:366 – 380, 2018.
347. L. Netsch. *A robust telephone-based speaker verification system.* Ph. D. dissertation, Electrical Engineering Dept., Univ. of Texas at Arlington, 1993.
348. R.T. Ng and A. Sedighian. Evaluating multidimensional indexing structures for images transformed by principal component analysis. In *Proc. of SPIE*, volume 2670, pages 50 – 61, Jan 1996.
349. K.N. Ngan and R.J. Clarke. Lowpass filtering in the cosine transform domain. In *Proc. of Intl. Conf. Commun.*, pages 37.7.1–37.7.5, 1980.
350. T. Nguyen and D. Marpe. Objective performance evaluation of the HEVC main still picture profile. *IEEE Transactions on Circuits and Systems for Video Technology*, 25(5):790–797, May 2015.

351. T. Nguyen and D. Marpe. Future video coding technologies: A performance evaluation of AV1, JEM, VP9, AND HM. In *Proc. of Picture Coding Symposium (PCS)*, volume 2670, pages 50 – 61, July 2018.
352. J. A. Nikara, J. H. Takala, and J. T. Astola. Discrete cosine and sine transforms-Regular algorithms and pipeline architectures. *Signal Processing*, 86(2):230 – 249, 2006.
353. V. Nikolajevic and G. Fettweis. Improved implementation of MDCT in MP3 audio coding. In *Proc of 5th Intl. Symp. on Multi-Dimensional Mobile Communications Proceedings. The Joint Conference of the 10th Asia-Pacific Conference on Communications*, volume 1, pages 309–312, Aug 2004.
354. P. Nirajan. *Discrete sine and cosine transforms on parallel processors.* M. S. thesis, Faculty of Computing and Electrical Engineering, Tempere University of Technology, 2015.
355. P. Norhashimah, H. Fang, and J. Jiang. Video extraction in compressed domain. In *Proc. of Conf. on Advanced Video and Signal Based Surveillance*, pages 321–326, July 2003.
356. AVS Workgroup of China. *Information technology - Advanced coding of audio and video, Part 2:Video, GB/T 20090.2-2006*, 2006.
357. K. Ogawa and G. Ohtake. Watermarking for HEVC/H.265 stream. In *Proc. of Intl. Conf. on Consumer Electronics (ICCE)*, pages 102–103, Jan 2015.
358. J. Ohm et al. Comparison of the coding efficiency of video coding standards-including high efficiency video coding (HEVC). *IEEE Transactions on Circuits and Systems for Video Technology*, 22(12):1669–1684, Dec 2012.
359. J.-R. Ohm and G.J. Sullivan. Invited talk: Versatile video coding-Towards the next generation of video compression. Picture Coding Symposium 2018, San Francisco, 2018.
360. P.A.M. Oliveira et al. Low-complexity image and video coding based on an approximate discrete Tchebichef transform. *IEEE Transactions on Circuits and Systems for Video Technology*, 27(5):1066–1076, May 2017.
361. T. M. Ono and K. Asai. *Ubiquitous technology, "hybrid video coding – MPEG-4 and H.264" (In Japanese, title translated).* Tokyo, Ohmsha Press, 2005.
362. T. Ooura. General purpose FFT (fast Fourier/cosine/sine transform) package. http://www.kurims.kyoto-u.ac.jp/~ooura/fft.html, 2017. [Online: accessed Dec. 2017].
363. J. Ou et al. A new image inpainting algorithm based on DCT similar patches features. In *Proc of. Intl. Conf. on Computational Intelligence and Security (CIS)*, pages 152–155, Dec 2016.
364. L.B. Ozer and W. Wolf. Real-time posture and activity recognition. In *Proc. of Workshop on Motion and Video Computing*, pages 133–138, Dec 2002.
365. N. Paramanandham and K. Rajendiran. Infrared and visible image fusion using discrete cosine transform and swarm intelligence for surveillance applications. *Infrared Physics and Technology*, 88:13–22, 2018.
366. J.-S. Park et al. 2-D large inverse transform (16x16, 32x32) for HEVC (high efficiency video coding). *Journal of Semiconductor Technology and Science*, 12:203–211, June 2012.
367. J.S. Park and T. Ogunfunmi. A new approach for image quality assessment: FSM (Frequency Similarity Method). In *Proc. of 7th Conf. on Industrial Electronics and Applications (ICIEA)*, pages 996–1000, July 2012.
368. S.Y. Park and P.K. Meher. Flexible integer DCT architectures for HEVC. In *Proc. of Intl. Symp. on Circuits and Systems (ISCAS)*, pages 1376–1379, May 2013.

369. G. Pastuszak. Hardware architectures for the H.265/HEVC discrete cosine transform. *IET Image Processing*, 9(6):468–477, June 2015.
370. D.-B. Perng and S.-H. Chen. Directional textures auto-inspection using discrete cosine transform. *Intl. Journal of Production Research*, 49(23):7171–7187, 2011.
371. G. Plonka. A global method for invertible integer DCT and integer wavelet algorithms. *Applied and Computational Harmonic Analysis*, 16(2):90 – 110, 2004.
372. G. Plonka and Tasche M. Integer DCT-II by lifting steps. In *Proc. of Advances in Multivariate Approximation*, pages 235–252, 2002.
373. G. Plonka and M. Tasche. Fast and numerically stable algorithms for discrete cosine transforms. *Linear Algebra and its Applications*, 49(9):309–345, Sept 2005.
374. N. Ponomarenko et al. On between-coefficient contrast masking of DCT basis functions. In *Proc. of 3rd Intl. Workshop on Video Processing and Quality Metrics for Consumer Electronics (VPQM)*, pages 1 – 4, 2007.
375. Ponomarenko, N. PSNR-HVS-M download page. https://github.com/Rolinh/VQMT/blob/master/inc/PSNRHVS.hpp, 2009. [Online: accessed July 2018].
376. U.S. Potluri et al. Multiplier-free DCT approximations for RF multi-beam digital aperture-array space imaging and directional sensing. *Measurement Science and Technology*, 23:1–15, 2012.
377. M.T. Pourazad et al. HEVC: The new gold standard for video compression: How does HEVC compare with H.264/AVC? *IEEE Consumer Electronics Magazine*, 1(3):36–46, July 2012.
378. S. Prattipati et al. A fast 8×8 integer Tchebichef transform and comparison with integer cosine transform for image compression. In *Proc. of 56th IEEE Intl. Midwest Symp. on Circuits and Systems (MWSCAS)*, pages 1294–1297, Aug 2013.
379. J. Princen and A. Bradley. Analysis/Synthesis filter bank design based on time domain aliasing cancellation. *IEEE Transactions on Acoustics, Speech, and Signal Processing*, 34(5):1153–1161, Oct 1986.
380. J. Princen, A. Johnson, and A. Bradley. Subband/Transform coding using filter bank designs based on time domain aliasing cancellation. In *Proc. of Intl. Conf. on Acoustics, Speech, and Signal Processing (ICASSP)*, volume 12, pages 2161–2164, April 1987.
381. Z. Qian et al. New framework of reversible data hiding in encrypted JPEG bitstreams. *IEEE Transactions on Circuits and Systems for Video Technology*, (Early access), 2018.
382. Z. Qian, W. Wang, and T. Qiao. An edge detection method in DCT domain. *Procedia Engineering, Intl. Workshop on Information and Electronics Engineering*, 29:344 – 348, 2012.
383. R.L. Queiroz. Processing JPEG-compressed images and documents. *IEEE Transactions on Image Processing*, 7(12):1661–1672, Dec 1998.
384. G. Quer et al. Sensing, compression, and recovery for WSNs: Sparse signal modeling and monitoring framework. *IEEE Transactions on Wireless Communications*, 11(10):3447–3461, Oct 2012.
385. M. Rabbat, J. Haupt, A. Singh, and et al. Decentralized compression and predistribution via randomized gossiping. In *Proc. of 5th Intl. Conf. on Information Processing in Sensor Networks (IPSN)*, pages 51–59, New York, NY, USA, 2006.
386. K.R. Rao. *Discrete transforms and their applications*. New York, NY: Van Nostrand Reinhold, 1985.
387. K.R. Rao, J.J. Hwang, and D.N. Kim. *High efficiency video coding and other emerging standards*. River Publishers, 2017.

388. K.R. Rao, D.N. Kim, and J.J. Hwang. *Fast Fourier transform-Algorithms and applications*. Springer, 2010.
389. K.R. Rao, D.N. Kim, and J.J. Hwang. *Video coding standards: AVS China, H.264/MPEG-4 part 10, HEVC, VP6, DIRAC and VC-1*. Springer, 2014.
390. K.R. Rao and P. Yip. *Discrete cosine transform: Algorithms, advantages, applications*. Academic Press, 1990.
391. S.K. Rao. Performance comparison of HEVC intra, JPEG, JPEG 2000, JPEG XR, JPEG LS and VP9 intra. Technical report, The University of Texas at Arlington, 2016.
392. D. Ravi et al. Semantic segmentation of images exploiting DCT-based features and random forest. *Pattern Recognition*, 52:260 – 273, 2016.
393. W. Ray and R. Driver. Further decomposition of the Karhunen-Loève series representation of a stationary random process. *IEEE Transactions on Information Theory*, 16(6):663–668, Nov 1970.
394. I.S. Reed and L.S Lan. A fast approximate Karhunen-Loève transform (AKLT) for data compression. *Journal of Visual Communication and Image Representation*, 5(4):304 – 316, Dec 1994.
395. H. Reitboeck and T.P. Brody. A transformation with invariance under cyclic permutation for applications in pattern recognition. *Information and Control*, 15(2):130 – 154, 1969.
396. H. Ren, P. Yu, and P. Zhang. Illumination invariant feature extraction using relative gradient difference. *Optik - Intl. Journal for Light and Electron Optics*, 126(23):4531 – 4533, 2015.
397. X. Ren and D. Fang. A sensor node compression algorithm based on H.264 integer DCT Transform in WSN. In *Proc. of IET Intl. Conf. on Wireless Sensor Network (IET-WSN)*, pages 214–217, Nov 2010.
398. I.E. Richardson. *The H.264 advanced video compression standard*. Wiley, 2010.
399. T. Richter. On the standardization of the JPEG XT image compression. In *Proc. of Picture Coding Symp. (PCS)*, pages 37–40, Dec 2013.
400. T Richter. On the integer coding profile of JPEG XT. In *Proc.of SPIE: Applications of Digital Image Processing XXXVII*, volume 9217, pages 9217 – 9217 – 19, 2014.
401. T. Richter. JPEG XT reference codec 1.31. http://www.jpeg.org/jpegxt/software.html, 2015. [Online: accessed July 2018].
402. T. Richter. Lossless coding extensions for JPEG. In *Proc. of Data Compression Conf. (DCC)*, pages 143–152, April 2015.
403. T. Richter. JPEG on steroids: Common optimization techniques for JPEG image compression. In *Proc. of Intl. Conf. on Image Processing (ICIP)*, pages 61–65, Sept 2016.
404. T. Richter. Error bounds for HDR image coding with JPEG XT. In *Proc. of Data Compression Conference (DCC)*, pages 122–130, April 2017.
405. T. Richter. libjpeg: A complete implementation of 10918-1 (JPEG). https://github.com/thorfdbg/libjpeg, 2017. [Online: accessed July 2018].
406. T. Richter, A. Artusi, and T. Ebrahimi. JPEG XT: A new family of JPEG backward-compatible standards. *IEEE MultiMedia*, 23(3):80–88, July 2016.
407. T. Richter et al. The JPEG XT suite of standards: status and future plans. In *Proc. of SPIE: Applications of Digital Image Processing XXXVIII*, pages 1–18, Aug. 2015.
408. T. Richter et al. Entropy coding and entropy coding improvements of JPEG XS. In *Proc. of Data Compression Conference (DCC)*, pages 87–96, March 2018.
409. B.A. Ringnyu, A. Tangel, and E. Karabulut. Implementation of different architectures of forward 4×4 integer DCT for $H.264/AVC$ encoder. In *Proc. of 10th Intl. Conf. on Electrical and Electronics Engineering (ELECO)*, pages 423–427, Nov 2017.

410. L.R. Roldan et al. Watermarking-based color image authentication with detection and recovery capability. *IEEE Latin America Transactions*, 14(2):1050–1057, Feb 2016.
411. S., Sreedevi and E., Sherly. A novel approach for removal of pectoral muscles in digital mammogram. *Procedia Computer Science*, 46:1724 – 1731, 2015.
412. M.A. Saad, A.C. Bovik, and C. Charrier. A DCT statistics-based blind image quality index. *IEEE Signal Processing Letters*, 17(6):583–586, June 2010.
413. M.A. Saad, A.C. Bovik, and C. Charrier. Blind image quality assessment: A natural scene statistics approach in the DCT domain. *IEEE Transactions on Image Processing*, 21(8):3339–3352, Aug 2012.
414. J.A. Saghri, A.G. Tescher, and J.T. Reagan. Practical transform coding of multispectral imagery. *IEEE Signal Processing Magazine*, 12(1):32–43, Jan 1995.
415. L. Sangkil and L. Insung. A low-delay MDCT/IMDCT. *ETRI Journal*, 35(5):935–938, 2013.
416. S. Saponara, L. Fanucci, and P. Terreni. Low-power VLSI architectures for 3D discrete cosine transform (DCT). In *Proc. of 46th Midwest Symp. on Circuits and Systems*, volume 3, pages 1567–1570, Dec 2003.
417. B.N.S. Sarvesh, Y.N. Mamatha, and M.K. Subramanian. Fast and efficient computation of MDCT/IMDCT algorithms for MP3 applications. *Intl. Journal of Scientific and Engineering Research*, 4(6):1121 – 1125, 2013.
418. A. Saxena and F.C. Fernandez. Jointly optimal intra prediction and adaptive primary transform. Technical report, ITU-T/ISO/IEC, 2010.
419. A. Saxena and F.C. Fernandez. CTVC-E125: CE7: Mode-dependent DCT/DST without full matrix multiplication for intra prediction. Technical report, ITU-T/ISO/IEC, 2011.
420. I.D. Schizas, G.B. Giannakis, and Z.Q. Luo. Distributed estimation using reduced-dimensionality sensor observations. *IEEE Transactions on Signal Processing*, 55(8):4284–4299, Aug 2007.
421. I.W. Selesnick, R.G. Baraniuk, and N.C. Kingsbury. The dual-tree complex wavelet transform. *IEEE Signal Processing Magazine*, 22(6):123–151, Nov 2005.
422. S.A. Selouani and D. O'Shaughnessy. Robustness of speech recognition using genetic algorithms and a mel-cepstral subspace approach. In *Proc. of Intl. Conf. on Acoustics, Speech, and Signal Processing (ICASSP)*, volume 1, pages 201–204, May 2004.
423. A. Sengupta et al. Low-cost obfuscated JPEG CODEC IP core for secure CE hardware. *IEEE Transactions on Consumer Electronics*, (Early access), 2018.
424. M. Shah. *Future of JPEG XT: Scrambling to enhance privacy and security*. M. S. thesis, Electrical Engineering Dept., The Univ. of Texas at Arlington, 2016.
425. Z. Shahid, M. Chaumont, and W. Puech. Fast protection of H.264/AVC by selective encryption of CABAC for I and P frames. In *Proc. of 17th European Signal Processing Conf.*, pages 2201–2205, Aug 2009.
426. Q. Shang et al. Single-port SRAM-based transpose memory with diagonal data mapping for large size 2-D DCT/IDCT. *IEEE Transactions on Very Large Scale Integration (VLSI) Systems*, 22(11):2423–2427, Nov 2014.
427. H.R. Sheikh and A.C. Bovik. Image information and visual quality. *IEEE Transactions on Image Processing*, 15(2):430–444, Feb 2006.
428. H.R. Sheikh, A.C. Bovik, and G. Veciana. An information fidelity criterion for image quality assessment using natural scene statistics. *IEEE Transactions on Image Processing*, 14(12):2117–2128, Dec 2005.

429. B.H. Shekar and B. Pilar. Discrete cosine transformation and height functions based shape representation and classification. *Procedia Computer Science*, 58:714 – 722, 2015.
430. B. Shen and S.K. Ishwar. Inner-block Operations on Compressed Images. In *Proc. of the Third ACM Intl. Conf. on Multimedia*, pages 489–498, New York, NY, USA, 1995.
431. S. Shen et al. A unified 4/8/16/32-point integer IDCT architecture for multiple video coding standards. In *Proc. of Intl. Conf. on Multimedia and Expo*, pages 788–793, July 2012.
432. D.I. Shuman et al. The emerging field of signal processing on graphs: Extending high-dimensional data analysis to networks and other irregular domains. *IEEE Signal Processing Magazine*, 30(3):83–98, May 2013.
433. T Sikora. Low complexity shape-adaptive DCT for coding of arbitrarily shaped image segments. *Signal Processing: Image Communication*, 7(4):381 – 395, 1995.
434. T. Sikora and B. Makai. Shape-adaptive DCT for generic coding of video. *IEEE Transactions on Circuits and Systems for Video Technology*, 5(1):59–62, Feb 1995.
435. H. Silverman. An introduction to programming the Winograd Fourier transform algorithm (WFTA). *IEEE Transactions on Acoustics, Speech, and Signal Processing*, 25(2):152–165, April 1977.
436. H. Silverman. Corrections and addendum to "An introduction to programming the Winograd Fourier transform algorithm (WFTA). *IEEE Transactions on Acoustics, Speech, and Signal Processing*, 26(3):268–268, June 1978.
437. P. Singh and G. Bhatnagar. A new robust watermarking system in integer DCT domain. *Journal of Visual Communication and Image Representation*, 53:86 – 101, 2018.
438. P. Singh and R.S. Chadha. A survey of digital watermarking techniques, applications and attacks. *Intl. Journal of Engineering and Innovative Technology (IJEIT)*, 2(9):165 – 175, 2013.
439. Digital Formats Web site. HD Photo, version 1.0 (Windows Media Photo). https://www.loc.gov/preservation/digital/formats/fdd/fdd000243.shtml, 2008. [Online: accessed July 2018].
440. P. Sjövall et al. High-level synthesis implementation of HEVC 2-D DCT/DST on FPGA. In *Proc. of Intl. Conf. on Acoustics, Speech and Signal Processing (ICASSP)*, pages 1547–1551, March 2017.
441. A. Skodras, C. Christopoulos, and T. Ebrahimi. The JPEG 2000 still image compression standard. *IEEE Signal Processing Magazine*, 18(5):36 – 58, Sept 2001.
442. B.C. Smith and L.A. Rowe. Algorithms for manipulating compressed images. *IEEE Computer Graphics and Applications*, 13(5):34–42, Sept 1993.
443. I.N. Sneddon. *Special functions of mathematical physics and chemistry.* New York, NY: Longman Inc, 1980.
444. J.A. Solomon, A.B. Watson, and A. Ahumada. Visibility of DCT basis functions: Effects of contrast masking. In *Proc. of Data Compression Conference (DCC)*, pages 362 – 370, March 1994.
445. G. Song and X. Jiang. All-zero block detection algorithm based on quantitative parameters. In *Proc of 3rd Intl. Conf. on System Science, Engineering Design and Manufacturing Informatization*, pages 296–299, Oct 2012.
446. X. Song, S. Wang, and X. Niu. An integer DCT and affine transformation based image steganography method. In *Proc. of 8th Intl. Conf. on Intelligent Information Hiding and Multimedia Signal Processing*, pages 102–105, July 2012.
447. H. Sorensen et al. On computing the discrete Hartley transform. *IEEE Transactions on Acoustics, Speech, and Signal Processing*, 33(5):1231–1238, Oct 1985.

448. D. Sridhar and I. M. Krishna. Brain tumor classification using discrete cosine transform and probabilistic neural network. In *Proc. of Intl. Conf. on Signal Processing , Image Processing Pattern Recognition*, pages 92–96, Feb 2013.
449. S. Srinivasan et al. Windows media video 9: Overview and applications. *Signal Processing: Image Communication*, 19(9):851 – 875, 2004.
450. T. Srisooksai et al. Practical data compression in wireless sensor networks: A survey. *Journal of Network and Computer Applications*, 35(1):37 – 59, 2012.
451. S. Srisuk. Robust face recognition based on texture analysis. *Intl. Journal of Advanced Robotic Systems*, 10(1):47, 2013.
452. D. Stanescu et al. Digital watermarking using Karhunen-Loeve transform. In *Proc. of 4th Intl. Symp. on Applied Computational Intelligence and Informatics*, pages 187–190, 2007.
453. G. Strang. Every unit matrix is a LULU. *Linear Algebra and its Applications*, 1997.
454. G. Strang. The discrete cosine transform. *SIAM Review*, 41(1):135–147, 1999.
455. G. Strang and T. Nguyen. *Wavelets and filter banks (Rev. ed)*. Wellesley-Cambridge Press, 1997.
456. P.-C. Su et al. A practical design of digital video watermarking in H.264/AVC for content authentication. *Signal Processing: Image Communication*, 26(8):413 – 426, 2011.
457. S. Subbarayappa. *Implementation and analysis of directional discrete cosine transform in H.264 for baseline profile*. M. S. thesis, Electrical Engineering Dept., The Univ. of Texas at Arlington, 2012.
458. B.N. Subudhi et al. Tumor or abnormality identification from magnetic resonance images using statistical region fusion based segmentation. *Magnetic Resonance Imaging*, 34(9):1292 – 1304, 2016.
459. N. Suehiro and M. Hatori. Fast algorithms for the DFT and other sinusoidal transforms. *IEEE Transactions on Acoustics, Speech, and Signal Processing*, 34(3):642–644, June 1986.
460. T. Sugimoto, Y. Nakayama, and T. Komori. 22.2 ch audio encoding/decoding hardware system based on MPEG-4 AAC. *IEEE Transactions on Broadcasting*, 63(2):426–432, June 2017.
461. M.A. Suhail and M.S Obaidat. Digital watermarking-based DCT and JPEG model. *IEEE Transactions on Instrumentation and Measurement*, 52(5):1640–1647, Oct 2003.
462. G.J. Sullivan et al. Overview of the high efficiency video coding (HEVC) standard. *IEEE Transactions on Circuits and Systems for Video Technology*, 22(12):1649–1668, Dec 2012.
463. C. Sun and E.H. Yang. An efficient DCT-based image compression system based on Laplacian transparent composite model. *IEEE Transactions on Image Processing*, 24(3):886–900, March 2015.
464. C.W. Sun and P. Yip. Split-radix algorithms for DCT and DST. In *Proc. of 23rd Asilomar Conf. on Signals, Systems and Computers*, pages 508–512, Nov 1989.
465. Z. Sun et al. Ordinal palmprint represention for personal identification. In *Proc. of Computer Society Conf. on Computer Vision and Pattern Recognition (CVPR)*, volume 1, pages 279–284, June 2005.
466. Z. Sun et al. Removal of non-informative frames for wireless capsule endoscopy video segmentation. In *Proc. of Intl. Conf. on Automation and Logistics*, pages 294–299, Aug 2012.

467. N. Surve and A. Kulkarni. Iris recognition using discrete sine transform and neural network. In *Proc. of Intl. Conf. and Workshop on Emerging Trends in Technology*, ICWET, pages 750–755, New York, NY, USA, 2010.
468. T. Suzuki and M. Ikehara. Integer DCT-based on direct-lifting of DCT-IDCT for lossless-to-lossy image coding. *IEEE Transactions on Image Processing*, 19(11):2958–2965, Nov 2010.
469. W. Sweldens. The lifting scheme: A construction of second generation wavelets. *SIAM Journal on Mathematical Analysis*, 29(2):511–546, Mar 1998.
470. D.L. Swets and J.J. Weng. Using discriminant eigenfeatures for image retrieval. *IEEE Transactions on Pattern Analysis and Machine Intelligence*, 18(8):831 – 836, Aug 1996.
471. W. Sze, M. Bugadavi, and G. J. Sullivan. *High efficiency video coding (HEVC): Algorithms and architectures*. Springer, 2014.
472. Y. Tachizaki, M. Fujiyoshi, and H. Kiya. A DCT coefficient sign-based background model for moving objects detection from motion JPEG coded movies. In *Proc. of Intl. Symp. on Intelligent Signal Processing and Communication Systems (ISPACS)*, pages 37–40, Jan 2009.
473. S.-C. Tai, Y.-G. Wu, and C.-W. Lin. An adaptive 3-D discrete cosine transform coder for medical image compression. *IEEE Transactions on Information Technology in Biomedicine*, 4(3):259–263, Sept 2000.
474. C. Tang, X. Yang, and G. Zhai. Noise estimation of natural images via statistical analysis and noise injection. *IEEE Transactions on Circuits and Systems for Video Technology*, 25(8):1283–1294, Aug 2015.
475. H. Tang, M. Fan, and L. Yu. Encoder-embedded temporal-spatial Wiener filter for video encoding. In *Proc. of Picture Coding Symp.*, pages 361–364, May 2012.
476. H. Tang and H. Shi. Fast mode decision algorithm for H.264/AVC based on all-zero blocks predetermination. In *Proc. of Intl. Conf. on Measuring Technology and Mechatronics Automation*, volume 2, pages 780–783, April 2009.
477. X. Tang and W.K Stewart. Texture classification using principal component analysis techniques. In *Proc. of SPIE*, volume 2315, pages 22 – 35, Sept 1994.
478. G. Tech et al. Overview of the multiview and 3D extensions of high efficiency video coding. *IEEE Transactions on Circuits and Systems for Video Technology*, 26(1):35–49, Jan 2016.
479. H. Tian et al. Robust watermarking of mobile video resistant against barrel distortion. *China Communications*, 13(9):131–138, Sept 2016.
480. Trellis Management Co. Ltd,. Software for JPEG XT. https://jpeg.org/jpegxt/software.html, 2015. [Online: accessed July 2018].
481. D. Tretter and C.A. Bouman. Optimal transforms for multispectral and multilayer image coding. *IEEE Transactions on Image Processing*, 4(3):296–308, Mar 1995.
482. B.D. Tseng and W.C. Miller. Comments on: An introduction to programming the Winograd Fourier transform algorithm (WFTA). *IEEE Transactions on Acoustics, Speech, and Signal Processing*, 26(3):268–269, June 1978.
483. B.D. Tseng and W.C. Miller. On computing the discrete cosine transform. *IEEE Transactions on Computers*, C-27(10):966–968, Oct 1978.

484. U. Tuna, S. Peltonen, and U. Ruotsalainen. Data estimation for the ECAT HRRT sinograms by utilizing the DCT domain. In *Proc. of Nuclear Science Symp. Conference Record (NSS/MIC)*, pages 5076–5080, Oct 2008.
485. U. Tuna, S. Peltonen, and U. Ruotsalainen. Interpolation for the gap-filling of the HRRT PET sinograms by using the slices in the direction of the radial samples. In *Proc. of Nuclear Science Symp. Conf. Record (NSS/MIC)*, pages 3273–3279, Oct 2009.
486. U. Tuna, S. Peltonen, and U. Ruotsalainen. Gap-filling for the high-resolution PET sinograms with a dedicated DCT-domain filter. *IEEE Transactions on Medical Imaging*, 29(3):830–839, March 2010.
487. P. Turcza, T. Zielinski, and M. Duplaga. Low complexity image coding algorithm for capsule endoscopy with Bayer color filter array. In *Proc. of Signal Processing Algorithms, Architectures, Arrangements, and Applications (SPA)*, pages 27–32, Sept 2007.
488. C. Turgay, O. Huseyin, and D. Hasan. Facial feature extraction using complex dual-tree wavelet transform. *Computer Vision and Image Understanding*, 111(2):229 – 246, 2008.
489. M.A. Turk and A.P. Pentl. Face recognition using eigenfaces. In *Proc. of Computer Society Conf. on Computer Vision and Pattern Recognition*, pages 586–591, Jun 1991.
490. K.-H. Tzou. Progressive image transmission: A review and comparison of techniques. *Optical Engineering*, 26(7):581–589, 1987.
491. P.T. Vanishree and A.M.V. Prakash. VLSI implementation of discrete cosine transform and intra prediction. In *Proc. of Intl. Conf. on Advances in Electronics Computers and Communications*, pages 1–6, Oct 2014.
492. A. Vashistha and A.M. Joshi. Fingerprint based biometric watermarking architecture using integer DCT. In *Proc. of Region 10 Conf. (TENCON)*, pages 2818–2821, Nov 2016.
493. V.D. Vaughn and T.S. Wilkinson. System considerations for multispectral image compression designs. *IEEE Signal Processing Magazine*, 12(1):19–31, Jan 1995.
494. V. Velisavljevic et al. Discrete multidirectional wavelet bases. In *Proc. of Intl. Conf. on Image Processing (Cat. No.03CH37429)*, volume 1, pages I–1025–8, Sept 2003.
495. S. Venkataraman et al. Discrete transforms via the Walsh-Hadamard transform. In *Proc. of 26th Midwest Symp. Circuits and Systems*, pages 74 – 78, Aug 1983.
496. S. Venkataraman et al. Discrete transforms via the Walsh–Hadamard transform. *Signal Processing*, 14(4):371–382, June 1988.
497. A. Vetro and D. Tian. Analysis of 3D and multiview extensions of the emerging HEVC standard. In *Proc. of SPIE: Applications of Digital Image Processing XXXV*, volume 8499, pages 8499 – 1 – 8499 – 7, 2012.
498. M. Vetterli and A. Ligtenberg. A discrete Fourier-cosine transform chip. *IEEE Journal on Selected Areas in Communications*, 4(1):49–61, Jan 1986.
499. M. Vetterli and H.J. Nussbaumer. Simple FFT and DCT algorithms with reduced number of operations. *Signal Processing*, 6(4):267 – 278, 1984.
500. Y. Voronenko and M. Püschel. Multiplierless multiple constant multiplication. *ACM Trans. Algorithms*, 3(2):1–39, May 2007.
501. G.K. Wallace. The JPEG still picture compression standard. *Communication of the ACM*, 34(4):30 – 44, April 1991.
502. R. Wallace et al. Cross-pollination of normalization techniques from speaker to face authentication using Gaussian mixture models. *IEEE Transactions on Information Forensics and Security*, 7(2):553–562, April 2012.

503. D. Wang et al. Perceptual differential energy watermarking for H.264/AVC. *Multimedia Tools and Applications*, 60(3):537–550, Oct 2012.
504. H. Wang, H. Du, and J. Wu. Predicting zero coefficients for High Efficiency Video Coding. In *Proc. of Intl. Conf. on Multimedia and Expo (ICME)*, pages 1–6, July 2014.
505. H. Wang et al. Early detection of all-zero 4 x 4 blocks in High Efficiency Video Coding. *Journal of Visual Communication and Image Representation*, 25(7):1784 – 1790, 2014.
506. H. Wang and S. Kwong. Hybrid model to detect zero quantized DCT coefficients in H.264. *IEEE Transactions on Multimedia*, 9(4):728–735, June 2007.
507. H. Wang and S. Kwong. Prediction of zero quantized DCT coefficients in H.264/AVC using Hadamard transformed information. *IEEE Transactions on Circuits and Systems for Video Technology*, 18(4):510–515, April 2008.
508. H. Wang, S. Kwong, and C.W. Kok. Analytical model of zero quantized DCT coefficients for video encoder optimization. In *Proc. of Intl. Conf. on Multimedia and Expo*, pages 801–804, July 2006.
509. H. Wang, S. Kwong, and C.W. Kok. Effectively detecting all-zero DCT blocks for H.264 optimization. In *Proc. of Intl. Conf. on Image Processing (ICIP)*, pages 1329–1332, Oct 2006.
510. H. Wang, S. Kwong, and C.W. Kok. Efficient prediction algorithm of integer DCT coefficients for H.264/AVC optimization. *IEEE Transactions on Circuits and Systems for Video Technology*, 16(4):547–552, April 2006.
511. H. Wang, S. Kwong, and C.W. Kok. Efficient predictive model of zero quantized DCT coefficients for fast video encoding. *Image and Vision Computing*, 25(6):922 – 933, 2007.
512. W. Wang, J. Yang, and W. Gao. Modeling background and segmenting moving objects from compressed video. *IEEE Transactions on Circuits and Systems for Video Technology*, 18(5):670–681, May 2008.
513. X.-F. Wang and De-B. Zhao. Performance comparison of AVS and H.264/AVC video coding standards. *Journal of Computer Science and Technology*, 21(3):310–314, May 2006.
514. Y. Wang et al. DCT inspired feature transform for image retrieval and reconstruction. *IEEE Transactions on Image Processing*, 25(9):4406–4420, Sept 2016.
515. Y. Wang et al. Parts for the Whole: The DCT norm for extreme visual recovery. *CoRR*, abs/1604.05451, 2016.
516. Y. Wang et al. DCT regularized extreme visual recovery. *IEEE Transactions on Image Processing*, 26(7):3360–3371, July 2017.
517. Y. Wang and M. Vilermo. Modified discrete cosine transform - Its implications for audio coding and error concealment. *J. Audio Eng. Soc.*, 51(1):52–61, Feb 2003.
518. Y. Wang, L. Yaroslavsky, and M. Vilermo. On the relationship between MDCT, SDFT and DFT. In *2000 5th Intl. Conf. on Signal Processing (ICSP). 16th World Computer Congress 2000 (WCC)*, volume 1, pages 44 – 47, Aug 2000.
519. Y.E. Wang et al. Some peculiar properties of the MDCT. In *Proc. of Intl. Conf. on Signal Processing (ICSP)*, pages 1–4, Aug 2000.
520. Z. Wang. A fast algorithm for the discrete sine transform implemented by the fast cosine transform. *IEEE Transactions on Acoustics, Speech, and Signal Processing*, 30(5):814–815, Oct 1982.
521. Z. Wang. Reconsideration of a fast computational algorithm for the discrete cosine transform. *IEEE Transactions on Communications*, 31(1):121–123, Jan 1983.

522. Z. Wang. Fast algorithms for the discrete W transform and for the discrete Fourier transform. *IEEE Transactions on Acoustics, Speech, and Signal Processing*, 32(4):803–816, Aug 1984.
523. Z. Wang. The discrete W transform. *Applied Mathematics and Computation*, 16:19–48, Jan 1985.
524. Z. Wang. Comments on a fast computational algorithm for the discrete sine transform. *IEEE Transactions on Communications*, 34(2):204–205, Feb 1986.
525. Z. Wang. Interpolation using the discrete cosine transform: Reconsideration. *Electronics Letters*, 29(2):198–200, Jan 1993.
526. Z. Wang et al. Image quality assessment: from error visibility to structural similarity. *IEEE Transactions on Image Processing*, 13(4):600–612, April 2004.
527. Z. Wang, E.P. Simoncelli, and A.C. Bovik. Multiscale structural similarity for image quality assessment. In *Proc. of 37th Asilomar Conf. on Signals, Systems Computers*, volume 2, pages 1398–1402, Nov 2003.
528. O. Watanabe, H. Kobayashi, and H. Kiya. Two-layer lossless HDR coding considering histogram sparseness with backward compatibility to JPEG. In *Proc. of Picture Coding Symp. (PCS)*, pages 1–5, June 2018.
529. J. Wei and G. Li. Automated lung segmentation and image quality assessment for clinical 3-D/4-D-computed tomography. *IEEE Journal of Translational Engineering in Health and Medicine*, 2(1):1–10, Dec 2014.
530. J. Wen et al. A compressive sensing image compression algorithm using quantized DCT and noiselet information. In *Proc. of Intl. Conf. on Acoustics, Speech and Signal Processing (ICASSP)*, pages 1294–1297, March 2010.
531. D. White and R. Jain. Similarity indexing: algorithms and performance. In *Proc. of SPIE*, volume 2670, pages 62–73, Jan 1996.
532. T. Wiegand. JPEG image coding standard. http://iphome.hhi.de/wiegand/, 2018. [Online: accessed July 2018].
533. M. Wien. *High efficiency video coding: Coding tools and specification*. Springer, 2014.
534. M. Wien. Versatile video coding – Video compression beyond HEVC: Coding Tools for SDR and 360° Video. University Lecture, 2018.
535. A. Willéme et al. Overview of the JPEG XS objective evaluation procedures. In *Proc. of SPIE: Applications of Digital Image Processing XL*, volume 10396, pages 103960N–1 – 103960N–18, 2017.
536. S. Wong et al. Radiologic image compression-A review. *Proc. of the IEEE*, 83(2):194–219, Feb 1995.
537. D. Wu et al. An adaptive thresholding technique for the detection of all-zeros blocks in H.264. In *Proc. of Intl. Conf. on Image Processing (ICIP)*, volume 5, pages V – 329–V – 332, Sept 2007.
538. J. K. Wu and R. E. Burge. Adaptive bit allocation for image compression. *Computer Graphics and Image Processing*, 19(4):392 – 400, 1982.
539. X.N. Wu and W. Sun. Robust copyright protection scheme for digital images using overlapping DCT and SVD. *Applied Soft Computing*, 13(2):1170 – 1182, 2013.
540. Y.-G. Wu. Medical image compression by sampling DCT coefficients. *IEEE Transactions on Information Technology in Biomedicine*, 6(1):86–94, March 2002.
541. Y.-G. Wu and S.-C. Tai. Medical image compression by discrete cosine transform spectral similarity strategy. *IEEE Transactions on Information Technology in Biomedicine*, 5(3):236–243, Sept 2001.
542. S. Xiang and J. He. Database authentication watermarking scheme in encrypted domain. *IET Information Security*, 12(1):42–51, Feb 2018.

543. Y. Xiang et al. Patchwork-based audio watermarking method robust to desynchronization attacks. *IEEE/ACM Transactions on Audio, Speech, and Language Processing*, 22(9):1413–1423, Sept 2014.
544. J.J. Xiao et al. Distributed compression-estimation using wireless sensor networks. *IEEE Signal Processing Magazine*, 23(4):27–41, July 2006.
545. Z. Xie et al. A general method for detecting all-zero blocks prior to DCT and quantization. *IEEE Transactions on Circuits and Systems for Video Technology*, 17(2):237–241, Feb 2007.
546. Z. Xin and Z. Wei. Fast inter prediction block mode decision approach for H.264/AVC based on all-zero blocks detection. In *Proc. of 8th Conf. on Industrial Electronics and Applications (ICIEA)*, pages 896–899, June 2013.
547. H. Xu, J. Xu, and F. Wu. Lifting-based directional DCT-like transform for image coding. *IEEE Transactions on Circuits and Systems for Video Technology*, 17(10):1325–1335, Oct 2007.
548. J. Xu, R. Joshi, and R.A. Cohen. Overview of the emerging HEVC screen content coding extension. *IEEE Transactions on Circuits and Systems for Video Technology*, 26(1):50–62, Jan 2016.
549. Z. Xuan, Y. Zhenghua, and Y. Songyu. Method for detecting all-zero DCT coefficients ahead of discrete cosine transformation and quantisation. *Electronics Letters*, 34(19):1839–1840, Sept 1998.
550. H.-Y. Yang et al. A geometric correction based robust color image watermarking scheme using quaternion exponent moments. *Optik - Intl. Journal for Light and Electron Optics*, 125(16):4456 – 4469, 2014.
551. J. Yang et al. Estimation of signal-dependent noise level function in transform domain via a sparse recovery model. *IEEE Transactions on Image Processing*, 24(5):1561–1572, May 2015.
552. J.-F. Yang, S.-C. Chang, and C.Y. Chen. Computation reduction for motion search in low rate video coders. *IEEE Transactions on Circuits and Systems for Video Technology*, 12(10):948–951, Oct 2002.
553. L. Yang and Y. Wan. Accurate image noise level estimation through DCT transformation and approximation by Chi-Square distribution. In *Proc. of 7th Intl. Conf. on Intelligent Human-Machine Systems and Cybernetics*, volume 1, pages 387–390, Aug 2015.
554. P. Yang and M.J. Narasimha. Prime factor decomposition of the discrete cosine transform and its hardware realization. In *Proc. of Intl. Conf. on Acoustics, Speech, and Signal Processing (ICASSP)*, volume 10, pages 772–775, April 1985.
555. X. Yang et al. Adaptive Wiener filter based chrominance up-sampling enhancement method for video coding. *IEEE Transactions on Consumer Electronics*, 57(4):1851–1856, Nov 2011.
556. Y. Yemini and J. Pearl. Asymptotic properties of discrete unitary transforms. *IEEE Transactions on Pattern Analysis and Machine Intelligence*, PAMI-1(4):366 – 371, Oct 1979.
557. C. Yeo et al. Mode-dependent transforms for coding directional intra prediction residuals. *IEEE Transactions on Circuits and Systems for Video Technology*, 22(4):545–554, April 2012.
558. P. Yip. Sine and cosine transforms. In P.D. Poularikas, editor, *Transforms and applications handbook*, pages 3–1 – 3–34. CRC Press, 2010.
559. P. Yip and B.P. Agrawal. Theory and applications of Toeplitz matrices. Technical report, CRL Report, Jan 1979.

560. P. Yip and K.R. Rao. A fast computational algorithm for the discrete sine transform. *IEEE Transactions on Communications*, 28(2):304–307, Feb 1980.
561. P. Yip and K.R. Rao. Fast DIT algorithms for DSTs and DCTs. *Circuits, Systems and Signal Processing*, 3(4):387–408, 1984.
562. P. Yip and K.R. Rao. DIF Algorithms for DCT and DST. In *Proc. of Intl. Conf. on Acoustics, Speech, and Signal Processing (ICASSP)*, volume 10, pages 776–779, April 1985.
563. P. Yip and K.R. Rao. The decimation-in-frequency algorithms for a family of discrete sine and cosine transforms. *Circuits, Systems and Signal Processing*, 7(1):3–19, 1988.
564. Y. Yokotani and S. Oraintara. Lossless audio compression using integer modified discrete cosine transform. In *Proc. of Intl. Symp. on Intelligent Signal Processing and Communication Systems*, pages 120 – 126, July 2003.
565. G. Yu and G. Sapiro. DCT image denoising: A simple and effective image denoising algorithm. *Image Processing - On Line*, 1(1):292–296, 2011.
566. L. Yu et al. Overview of AVS-video: Tools, performance and complexity. In *Proc. of SPIE: Visual Communications and Image Processing*, volume 5960, pages 679 – 690, July 2006.
567. F. Zargari and S. Ghorbani. Fast calculation of 8×8 integer DCT in the software implementation of H.264/AVC. In *Proc. of 7th Intl. Conf. on Application of Information and Communication Technologies*, pages 1–4, Oct 2013.
568. B. Zeng et al. Perceptual encryption of H.264 videos: Embedding sign-flips into the integer-based transforms. *IEEE Transactions on Information Forensics and Security*, 9(2):309–320, Feb 2014.
569. B. Zeng and J. Fu. Directional discrete cosine transforms for image coding. In *Proc. of Intl. Conf. on Multimedia and Expo*, pages 721–724, July 2006.
570. B. Zeng and J. Fu. Directional discrete cosine transforms - A new framework for image coding. *IEEE Transactions on Circuits and Systems for Video Technology*, 18(3):305–313, March 2008.
571. Y. Zeng et al. Integer DCTs and fast algorithms. *IEEE Transactions on Signal Processing*, 49(11):2774–2782, Nov 2001.
572. G. Zhai et al. Bayesian error concealment with DCT pyramid for images. *IEEE Transactions on Circuits and Systems for Video Technology*, 20(9):1224–1232, Sept 2010.
573. C.-X. Zhang et al. The technique of pre-scaled integer transform. In *Proc. of Intl. Symp. on Circuits and Systems*, volume 1, pages 316–319, May 2005.
574. D. Zhang et al. Online palmprint identification. *IEEE Transactions on Pattern Analysis and Machine Intelligence*, 25(9):1041–1050, Sept 2003.
575. H. Zhang and Z. Ma. Fast intra prediction for high efficiency video coding. In *Proc. of 13th Pacific-Rim Conf. on Advances in Multimedia Information Processing*, pages 568–577, Berlin, Heidelberg, 2012.
576. J. Zhang et al. Robust video watermarking of H.264/AVC. *IEEE Transactions on Circuits and Systems II: Express Briefs*, 54(2):205–209, Feb 2007.
577. J. Zhang and A.T.S. Ho. An efficient digital image-n-image watermarking algorithm using the integer discrete cosine transform (IntDCT). In *Proc. of Fourth Intl. Conf. on Information, Communications and Signal Processing, 2003 and the Fourth Pacific Rim Conf. on Multimedia. Proc. of the 2003 Joint*, volume 2, pages 1163–1167, Dec 2003.
578. T. Zhang et al. Signal dependent transform based on SVD for HEVC intracoding. *IEEE Transactions on Multimedia*, 19(11):2404–2414, Nov 2017.

579. W. Zhang and W.K. Cham. Hallucinating face in the DCT domain. *IEEE Transactions on Image Processing*, 20(10):2769–2779, Oct 2011.
580. Y. Zhang. Digital watermarking technology: A review. In *Proc. of ETP Intl. Conf. on Future Computer and Communication*, pages 250–252, June 2009.
581. Y. Zhang and R. Hu. Scalable audio coding based on integer transform. In *Proc. of 1st. Intl. Conf. on Communications and Networking in China*, pages 1–5, Oct 2006.
582. Z. Zhang et al. Focus and blurriness measure using reorganized DCT coefficients for an autofocus application. *IEEE Transactions on Circuits and Systems for Video Technology*, 28(1):15–30, Jan 2018.
583. W. Zhao, T. Onoye, and T. Song. High-performance multiplierless transform architecture for HEVC. In *Proc. of Intl. Symp. on Circuits and Systems (ISCAS)*, pages 1668–1671, May 2013.
584. X. Zhao et al. Joint separable and non-separable transforms for next-generation video coding. *IEEE Transactions on Image Processing*, 27(5):2514–2525, May 2018.
585. Y.Q. Zhao et al. Tampered region detection of inpainting JPEG images. *Optik - Intl. Journal for Light and Electron Optics*, 124(16):2487 – 2492, Aug 2013.
586. Y.W. Zhao et al. The optimal digital filters of sine and cosine transforms for geophysical transient electromagnetic method. *Journal of Applied Geophysics*, 150:267 – 277, March 2018.
587. D. Zhong and I. Defee. Pattern recognition by grouping areas in DCT compressed images. In *Proc. of 6th Nordic Signal Processing Symposium, (NORSIG)*, pages 312–315, June 2004.
588. G. Zhong, L. Lu, and N. Jiang. Fast mode decision based on all-zero block in H.264/AVC. In *Proc. of 2nd Intl. Conf. on Digital Manufacturing Automation*, pages 535–538, Aug 2011.
589. M. Zhou et al. HEVC lossless coding and improvements. *IEEE Transactions on Circuits and Systems for Video Technology*, 22(12):1839–1843, Dec 2012.
590. C. Zhu et al. Multi-level low-complexity coefficient discarding scheme for video encoder. In *Proc. of Intl. Symp. on Circuits and Systems (ISCAS)*, pages 5–8, June 2014.
591. S. Zhu, S.K.A. Yeung, and B. Zeng. In search of better-than-DCT unitary transforms for encoding of residual signals. *IEEE Signal Processing Letters*, 17(11):961–964, Nov 2010.
592. N.A. Ziyal. Image representation, recovery and analysis using principal component analysis. In *Proc. of ICSPAT*, pages 321–324, Sept 1997.

List of Authors

Aas, J.
Abbas, H.M.
Abdelfattah, E.
Acton, S.
Adams, M.
Agbinya, J.I.
Agrawal, B.P.
Ahaitouf, A.
Ahmad, M.O.
Ahmed, A.
Ahmed, N.
Ahonen, T.
Ahuja, N.
Ahumada, A.
Alattar, A.M.
Alexander, P.P.
Alfalou, A.
Alkachouh, Z.
Alshin, A.
Alshina, E.
Alvarado, R.G.M.
Amani, A.
Amornraksa, T.
Ansari, A.E.
Arayacheeppreecha, P.
Arriens, H.J.L.
Artusi, A.
Asai, K.
Asikuzzaman, M.
Astola, J. T.
Atta, R.
Avsgroup

Baali, H.
Babu, S.H.
Bacchelli, S.
Bae, S.H.
Bai, H.
Baig, M.Z.
Bamberger, R.H.
Baraniuk, R.G.
Baron, D.
Basiri, M.M.A.
Bayer, F.M.
Beghdadi, A.
Bellanger, M.G.
Bertrand, A.
Bhatnagar, G.
Bi, G.
Bigun, J.
Bisherurwa, E.J.K.
Bjontegaard, G.
Bloom, A.
Bora, P.K.
Bouguezel, S.
Bouman, C.A.
Bouridane, A.
Bovik, A.C.
Bozinovic, N.
Bradley, A.
Bradley, A.B.
Brahimi, N.
Britanak, V.
Brody, T.P.
Brown, L.G.
Budagavi, M.
Bugadavi, M.
Burge, R. E.
Burl, J.B.

Cai, Q.
Calderbank, A.R.
Cao, Z
Chadha, R.S.
Chaitanya, C.K.
Chakrabarti, N.B.
Cham, W.-K.
Cham, W.K.

Index